T0138252

Probing the Sky with Radio Waves

# Probing the Sky with Radio Waves

## FROM WIRELESS TECHNOLOGY TO THE DEVELOPMENT OF ATMOSPHERIC SCIENCE

Chen-Pang Yeang

The University of Chicago Press CHICAGO & LONDON

CHEN-PANG YEANG is associate professor in the Institute for the History and Philosophy of Science and Technology at the University of Toronto.

The University of Chicago Press, Chicago 60637
The University of Chicago Press, Ltd., London
© 2013 by The University of Chicago
All rights reserved. Published 2013.
Printed in the United States of America

22 21 20 19 18 17 16 15 14 13    1 2 3 4 5

ISBN-13: 978-0-226-01519-4 (cloth)
ISBN-13: 978-0-226-03481-2 (e-book)

Library of Congress Cataloging-in-Publication Data
Yeang, Chen-Pang.
Probing the sky with radio waves : from wireless technology to the development of atmospheric science / Chen-Pang Yeang.
pages cm
Includes bibliographical references and index.
ISBN 978-0-226-01519-4 (cloth : alk. paper)—ISBN 978-0-226-03481-2 (e-book)
1. Radio waves. 2. Radio waves—Research—History. 3. Atmospheric physics—History. 4. Amateur radio stations—History. I. Title.
QC676.4.Y43 2013
538′.767—dc23

2013009936

♾ This paper meets the requirements of ANSI/NISO Z39.48-1992 (Permanence of Paper).

*To my parents, wife, and brother*

**CONTENTS**

# ACKNOWLEDGMENTS

This book originated from a paper I wrote when I was a graduate student at MIT Science, Technology, and Society (STS) Program in the early 2000s. In that paper, I attempted to investigate a turn-of-the-century debate regarding whether the electromagnetic waves of Marconi's first transatlantic wireless signals crept along the sea or bounced between the earth and the sky, a debate I had heard about in my undergraduate electromagnetics course. Over the following decade, the paper expanded into ten talks and conference presentations, two journal articles, and eventually a monograph. And my initial curiosity about the surface-wave-vs.-sky-wave controversy led to an exploration of the epistemic status of scientific theories, the activities of radio amateurs, the discovery of the ionosphere, the nature of field experiments, the microphysics of radio wave propagation in ionic media, and the significances of active sensing in general. It is fair to state that the book project has grown with and spanned my humble academic path in the history of science and technology. Thus, I would like to thank not only those who helped me directly with the book, but also those who helped me with my career.

Above all, I am grateful to my mentors at graduate school: Jed Buchwald, George Smith, and David Mindell. Jed was my eye opener to the history of science. From his teaching, I learned the value and pleasure of parsing out technical details with the reasoning of historical actors themselves. He also gave me intellectually stimulating and pragmatically important suggestions at every stage of my book project. I am very fortunate to have him as an adviser. George guided me to the world of philosophy. His enlightening pedagogy and crystal-clear explications refreshed my understanding of the philosophy of science. He taught me one thing that I consider perhaps the most important research guideline: think hard of the nature of questions and their answers. David was a great adviser. He urged me to pay close attention to the coherence of narratives and the arts of studying and writing the history of technology. He also coached me in various aspects of my professional preparation.

In addition, I am indebted to David Kaiser's tremendous help. He read carefully an earlier work on this project and a draft of several chapters of this book. His comments and critiques on the structures, concepts, and contents of my writings were always insightful. I also appreciate that he brought my book project to the attention of the University of Chicago Press.

I developed and wrote this book at six institutions: MIT STS Program; the Division of Humanities and Social Sciences at California Institute of Technology; the former Dibner Institute for the History of Science and Technology in Cambridge, Massachusetts; the Max Planck Institute for the History of Science in Berlin; the Institute for the History and Philosophy of Science and Technology (IHPST) at the University of Toronto; and the Institute for Modern History at Academia Sinica in Taipei. I was fortunate to befriend the faculty, fellows, students, and staff at these institutions and to receive their generous support in various ways. At MIT, Sam Schweber, Evelyn Fox Keller, Merritt Roe Smith, and Larry Bucciarelli mentored me on the history of physics, philosophy of science, history of technology, social studies of engineering, and the nuts and bolts of academia. I also thank my fellow graduate students for their warm friendships: Tim Wolters, David Lucsko, Shane Hamilton, Sandy Brown, Wen-Hua Kuo, Rob Martello, Rachel Prentice, Eden Medina, Rebecca Slayton, Anne Pollock, and Kaushik Sunder Rajan. Tim read an early draft of this book's chapters 2–4 and provided brilliant comments. He and his wife Karen nicely offered me lodging at their house when I visited Washington, D.C., for archival research. My office mate David (Lucsko) helped me considerably with my writing. I also benefited from the writing workshops organized by David (Kaiser), Susan Silbey, Rachel, Rebecca, and Kaushik.

At the Dibner Institute, I received invaluable advice from senior fellows including Bruce Sinclair, Deborah Cramer, Conevery Bolton Valencius, Edith Sylla, David Cahan, Karine Chemla, Bruce Belhoste, Tom Archibald, Olival Freire, Giora Hon, James Voelkel, Sara Wermiel, and Emily Thompson. Bruce discussed my research project with me, read an earlier work on this project, and gave me useful and detailed suggestions on how to turn it into a book. Deborah helped me to craft my book proposal and book plan. Conevery gave me fabulous encouragement. In addition, I enjoyed the company of my cohort postdoctoral fellows: David Pantalony, Jeremiah James, Elizabeth Cavicchi,

Takashi Nishiyama, Claire Calcagno, Martin Niss, Dane Daniel, Yunli Shi, Arne Hessenbruch, Mathew Harpster, Ben Marsden, Kristine Harper, Peter Bokulich, Sandro Caparrini, Alberto Martinez, and Gerard Fitzgerald. In particular, David was kind enough to proofread many of my writings. We all miss the beautiful, cozy, and productive environment of the Dibner Institute, and deeply regret that it no longer exists.

While my stays at Caltech, the Max Planck Institute, and Academia Sinica were shorter, they too were fruitful and critical to my book project. I am grateful to Jed Buchwald, Otto Sibum, and Yung-fa Chen for arranging my visits at these three institutions, respectively. At Caltech, Diana Kormos-Buchwald and James Lee talked with me about my career plans in general. At the Max Planck Institute, David Bloor read a draft of my book's chapter 5 and provided me with excellent suggestions.

At the IHPST, I am privileged to have wonderful colleagues: Trevor Levere, Bert Hall, Janis Langins, Craig Fraser, Marga Vicedo, and Mark Solovey, who offered me their advice and shared with me their experiences in research, teaching, professional development, and the details of book publishing. In Toronto, I also enjoy the company and suggestions of Edward Jones-Imhotep, who shares my interest in the history of radio ionospheric propagation studies.

I would also like to thank the staffs of the six institutions for their unreserved support: Chris Bates, Deb Fairchild, Debbie Meinbresse, May Maffel, Shirin Fozi, Kris Kipp, and Judy Spitzer at MIT STS; Rita Dempsey, Carla Chrisfield, Trudy Kontoff, Bonnie Edwards, and Evelyn Simha at the Dibner Institute; Gail Nash and Susan Davis at Caltech; Nina Ruge at the Max Planck Institute; Hsiuchuan Lin at Academia Sinica; and Muna Salloum and Denise Horsley at the IHPST.

As an engineer turned historian, I owe much to my former advisers and friends in electrical engineering. The training and mentorship they provided enabled me to engage the technical history of radio ionospheric propagation, and their interest in my project motivated me to finish the book. I would like to express my gratitude to four faculty members at MIT Department of Electrical Engineering and Computer Science. Hermann Haus was a reader of my earlier work in the history of science before he passed away. He was most concerned with my writing. I am grateful that he went through my chapters line by line with corrections and gave me useful critiques on my English. The late Jin Au Kong was the supervisor for my Master of Science in electrical engineering. He taught me the beauty and intricacy of electromagnetic theory, which I needed to write this book. I really wish they could have lived to see this book in print. Jeffrey Shapiro was the supervisor of my ScD in electrical engineering, and Gregory Wornell supervised my engineering postdoctoral research. I appreciate their open-mindedness to my digression to history and their curiosity about my book project. In addition, I presented several chapters of this book at David Rutledge and William Bridges's group meeting at Caltech and Frank Kschischang's group meeting at the University of Toronto.

I am very thankful to three referees who reviewed my book manuscript; I came to know later that they were David Wunsch, Sungook Hong, and Suman Seth. Their frank but positive comments were extremely useful for me in reshaping and revising my manuscript. I also appreciate the enormous help of my editors, Jennifer Howard and Karen Darling at the University of Chicago Press. Jennifer launched the review process for my manuscript; Karen has taken over since Jennifer left the press. Without their efforts, this book would not exist.

I would also like to thank my research assistants Chris Conway and Nicolas Sanchez Guerrero at the University of Toronto and my copyeditor John Parry. Chris and Nicolas helped me to prepare the figures and tables and to acquire copyright permissions. John went through my manuscript relentlessly and made a lot of stylistic modifications. While I hold responsibility for any stylistic awkwardness, he contributed to making this book much more readable than it was. In addition, I express my gratitude to Michael Koplow, a manuscript editor at the University of Chicago Press, for his work on my book.

My book project could not be done without the generous assistance of the staff at the archives and libraries I had accessed: American Radio Relay League Archives, Newington, CT; Arnold Sommerfeld Papers (NL 056) and Jonathan Zenneck Papers (NL 053), Deutsches Museum Archives, Munich; AVIA Records and DSIR Records, Public Record Office, London; Bureau of Ships Records (RG 19), U.S. National Archives I and II, Washington and College Park, MD; Edward Appleton Papers (H37), Special Collections, University of Edinburgh Libraries; General Collection, Philips Research Public Relations Department, Eindhoven; George Neville Watson Papers, Special Collections, University of Birmingham Libraries; Hiram Percy Maxim Papers, RG 69:12, Connecticut State Library, Hartford; Historical Archives, Naval Research Laboratory, Anacostia, Washington; Historical Records, Rutherford Appleton Laboratory (Space Science Department), Didcot, England; Ionospheric Section Records, 1927–59, Department of Terrestrial Magnetism Archives, Carnegie Institution, Washington; Joseph Larmor Papers, Special Collection, St. John's College Library, Cambridge University; Merle Tuve Papers, Library of Congress, Washington; Oliver Heaviside Papers (UK0108 SC MSS), Institution of Electrical Engineers (IEE) Archives, London; Radio Pioneers, Columbia University Oral History Collection, New York; Rayleigh Papers, Rare Book Collection, Air Force Research Laboratory, Hanscom Air Force Base, Lexington, MA; Service Historique de l'Armée de Terre, Paris; Special Collection, Libraries of Radio France, Paris; Special Collection, University of Aberdeen Libraries, Aberdeen. I am especially grateful to Chris Davis at the Rutherford Appleton Laboratory, Sylvain Alzial at Radio France, Wilhelm Füßl at Deutsches Museum, William Perry at the American Radio Relay League, Marjery Ciarlante and Tab Lewis at the National Archives II, the late David van Keuren at the Naval Research Laboratory, Sheila Noble at the Edinburgh University Libraries, and Shaun Hardy at the Carnegie Institution.

My research and writing for this book project were sponsored by the MIT STS fellowship (1999–2001), fellowship at the Caltech History Program (2002), the Dibner Graduate Fellowship (2002–4), the Dibner Sloan Project for the History of Recent Sciences and Technologies (2002–4), the MIT Kelly-Douglass Fund (2001–3), the MIT Siegel Prize (2003), the Dibner Postdoctoral Fellowship (2004–5), the IEEE Life Member Fellowship in Electrical History (2004–5), the postdoctoral fellowship at the Max Planck Institute for the History of Science (2006), the Connaught Start-Up Fund at the University of Toronto (2006–8), and the Connaught New Staff Matching Fund at the University of Toronto (2007–10). Jed Buchwald arranged the Caltech fellowship; Buchwald and David Mindell arranged the grants from the Dibner Sloan Project; Otto Sibum arranged the fellowship at the Max Planck Institute.

Portions of this book have appeared previously in two articles. Sections of chapters 2 and 4 originally appeared in Chen-Pang Yeang, "The study of long-distance radio-wave propagation: 1900–1919," *Historical Studies in the Physical and Biological Sciences*, 33:2 (2003), 369–403. Most of chapter 3 originally appeared in Chen-Pang Yeang, "Scientific fact or engineering specification? The U.S. Navy's experiments on long-range wireless telegraphy circa 1910," *Technology and Culture*, 45:1 (2004), 1–29. I thank the editors of each journal for permission to use the material here.

Above all, I owe great debt to my family: my father Chia-Chiu Yeang, my mother Lan-Chun Hsu, my sisters Hui-Shan and Hui-Chi, my brother Chen-Hsiang, who accompanied me during my long life as a graduate student in Cambridge, Massachusetts, and my wife, Wen-Ching Sung, who has given me love and support all the time.

CHAPTER 1

# Introduction: From Propagation Studies to Active Sensors

On 26 October 2004, the *Cassini* Orbiter had its first close flyby of Titan, Jupiter's largest moon. Under the control of the Jet Propulsion Laboratory (JPL) at California Institute of Technology in Pasadena, the spacecraft had undertaken a seven-year cosmic odyssey since its launch from the Kennedy Space Center in Florida. The JPL staff and participating researchers around the world had waited for this day, since one of the project's major missions—arguably its primary task—was to explore Titan. As an exploratory platform, the *Cassini* boasted a dozen detectors, including two state-of-the-art cameras that had captured stunning images of Saturn and its rings and a cadre of spectrometers to monitor the chemical composition of any radiating celestial body.

But Titan posed a particular challenge to the instrument designers. It is the only satellite of the solar system possessing an atmosphere, and a thick, yellow haze of hydrocarbons almost perpetually blocks it to cameras and spectrometers. Even the onboard *Huygens* probe—the landing unit that the European Space Agency had made for exploring Titan's surface—was not the complete solution, as it could take measurements only near its landing site. To reveal the yellow moon's macroscopic geological characteristics, the *Cassini* team rested its hopes on imaging radar. Unlike cameras and spectrometers, which received light, energy, or particles that emanated from the observed object, the radar bounced microwaves off the object and timed their return. This apparatus did not disappoint the JPL staff. During the flyby, it scanned 1 percent of Titan's overall surface and relayed the echoed signals back to earth. Three days later, JPL's Media Relations Office proudly displayed the first radar images of Titan, which unveiled such novel features as the active surface, complex terrains, and the possible existence of lakes. The Titan radar had made its début.[1]

The mapping of Titan's geology signified a mode of seeing that has permeated our world, ranging from the spectacular weather-radar images of ash out of the recently erupting Eyjafjallajökull volcano in Iceland and the underwater sounding of the *Titanic*'s debris in the North Atlantic, through the mundane altimeters every aircraft now carries and acoustic pulse-echo devices popular among oil-drilling stations, to the ubiquitous magnetic-resonance imaging (MRI), ultrasound, and X-ray machines that adorn modern hospitals. In none of these endeavors do instruments passively observe and measure objects in a nonintervening manner; rather, an acoustic wave, electronic beam, electromagnetic wave, or other form of energy or particle flow "pokes" the objects and then reconstructs their properties from their modification of the flow. This is the principle of the *active sensor*, one of the most powerful scientific instruments since the early twentieth century.

Instruments are never "just" instruments. Introducing a new instrument is not simply the addition of more advanced hardware to enhance human capacity. As history shows, it often accompanies a sea change of understanding and doing things: the telescope initiated the Scientific Revolution; the air pump nurtured laboratory science; the thermometer pioneered quantitative experimentation; the microscope redefined diseases; the particle accelerator made "big science"; the polymerase chain reaction heralded the genetic worldview. Likewise, the employment of active sensors represents a distinct approach to probing nature, the body, and artifacts that involves not only instrument design but also the making of theories and experimentation.

How did the approach of active sensing come into existence? What characterized this approach as it was developing? How did such a novel mode of seeing change the meanings of experimentation and the patterns of experimental practice? How did it affect the standard of legitimate evidence? How did theories of wave or particle propagation help form and refine active sensing? What kinds of epistemic functions did these theories aim to undertake? Why did this mode of seeing prevail?

While the complete answers to these questions require an overwhelming

1. Jet Propulsion Laboratory news releases, "First close encounter of Saturn's hazy moon Titan" (25 Oct. 2004), "Cassini's radar shows Titan's young active surface" (29 Oct. 2004), "Cassini radar sees bright flow-like features on Titan" (9 Nov. 2004), all from Cassini Equinox Mission homepage (http://saturn.jpl.nasa.gov/news/newsreleases/, last accessed on 3 December 2012).

comparative analysis and synthesis of all active sensors that easily go beyond the scope of a monograph, investigation of an informative case may shed some light on such vexing puzzles. This book examines research on radio ionospheric propagation between 1900 and 1935. It is a story of mutual shaping between wireless technology and atmospheric science. After Guglielmo Marconi's first successful transatlantic test in 1901, scientists were curious about why and how radio waves could propagate over such a long distance without the earth's blocking them. From 1901 to 1925, European theoreticians and American engineers grappled with this problem. Its solution led to the discovery of an electrically active region in the upper atmosphere, which they named the "ionosphere."

This revelation opened a new field in earth sciences, and, with the assistance of propagation studies, initiated a novel method of experimentation based on manipulating waves: sending radio waves to the ionosphere and detecting their return. Known as "radio sounding," this method transformed atmospheric studies from passive observation to active experimentation, undercutting the traditional distinction between field and laboratory sciences. From wireless to geophysics, the emergence of studies of radio ionospheric propagation occupies a significant position in the history of active sensing: it began this mode of seeing with electromagnetic waves and led directly to radar during World War II and various sensors in space exploratory programs since *Sputnik* and *Apollo*.

## FROM PROPAGATION STUDIES TO ACTIVE SENSING: EXPERIMENT AND THEORY

Similar to the emergence of some other active sensors, the history of radio ionospheric propagation displays a transformation from studies of wave propagation to development of active-sensing systems. Looking at images or data from a lidar, radar, seismic sounder, sonar, or X-ray machine, we may assume that the stream of energy that the instrument sends to observed objects is a transparent medium that merely helps to illuminate the invisible, like a spotlight on a dark stage. But that is not the case. Far from being transparent, that medium is usually complex, entangles itself with imaging and measuring, attracts researchers' attention for its own sake, and thus has a rich history.

The origin of radio ionospheric sounding attests to the importance of wave-propagation research. What spurred radio echo-sounding probes of the ionosphere in the mid-1920s was not geoscientists' pressing need to measure the upper atmosphere, but physicists and engineers' desire to understand how radio waves propagated over long distances above the earth. Only after the discovery of the ionosphere and the invention of the sounding-echo scheme during wave-propagation research did scientists refocus from the waves to the upper atmosphere.

Along this axis of transformation, the history of radio ionospheric propagation epitomizes the challenges that active sensing has brought to our understanding of modern science and technology. In its first thirty years—from Marconi's wireless test to the establishment of the so-called magneto-ionic theory—which constitutes the scope of this book, such a history raises at least three major issues in experiment and theory: Is it possible to experiment outside laboratories? How do we define direct evidence? What role does theory play at different stages of research?

In a subsequent book, I will examine the development and ramification of the automatic ionospheric sounders in the 1930s based on the theoretical and experimental work on radio propagation, and the establishment of radio ionospheric forecasting services around the world during World War II. This forthcoming work will address more closely the issues of instrumentation and technology in radio ionospheric research.

The issues of experiment and theory raised in the development of active sensing were embedded in a broader context of changing senses of reality at the turn of the twentieth century. Historians have found that scientists during this period were increasingly concerned with the epistemic ground of various experiments, observations, and instruments that promised to make the invisible visible: Do scientific instruments uncover phenomena, or create them? What is the role of sensory experiences in the process of generating empirical knowledge? How does one make claims about microscopic or hidden entities based on macroscopic or observable effects? While the scientists' views on these questions diverged, they were all aware of the instrument-mediating character of scientific evidence, and the shaping force of instruments on experiment and theory.[2]

### *Field Experiments and Direct Evidence*

Above all, studies of radio ionospheric propagation in the early twentieth century broaden our historical understanding of experiment. The empiri-

2. See Sibum, "Science and the changing senses of reality circa 1900" (2008), 295–97, and the special issue *Studies in History and Philosophy of Science A*, 39:3 (2008). (Sibum's article is the introduction to the issue's theme.)

cal investigations on wave propagation and the ionosphere, like the research and development relating to many other active sensors, had to take place outdoors. The scale could be as large as several thousand miles, and the objects of interest were geophysical in nature. These outdoor measurements and tests were by no means feasible in any laboratory. Therefore, we may not be able to understand them in terms of the laboratory studies that historians have explored in the past. For example, measurements of wave propagation and radio sounding of the ionosphere hardly followed what historians Steven Shapin and Simon Schaffer have called the "laboratory form of life":[3] Control and manipulation of material conditions and relevant variables were often very challenging; replication was usually almost impossible; authoritative eyewitnesses of results were rare; and investigators aimed not so much to generate novel "matters of fact" or "scientific effects" as to figure out how those scientific effects interacted with large-scale nature.

Rather, the empirical work in this story resembled more the tradition of field sciences such as astronomy, botany, geology, geodetics, meteorology, and zoology. What characterized these sciences and radio ionospheric propagation alike were comprehensive and extensive fieldwork, careful preparation for expeditions, meticulous collection of data, and precise instrumentation for observations. The "Humboldtian approach" marked an apex of efforts to turn natural history into integrated, modern field science.[4]

Nevertheless, calling radio ionospheric propagation Humboldtian may downplay its experimental features. Throughout the first half of the twentieth century, the physicists and engineers measuring wave propagation and sounding the ionosphere frequently called their activities "experiments." Much of their practice closely resembled experimentation rather than field observation: their instruments, not nature, produced the radio waves. Although they could not control the macroscopic geophysical structures that shaped the propagation of radio waves over distance, they could manipulate radio waves, including their frequencies, power, polarizations, and waveforms.

Such delicate control encouraged them to tinker with devices, redesign procedures, coordinate measurements, and manipulate signals. For example, military engineers tested wireless equipment between warships, radio amateurs demonstrated long-range radio communications with coordinated voluntary actions all over Europe and the Americas, physicists explored scientific effects between ground and sky at particular experimental sites, and

3. Shapin and Schaffer, *Leviathan and the Air Pump* (1989), 22.

4. Humboldt, *Cosmos*, vol. 1, (1997), vii–xxxv, 7–12, 23–34.

geoscientists acquired data from networks of observing stations (which resembled labs) and interpreted them. These were not laboratory but rather field experiments in our eyes and in theirs. Instead of working indoors, they experimented outdoors and turned nature into a laboratory.

How credible was their empirical evidence? While wave propagation studies had suggested the possibility of an electrically active upper atmosphere, general acceptance of the ionosphere's existence occurred only after the sounding-echo experiments in the 1920s. Why? Many scientists believed that the sounding-echo experiments generated "direct" evidence for the ionosphere. But what was direct about the evidence produced by this particular approach? Control in field experiments, I believe, offers the answer: unlike propagation experiments, which only changed the transmitting radio waves' power and frequency, the sounding-echo tests relied on more elaborate control of waveforms. Instead of sinusoids modulated by Morse code dots and dashes, radio waves now could be chirps, pulses, or other patterned undulations, which scientists designed so that their return, scattering, or deflection from an unknown entity would exhibit their properties more clearly.

Here Nancy Cartwright and Ian Hacking's concept of entity realism may help us: a scientific object is real if we can manipulate it.[5] Sounding-echo experimenters in the 1920s could not modify the ionosphere, yet they could tinker with the transmitting radio waveforms as malleable signals and observe the corresponding changes at the receivers. The introduction of waveform control in propagation experiments made the ionosphere seem more "real" and transformed propagation studies into active sensing.

### *Epistemic Status of Theories*

A central desideratum of research on radio ionospheric propagation was to understand how radio waves traveled above the ground, across water, within the atmosphere, around geographical obstacles, or in any other open environment on the earth. Through the first half of the twentieth century, scientists and engineers proposed, elaborated, and fought over several theories: surface diffraction, atmospheric reflection, ionic refraction, and magneto-ionic refraction. The contest between these theories spurred studies of radio ionospheric propagation, especially up to 1930.

However, it is misleading to interpret the story as one theory replacing

5. Cartwright, *How the Laws of Physics Lie* (1983), 1–20; Hacking, *Representing and Intervening* (1983), 262–75.

another, like Hempel-style expansion of covering laws or Kuhnian paradigm change. Scientists devised these theories for different purposes, and they served different functions. Although they did generate mutually incompatible predictions on some empirical questions, and although researchers fiercely debated such forecasts, more often they operated within their own realms and were either irrelevant or marginal to others.

This plurality becomes clear as we examine the epistemic status of wave-propagation theories. At least six questions are germane here. What was a theory's aim and function? What was its most important intellectual virtue? What was the empirical knowledge essential to the theory? What were the central questions it meant to answer? What was its method for tackling these questions? What sort of answers did scientists expect?

The two dominant theories of wave propagation between 1900 and 1920, for instance, differed in nature, even though they both attempted to explain the possibility of long-distance radio. Consistent with Pierre Duhem's two types of scientific theories,[6] the hypothesis of surface diffraction aimed at *formal representation* of an empirical fact, whereas atmospheric reflection proposed *causal explanations* for a broader set of wireless phenomena.

Mathematical physicists worked on surface diffraction and sought a mathematical model to represent long-distance propagation of radio waves along the earth's curvature. Their model comprised a wave equation and a simple boundary condition and gained a life of its own. It became more and more a platform to develop approximating techniques in solving differential equations instead of a reference point for empirical observations. In other words, mathematics was replacing physics.

By contrast, radio engineers were the main explorers of atmospheric reflection. The theory's mathematical structure was much cruder and simpler than surface diffraction before 1919. It boasted no differential equations, no Bessel functions, no asymptotic approximations; it worked with just naïve ray tracing and geometric optics. But formal refinement was never the point. Rather, radio engineers sought to explain numerous wireless phenomena from daily practice—not only long-distance propagation, but also diurnal, geographical, and seasonal variations of ambient noise from the atmosphere. Even though the explanations that they generated were only partly quantitative, their broad but rough theory of atmospheric reflection explained their field observations much better than the precise but narrow theory of surface diffraction.

The magneto-ionic theory, which evolved from the model of atmospheric

6. Duhem, *The Aim and Structure of Physical Theory* (1982), 1–54.

reflection, originally served to explain new short-wave propagation phenomena in the early 1920s. As wave-propagation studies were evolving into ionospheric sounding in middecade, however, the epistemic status of the magneto-ionic theory was changing, too. The new radio echo sounding absorbed this theory (thanks to its dependence on the atmospheric structure), which came to be neither strictly mathematical nor exclusively causal.

Recent horizon-expanding literature on scientific theories sheds light on the magneto-ionic model after the 1920s. Historians and philosophers have identified other functions for theories than the conventional ones such as facilitating hypothetical deduction, establishing natural laws, and providing ontological assurance. Andrew Warwick, Ursula Klein, and David Kaiser have stressed theories' use as "paper tools" for computation and supplying information to experimenters. Similarly, Suman Seth has characterized German physicist Arnold Sommerfeld's approach as "physics of problems," which used certain core mathematical techniques to (somehow opportunistically) formulate and solve various physical problems. George Smith has highlighted their roles in building a recursive convergence between predictions and empirical observations and thus achieving "more secure arguments." Sylvain Bromberger has pinpointed their utility as generators of unexpected puzzles and solutions.[7] The magneto-ionic model, it turned out, became "theoretical machinery" that helped radio sounders generate systematic information about the ionosphere as they used echoing radio signals to infer the upper atmosphere's electron density. From the mid-1920s on, the theory's utilitarian value dominated over other goals. Scientists worried less about explaining radio-wave propagation while imposing the ionosphere as part of that explanation; they looked more to unveiling the ionosphere's nature by employing radio-wave propagation. Means and end swapped places.

That conclusion nonetheless does not imply that all scientists after the mid-1920s treated the magneto-ionic theory as a totally black-boxed tool and ceased to think about its accuracy or physical foundation. In the 1920s and 1930s, some physicists still paid attention to the microscopic basis of the magneto-ionic theory and tried to revise it according to such microphysical arguments. These attempts did not succeed, for one reason or another. Yet

7. Warwick, *Masters of Theory* (2003); Klein, "Paper tools in experimental cultures" (2001), 265–302; Kaiser, *Drawing Theories Apart* (2005); Seth, *Crafting the Quantum* (2010); Smith, "The methodology of the *Principia*" (2002), 138–73; Bromberger, *On What We Know We Don't Know* (1992).

they remind us of the ambiguity and ontological complexity of theoretical machinery or paper tools.

## WHERE THIS BOOK FITS IN

An interdisciplinary subject connotes a pluralistic historiography. Examining research on radio ionospheric propagation as the emergence of a mode of seeing via active sensing can bring new perspectives to the existing literature on the history of physics, history of radio, and history of geoscience. In the history of physics, scholars usually assume a massive change about 1900. Before then, there were elaborations and extensions of Newton's doctrines in mechanics and centuries-long inquiries into light, electricity, and magnetism, culminating in Maxwell's theory. Afterwards, the revolutions of relativity and quantum mechanics dominated modern physics.[8]

In electromagnetism, a subarea of physics, historians have rarely considered anything after 1900 interesting unless it related to quantum mechanics or relativity. Thus, most historical writing has concentrated either on pre-1900 optics, electricity, and magnetism—as in the pivotal works of Jed Buchwald, Bruce Hunt, Daniel Siegel, and Crosbie Smith and Norton Wise[9]—or on quantum and relativistic electrodynamics—as with the renowned studies by Olivier Darrigol, Gerald Holton, and Sylvan Schweber.[10] The most recent overview of electrodynamics, Darrigol's *Electrodynamics from Ampère to Einstein*, has stopped at Einstein, too.[11] This book attempts to remedy exactly this historiographical blind spot. It examines research on electromagnetic waves, but not in relation to quantum physics or relativity, in the early twentieth century. It is clear from this case that novelty in "classical" electromagnetism did not end with Maxwellians, nor were quantum mechanics and relativity the only noteworthy developments of twentieth-century physics. The discoveries regarding wave propagation in the ionosphere formed a breakthrough in

8. Recently, some historians have challenged this clear-cut demarcation between "classical physics" and "modern physics" and started to examine the historical orgin of the discourses on this demarcation. For example, see Staley, *Einstein's Generation* (2008), 345–422.

9. Buchwald, *From Maxwell to Microphysics* (1985), *The Creation of Scientific Effects* (1994), and *The Rise of Wave Theory of Light* (1989); Hunt, *The Maxwellians* (1991); Siegel, *Innovation in Maxwell's Electromagnetic Theory* (1992); and Smith and Wise, *Energy and Empire* (1989).

10. Darrigol, *From C-Numbers to Q-Numbers* (1992); Holton, *Thematic Origins of Scientific Thoughts* (1988); Schweber, *QED and the Men Who Made It* (1994).

11. Darrigol, *Electrodynamics from Ampère to Einstein* (2000).

twentieth-century physics. And the history of these findings is more interdisciplinary and involves broader contexts than the above works.

Most literature on the history of radio has focused on devices and their technological systems, not on the effects of propagation of immaterial waves on devices and systems. In recent writing about the technical development of the key devices in radio technology, Hugh Aitken and Sungook Hong have offered the most comprehensive accounts. They have traced Marconi and others' invention of tuning, the transformation of wireless from laboratory apparatus to powerful machinery, and the emergence of electronic tubes. While they have connected the early development of wireless to physicists' research on electromagnetic waves, they have not addressed radio-wave propagation after 1900, which, as this volume shows, helped shape wireless technology.[12] Among the abundant cultural, political, and social histories of radio broadcasting, Susan Douglas and Linwood Howeth's works have stood out, as they have stressed the institutional processes that transformed radio from novel machines into a prevalent technological system. While Howeth has discussed the U.S. Navy's adoption of radio, Douglas has explored inventor-entrepreneurs, the navy, and radio amateurs as incubators of American broadcasting.[13] They have focused on social dimensions of the construction of *hardware* and *applications*—how these people invented devices, improved machines, promoted new services, and established organizations for these purposes. By contrast, this book looks at the immaterial agent of radio technology—electromagnetic waves—as it examines how these parties worked together on radio-wave propagation.

Writings on the history of geoscience have been fewer than those on electromagnetism and radio. Recently, growing interest in environmental history and awareness of big science's interdisciplinary nature have inspired more research on this subject. One common trend is emphasis on how the methods of laboratory physical sciences diffused throughout and changed the practices of geoscience in the last two centuries. Naomi Oreskes's *Rejection of Continental Drift* (on geology) and Helen Rozwadowski's *Fathoming the Ocean* (on oceanography) are the leading products of such a trend.[14] *Probing*

12. Aitken, *Syntony and Spark* (1976), and *The Continuous Wave* (1985); Hong, *Wireless* (2001).

13. Douglas, "Technological innovation" (1985), 117–73, and *Inventing American Broadcasting* (1987); Howeth, *History of Communications-Electronics* (1963).

14. Oreskes, *The Rejection of Continental Drift* (1999); Rozwadowski, *Fathoming the Ocean* (2005).

*the Sky with Radio Waves* complements these books by concentrating on how the introduction of active sensing facilitated the rise of ionospheric science.

Of course, this is not the only historical work on the subject. Decades ago, Stewart Gillmor and Bruce Hevly wrote on different aspects of the field's history, including the discovery of the ionosphere, the mathematical formulation of the magneto-ionic theory, and the institutional framework of early radio ionospheric research. Recently, Edward Jones-Imhotep and Dominique Pestre have studied ionospheric research during and after World War II. Although all these and other writings provide valuable insights, they deal with only part of the story: either conspicuous discoveries (the early works), or the institutional and political contexts of ionospheric studies *after* their major development (the later efforts). An exception is Aitor Anduaga's *Wireless and Empire*, perhaps the most comprehensive examination on the topic in recent years. It offers a careful and detailed account of interwar research on radio ionospheric propagation in Britain, Australia, Canada, and New Zealand, with the intention of illuminating the geopolitical and commercial structure of the British Empire that nurtured these endeavours. It therefore does not touch on developments in France, Germany, and the United States, or the pioneer era 1900–1920.[15]

This volume presents a longue-durée conceptual, epistemic, and technical history of radio ionospheric propagation and situates it in the emergence of a new mode of sensing the world.

## OUTLINE OF CHAPTERS

I have divided this book into three parts. Part I (chapters 2–4) concerns the early studies of long-distance radio-wave propagation between 1901 and 1919. The success of Marconi's transatlantic wireless trial in 1901 immediately raised a question: how did radio waves traverse the earth's curvature? Chapter 2 is about the introduction of the first theory to account for long-distance propagation of waves—the theory of "surface diffraction" contended that radio waves crept along the earth's surface like sound or light flowed around an obstacle. Hector Macdonald at Cambridge first proposed this model in 1901.

15. Gillmor, "The big story" (1994), 133–41, "Threshold to space" (1981), 101–14, and "Wilhelm Altar, Edward Appleton, and the magneto-ionic theory" (1982), 395–440; Hevly, "Building a Washington network for atmospheric research" (1994), 143–48; Jones-Imhotep, "Nature, technology, and nation" (2004), 5–36; and Pestre, "Studies of the ionosphere and forecasts for radiocommunications" (1997), 183–205; Anduaga, *Wireless and Empire* (2009).

In the decade that followed, the French mathematician Henri Poincaré, Macdonald's colleague John Nicholson, the German engineer Jonathan Zenneck, and the German physicist Arnold Sommerfeld revised Macdonald's model and turned it into a tour de force of mathematical exercises.

Yet those proponents of the theory lacked quantitative empirical evidence. Chapter 3 deals with the U.S. Navy's construction of the first empirical formula for radio-wave propagation. From 1909 to 1913, the U.S. Naval Radiotelegraphic Laboratory in Washington, directed by Louis Austin, performed a series of transatlantic wireless measurements to test the navy's new high-power transmitting station, and these engineering tests became scientific experiments. Austin and his colleague Louis Cohen's resulting Austin-Cohen formula constituted substantial empirical evidence for propagation theorists. Its wavelength dependence, however, contradicted the surface-diffraction theory.

Chapter 4 looks at how scientists and engineers resolved this discrepancy. When Macdonald developed his theory, the English physicist Oliver Heaviside and the American engineer Arthur Kennelly proposed "atmospheric reflection" to explain long-distance propagation, whereby radio waves bounced back and forth between a conducting layer in the upper atmosphere and the earth. This simple model was not even quantitative, let alone capable of making predictions consistent with Austin-Cohen. In the mid-1910s, the British radio experimenter William Eccles revised the model by suggesting that the upper layer consisted of free electrons that emerged as sunlight ionized the atmosphere. The ionized layer did not directly reflect radio waves but rather "refracted" them by deflecting their trajectories, just as water deflected light from air. Although Eccles's theory provided quantitative predictions on wave propagation, the results did not match the empirical formula either. Predictions and data eventually fitted each other in 1919, when the Cambridge mathematician George Watson built a formal theory of atmospheric reflection that could reproduce the Austin-Cohen formula. A long-distance radio experiment by the Marconi Wireless Company corroborated the theory and ended the debate.

Austin-Cohen's most valuable prediction was that radio waves traveled farther at longer wavelengths (i.e., lower frequencies). Before 1920, therefore, all the military and commercial long-range wireless systems used long wavelengths. During the early 1920s, however, the situation changed entirely: short-wave systems marginalized long-wave radio and came to dominate long-range wireless communications, which led to discovery of the ionosphere—the focus of part II (chapters 5–8). Chapter 5 delves into a major finding by radio amateurs: contrary to Austin-Cohen, radio waves shorter than 300 meters could travel up to several thousand miles. At the time, laws restricted radio amateurs to operating at "useless" short wavelengths, a disadvantage that turned them into pioneers. In the early 1920s the amateur American Radio Relay League undertook large-scale wireless experiments that involved hundreds of hobbyists on both sides of the Atlantic. The results showed that waves as short as 100–200 meters could cross the ocean. This launched a "gold rush" to explore shorter waves.

The short-wave bandwagon led to identification of another aspect of propagation and modification of the theory of ionic refraction, as we see in chapter 6. Among the explorers was the U.S. Naval Research Laboratory. Albert Taylor, the head of its radio department, collaborated with American amateurs to experiment with short-wave propagation. They kept on decreasing the wavelength and measured the change in the range of propagation. It turned out that radio waves shorter than 50 meters behaved strangely: instead of diminishing steadily with distance, their signal strength became zero in some intermediate region but then rose again, as if they "skipped" over this region. These three phenomena—transatlantic short waves, the skip zone, and the variation of propagating range with wavelength—all called for explanation. Eccles's model (ionic refraction) explained them better than Watson's (earth-sky reflection). In 1924, the Cambridge physicist Joseph Larmor reformulated Eccles's theory. Then Taylor and E. O. Hulburt at the Naval Research Laboratory modified ionic refraction and explained the skip zones. Finally, Edward Appleton of the Cavendish Laboratory and Harold Nichols and John Schelleng of American Telephone and Telegraph (AT&T) incorporated geomagnetism into ionic refraction to account for some features in the range-wavelength data. All these experimental and theoretical efforts led to the conclusion that short radio waves propagated over long distances because an ionized layer in the upper atmosphere refracted them and the earth's magnetic field affected their velocity, intensity, and polarization.

Nevertheless, no one had actually seen this layer. Was it a hypothetical entity to explain some observations, or a real object? Again, studies of radio-wave propagation offered answers. Between 1924 and 1928, as chapter 7 details, wireless experiments by British physicists yielded what seemed to be direct evidence for the ionosphere. In 1920, the English engineer Thomas Eckersley suggested, from his experience in the British army with wireless direction finders, that radio waves often traveled great distances, going into the sky and returning to the ground. The existence of such sky waves indicated to him the existence of a wave-deflecting boundary in the sky, without evoking specific propagation theories. And one could determine the height of this boundary by measuring sky waves, whose detection became the primary route to direct evidence for the ionosphere. This endeavor preoccupied three groups of researchers—two in Britain and one in the United States. Reginald Smith-Rose and R. H. Barfield of Britain's National Physical Laboratory followed Eckersley's approach of looking for sky waves by tracing the polarizations of incoming waves. Before they could obtain positive results, however, Appleton and his assistant, Miles Barnett, under the aegis of Britain's Radio Research Board, claimed success in 1924. In contrast to the polarization method, they changed radio signals' frequencies to produce interference between sky waves and waves propagating from transmitter to receiver along the ground. They could also easily determine the ionosphere's height from their measurements of interference.

The four British researchers' radio experiments triggered a series of investigations that harnessed echo sounding to probe the upper atmosphere—the subject of chapter 8. About the same time as Appleton and Barnett's frequency-change trials, Gregory Breit and Merle Tuve at the Carnegie Institution of Washington began sending narrow radio pulses to the sky and observing whether they bounced back, and they used the time delay between a pulse and its echo to determine the ionosphere's height. Unlike previous propagation experiments that passively measured radio-wave intensity, Appleton-Barnett and Breit-Tuve actively changed signal patterns at transmitters and inferred properties from corresponding changes at receivers. In so doing, they generated evidence for the ionosphere that contemporaries considered direct. Furthermore, their method of active experiments using radio waves allowed scientists to explore the physical state of the ionosphere. For example, Appleton soon claimed that it possessed a second layer. Propagation studies were revolutionizing atmospheric science.

The discovery of the ionosphere offered new clues to some long-standing problems in geoscience, and radio waves became a tool to help solve these problems. Yet turning radio measurements into ionospheric data required better understanding of radio-wave propagation. Part III (chapters 9–10) examines physicists and engineers' efforts between the world wars to explain wave propagation in the ionosphere. Chapter 9 concentrates on the incorporation of the geomagnetic effect into ionic refraction. Extending Appleton, Nichols, and Schelleng's earlier work, a few Germans, an Austrian, and a few English associated with Appleton deduced and elaborated a generalized magneto-ionic theory (the Appleton-Hartree formula), which gained empirical support from the Cavendish Laboratory's polarization measurements in England and Australia. The English researchers also developed computational methods to turn the general theory into a paper tool to help echo-sounding experimenters uncover the structural characteristics of the ionosphere.

Domesticating the magneto-ionic theory did not stop scientists from worrying about its physical foundation. Chapter 10 addresses physicists' interwar debates on the theory's ontological—or, more precisely, microscopic—status. American, British, Danish, and French researchers contended over whether to include two extra terms in Appleton-Hartree because of intermolecular interactions: the Lorentz correction, proposed by the Cambridge mathematician Douglas Hartree, and the quasi-elastic force, advocated by the French physicist Camille Gutton and his disciples in Nancy. Although these microphysical debates did not get settled until the 1950s and did not much affect practical application of the theory, they reflected physicists' concern about the ultimate basis of a theory and the technical difficulty of exploring that conundrum.

Chapter 11 concludes the book. It sketches later research in radio ionospheric propagation, summarizes key themes and crucial developments between 1900 and 1930, and compares this story with other ones relating to the rise of active sensing.

## A NOTE ON UNITS

Scientists and engineers in the first half of the twentieth century did not settle on a single system of units for measurement. While those in Continental Europe had adopted the metric units of meter and kilometer for length and gram and kilogram for weight, the Anglo-American researchers stayed with the imperial units of inch, foot, mile, and pound. In electromagnetics, a widely used system of units, which is the only one most scientists and engineers know today, was the rationalized MKS system comprising meter, kilogram, second, coulomb (for electric charge), ampere (for electric current), and other common technical units. At the time, however, another system of units was also popular: the Gaussian system comprising CGS (centimeter, gram, second), the electrostatic unit (esu) of charge, and the electromagnetic unit (emu) of current. And there were at least two additional systems: the Heaviside-Lorentz system that eliminated the constants of permittivity and permeability in Maxwell's equations, and the "natural" units that equaled the speed of light and reduced Planck constant to 1.[16] In this book, I do not convert all the physical units into today's standard system such as metric or rationalized MKS. Rather, I follow historical actors' own use of units. When a scientist or engineer in the story chose to express physical quantities in terms of a specific system of units, I present the same figures and units.

16. Panofsky and Phillips, *Classical Electricity and Magnetism* (1962), 459–69.

# * 1 *

# *Conceiving Long-Range Propagation: 1901–19*

CHAPTER 2

# Theorizing Transatlantic Wireless with Surface Diffraction

It was a chilly winter day. The freezing Arctic winds blew on the rocky coast, not pleasant for the men on the hill, but good for flying kites, which they were doing. The young *signor* sat in a room of an abandoned hospital nearby, with drums, knobs, tubes, and wires on the table. Outside with the local laborers were his two moustached, hard-working assistants. The signor had planned for his "big thing" for years; this expedition was its culmination. Before departing for Newfoundland, the assistants shipped six 500-foot-long kite antennae, two balloons, and various receiving apparatuses. On arrival, the signor cabled his experimental station in England to transmit the Morse code of "SSS" (dot-dot-dot) to him every day between 11:30 a.m. and 2:30 p.m., his time. Keeping their real goal secret, team members pretended that they were just studying the influence of rocks on Hertzian waves.

The historical moment came at 12:30 p.m. The signor was the first to hear the three sharp clicks from the earphone. He then fetched one of his assistants, who heard the same thing. The pattern repeated itself at 1:10 and 2:20 p.m. and the next day at 1:38 p.m. Deteriorating weather forbade further testing. But that was enough. The signor informed his business partner in London and issued a brief press statement. Two days later, the *New York Times* featured the story: "St. John's, Newfoundland, December 14 [1901]—Guglielmo Marconi announced tonight the most wonderful scientific development of recent times. He stated that he had received electric signals across the Atlantic Ocean from his station in Cornwall."[1] The world beheld the wireless.

1. Dunlap, *Marconi* (1937), 99; anonymous, "Wireless signals across the ocean" (1901). There are many accounts of Marconi's transatlantic wireless test of 1901. For example, see Dun-

Like all telecommunications technologies, wireless has been a project of conquering space. When Heinrich Hertz discovered the spark-induced electric waves in 1886–88, the phenomenon was a curiosity interesting only to physicists. A series of efforts by scientists and engineers in the 1890s turned this scientific effect into a workable means of telegraphy. A major development took the experimental sets outside the laboratory. In Hertz's original experiment in Karlsruhe, the whole apparatus was indoors, and the maximum distance to detect electric waves was 12 m. In 1891, the Briton George Minchin detected the waves at 130 feet (39.65 m). In 1894, the Maxwellian physicist Oliver Lodge claimed success in detecting electric waves between a spark-gap transmitter in the Clarendon Laboratory and a coherer receiver 180 feet (54.9 m) away in the Oxford Museum. From that time on, experimenters no longer stationed the transmitter and receiver in the same room, and the development of wireless became a race of expanding distances: in 1895–96, Captain Henry Jackson of the British Royal Navy reached 100 yards (91.4 m). In 1896, the New Zealander Ernest Rutherford reached 0.5 miles (800 m) at Cambridge's Cavendish Laboratory. In 1894, the Italian-Briton inventor Guglielmo Marconi reached 2 miles (3.2 km) in Bologna, and his reproduction of this result in 1896 at a demonstration to the British Post Office marked the beginning of his enterprise in wireless.[2]

Marconi's career in the late 1890s rose with his further increase of the Hertzian waves' communications range. From May to November 1897, he experimented across the Bristol Channel and achieved first 3.3 miles (5.28 km) and later 7.3 miles (11.68 km). At the same time, his demonstration for the Italian navy reached 12 miles (19.2 km). In March 1899, he sent wireless signals across the English Channel, between the South Foreland Lighthouse near Dover and Wimereux near Boulogne. The distance was about 30 miles (48 km), at which the receiver was significantly below the transmitter's earth horizon.[3]

Could the traveling distance of the Hertzian wave increase indefinitely? What was the spatial limit of wireless telegraphy? Was it possible to transmit signals across, say, the entire ocean? To Marconi, these were technological

lap, *Marconi*, 87–102; Bussey, *Marconi's Atlantic Leap* (2000), 43–52; Hong, *Wireless* (2001), 77–80; and Marconi, *My Father Marconi* (1962), 111–20. For the historiography of this event, see Hong, *Wireless* (2001), 213.

2. Aitken, *Syntony and Spark* (1976), 118–20; Buchwald, *The Creation of Scientific Effects* (1994), 297; Hong, *Wireless*, 5, 14, 16, 20, 34.

3. Hong, *Wireless* (2001), 53–58; Jolly, *Marconi* (1972), 33–67.

rather than scientific questions. After the successful test across the Channel, he was confident that elevating the power of spark-gap machines, refining the sensitivity of detectors, and improving antennae could overcome distance, regardless of any laws of wave propagation. So he aimed at a much more ambitious range: the 2,000 miles (3,200 km) of the North Atlantic between Ireland and Newfoundland.

Marconi clandestinely prepared for the transatlantic experiment throughout 1900 and 1901. With the help of John Ambrose Fleming, a physics professor at University College, London, he established a high-power transmitting station in Poldhu near Cornwall, England. In September 1901, a test between there and a receiver 200 miles (320 km) away in Crookhaven, Ireland, was successful. Marconi considered several receiving sites in North America and first picked Cape Cod, Massachusetts. But a storm destroyed the aerial mast there, which forced him to move the effort to Newfoundland.[4] The rest of the story is familiar: Marconi achieved the first wireless communication across the Atlantic on 12 December 1901. Some experts reasonably suspected the reliability of the 1901 test—only Marconi and his assistant George Kemp witnessed SSS, the signals appeared just four times, and the result was not repeatable. But the doubt evaporated in February 1902, as Marconi, his assistants, and their apparatuses boarded SS *Philadelphia* in Southampton, England, to sail to New York. Their ability to receive signals from Poldhu in the course of the voyage demonstrated the actuality of long-range wave transmission. By the end of 1902, Marconi had managed to enact a receiving station in Glace Bay, Nova Scotia, and launched a regular transatlantic service.[5]

Marconi's transatlantic wireless tests caught the public's attention for their technological implications and also raised a scientific question about how electric waves traverse space. Despite the enthusiastic late-century competition to increase the effective range of Hertzian waves, attempts to understand the variation of the electric-wave intensity with distance were scarce. Hertz's work remained the gold standard. In 1889, he proposed a theory for his spark-gap experiment: he modeled the spark gap as a tiny radiating dipole source and solved Maxwell's equations under spherical symmetry. The solution was

4. Jolly, *Marconi* (1972), 103.

5. For the reception of Marconi's 1901 test, see Hong, *Wireless* (2001), 79–80. For Marconi's experiment on SS *Philadelphia* and the creation of the Glace Bay station, see Bussey, *Marconi's Atlantic Leap* (2000), 71–87. There has been a controversy regarding whether Marconi "really" received the letters "SSS" transmitted from Britain in December 1901. For a skeptical view about the reality of Marconi's claimed success in receiving "SSS," see Sarkar et al., *History of Wireless* (2006), 392–94.

classical: if the whole space contained nothing but the source, then the radiated field intensity $E$ observed far from the source attenuated with distance $r$; in other words, $E \sim 1/r$.[6]

In Hertz's theory, the electric-wave intensity decreased with distance. And the decreasing rate of $1/r$ was identical to that of a point source spreading energy spherically to the surrounding space, which was consistent with the picture of the pointlike Hertzian dipole. So an electromagnetic wave *spread out rectilinearly* from the radiating source, and therefore Hertz applied the German *Ausbreitung* (spreading out); the English *propagation* connotes the same meaning.

Hertz's theory claimed that electromagnetic waves propagated along straight lines. But in wireless communications farther than 20 km (such as Marconi's 1899 trial across the English Channel), the transmitter and receiver were below each other's horizons, and the earth blocked line-of-sight propagation. What mechanism delivered electromagnetic energy in those cases? Marconi suspected that the ground, like a huge cable, carried the electric signals. The Serbian-American inventor Nikola Tesla held the same belief. The idea of ground-creeping Hertzian waves did not receive much attention in the 1890s; the effect seemed minor for the distances of experimentation, at which Hertzian radiation still behaved like light. Nor were Marconi and his fellow technologists looking for scientific explanations. For a while, how the Hertzian waves propagated remained a not particularly interesting question.

The success of the transatlantic wireless in 1901 altered the nature of this question. The distance of 3,200 km was much beyond the line of sight. Why could the signals travel this far? To traverse one-sixth of the earth's perimeter, the waves had to conform to the curving surface of the earth. Why did they, unlike optical and acoustic waves, not follow rectilinear trajectories? Why did the earth's curvature not block their propagation? After Marconi's triumph, scientists could no longer ignore this problem or pretend it was the result of some minor physical effect.

Starting in 1901, the theoretical physicists and mathematicians in Britain, France, and Germany took on the theoretical problem of long-range propagation of radio waves, the new name for the Hertzian, or electric, waves.[7] They

6. Hertz, "The forces of electric oscillations" (1900), 137–59; Buchwald, *The Creation of Scientific Effects* (1994), 312–14.

7. The term "radio" began to replace the scientific "Hertzian effect" and technological "wireless" after about 1905. In the rest of the book, I follow today's convention by using "radio waves" to refer to electromagnetic waves longer than 1 mm.

entertained a model that resembled Marconi and Tesla's ground-creeping hypothesis, but with a much more specific physical connotation: the waves traversed long distances by *diffraction* along the earth's surface. They produced formal representations of this model, thereby converting transatlantic wave propagation into a mathematical problem, and developed analytical techniques to solve this problem. They argued fiercely over the legitimacy, rigor, and consistency of such techniques, all of which seemed unable to explain Marconi's experimental result.

## *European Theoretical Physicists and Wave Propagation*

European theoretical physicists were the first group to investigate radio-wave propagation. Although mathematical theorization had been a hallmark of physics since its split from natural philosophy after the Scientific Revolution, theoretical, or mathematical physics became an independent discipline only after 1850. Three academic traditions—British, French, and German—contributed to its emergence. One was the teaching of physics-laden mathematics at Oxbridge that dated back to Newton. Another was the school training for rational, mathematized engineering in post-Revolutionary France. The most recent one was the rise of physics research laboratories at German universities after the Napoleonic Wars and their demand for in-house theoretical calculations.

At the end of the nineteenth century, these centers had cultivated certain common epistemic values in their research and pedagogy. The theoretical physicists, mathematicians, and engineers there held an ideal of how to study a physical problem: it should correspond to a model that was simple enough to be reducible to a set of mathematical equations, usually of the ordinary or partial differential sort. A boundary condition or an initial condition relating to the equations should represent the physical circumstance of the problem. The solution should be elegant and closed, with physical meanings directly identifiable, rather than consisting of tedious, messy numerical terms, even though approximations might replace decimal accuracy. The solution should be general enough so that the same procedure, with slightly modified numerical coefficients, boundary or initial conditions, or analytical approximations, could solve other physical problems.

Just as theoretical physicists were consolidating this view, the French scientist-philosopher Pierre Duhem proposed two types of physical theory—*causal explanation* of empirical phenomena and *formal representation* of experimental regularities—and contended that the former was more important than the latter.[8] If Duhem's view reflected his contemporaries' collective mentality, it's no wonder that the new mathematical physics tended to offer formal expressions of facts rather than explaining them and to focus on empirical phenomena that they could reduce to mathematical representations.

Transoceanic propagation of radio waves became one such theoretical problem just after 1900. It was a natural extension of the electrodynamic questions that had preoccupied Helmholtz, Hertz, Maxwell, and their peers. To many physicists, wave propagation above the ground resembled the flow of electromagnetic energy along a conductor, the passing of sound over a surface, and the travel of light around an obstacle, all suggesting the possibility of a ready-made mathematical representation, and perhaps even a formal solution, for the phenomenon. Highly mathematicized research on long-range propagation nonetheless entailed a direction that marginalized the original problem. As theoretical physicists and mathematicians delved deeper and deeper into the complexity of the analytical apparatuses, they found Marconi's experimental result, as well as other wireless phenomena, less and less relevant. Their undertakings looked more and more like a mathematical tour de force for its own sake.

This chapter explores efforts by mathematical physicists to develop theories about surface diffraction early in the twentieth century. We look at how they translated Marconi's experimental findings on wave propagation into a simple model consisting of a radiator and the earth; at how some dons in Cambridge, physicists in Munich, and a mathematician in Paris attacked the problem with various approaches and analytical techniques; and at how their debate narrowed itself down to mathematical technicalities.

## SURFACE DIFFRACTION THEORY IN BRITAIN AND FRANCE

Although wireless telegraphy was a brand-new technology, radio-wave propagation was not entirely novel. The distribution and change of electromagnetic energy in various media and spatial configurations (of which propagation of radio waves above the ground was a special case) had for several decades obsessed researchers on electrodynamics, disciples of Helmholtz, Maxwell, and Weber alike.[9] Moreover, the picture underlying Marconi's transatlantic test—the path of waves bending along a curving surface—was familiar in acoustics,

8. Duhem, *The Aim and Structure of Physical Theory* (1982), 7.

9. Buchwald, *The Creation of Scientific Effects* (1994), 7–24.

geomechanics, and physical optics since the first half of the nineteenth century. If one considered the effect of diffraction—waves' ability to diffuse into the shadow region behind an obstacle—then the fact that the earth, which blocked the line of sight, did not obstruct the propagation of radio waves might not be so counterintuitive.

The primary challenge for any theory of long-distance propagation of radio waves was thus to calculate how much of the electromagnetic energy that the earth diffracted (or, more generally, scattered) was present at a given point. Analytical tools that scientists developed to treat diffracting and scattering problems for light, sound, and seismic waves inevitably could assist vis-à-vis radio waves. An example is the well-known treatise of the British physicist John William Strutt (third Baron Rayleigh and former director of the Cavendish Laboratory), *The Theory of Sound*, which provided abundant mathematical techniques for solving problems of wave scattering.[10]

In the early twentieth century, two groups of mathematical physicists and mathematicians sought diffraction theories for the bending of radio waves along the earth. One consisted of scholars who had trained and worked at Cambridge University plus the French mathematician Henri Poincaré. The other comprised a German school under the theoretical physicist Arnold Sommerfeld at the University of Munich and the electrical engineer Jonathan Zenneck. The investigation began in Cambridge.

## *Macdonald's Initiative*

Cambridge University had been a powerhouse of theoretical physics for three hundred years. The key to its success, as historian Andrew Warwick has indicated, was a distinctive cultural institution—the Mathematical Tripos exam—for training in mathematics. In the late eighteenth century, several colleges transformed the previous graduation examination for bachelor's students into this intensive, days-long written test on applied mathematics and mathematical physics. The Tripos helped shape instruction at Cambridge. The formal curriculum in the hands of university professors and college lecturers yielded to private tutoring, which prepared students for their ultimate test. This system produced an army of elite mathematicians and theoretical physicists—the famous Cambridge "wranglers," those who achieved top rank in the Tripos—and shaped their research. To many of them, attacking a physical problem was similar to tackling a tough but answerable exam question:

10. Rayleigh, *The Theory of Sound* (1945).

simplified modeling, reduction into differential equations, sophisticated but familiar techniques from analytical mechanics, complex analysis, harmonic analysis, special functions, adjustable boundary and initial conditions, and handy generalization into other problems—in brief, marshaling theory's form of representation and the techniques for doing so.[11]

The first person to examine long-range radio-wave propagation was a typical Cambridge wrangler. Born in Edinburgh, Hector Munro Macdonald studied at the University of Aberdeen before entering Clare College, Cambridge, where he became fourth wrangler in 1889 and stayed as a research fellow until 1905. The intellectual atmosphere and pedagogical style of the university's leading physicists, such as George Darwin, George Gabriel Stokes, and John Joseph Thomson, influenced him strongly, even though he had trained as an applied mathematician. Like other Cambridge mathematicians in the 1890s, he possessed a working knowledge of crucial techniques for solving differential equations, including complex and harmonic analysis. Unlike some of his colleagues, he could also work with Bessel functions, useful in solving wave equations with spherical or cylindrical symmetry. His first published research papers concerned the zeros of Bessel functions.[12]

Macdonald took up radio-wave propagation while competing for the Adams Prize, a mathematics award at St John's College since 1848. In 1899, the prize committee announced the subject as the improvement of knowledge with respect to the following topics: "modes and periods of electric vibrations in charged bodies and the radiation from them"; "propagation of electric waves under the influence of conducting wires and conducting or dielectric bodies"; "electric resonance of linear conducting circuits"; "maintenance of forced vibrations"; and "theory of wireless telegraphy." In short, the competition's goal was to forge a theoretical foundation for the wireless experiments from Hertz to Marconi.

Macdonald submitted a long two-part paper to the committee and won the prize in 1901. In the first part, he challenged a Maxwellian convention that represented electromagnetic energy in terms of magnetic field and proposed instead to use Maxwell's older form, which expressed energy in terms of electric current. In so doing, he attributed the source of energy to charge and current rather than to electric and magnetic fields. In the second part,

11. Warwick, *Masters of Theory* (2003), 1–48.

12. Edmund Taylor Whittaker, "Hector Munro Macdonald" (1935), 551–52. For Macdonald's life, also see "As a tribute to the memory of the late Professor Hector Munro Macdonald," undated obituary pamphlet, Special Collection, University of Aberdeen Libraries.

Macdonald applied his formulation to various problems of wave propagation, especially the transmission of electric waves surrounding a conductor in open space (such as Marconi's antenna) and that between a closed chamber and a conductor within (such as an antenna in a cavity). In his own words, "the complete determination of the circumstance of propagation of waves . . . can be reduced to the solution of linear differential equations involving one independent variable" at least for the latter case. The long paper appeared as a monograph, *Electric Waves*, in 1902.[13]

This research provided Macdonald with the conceptual and technical basis for tackling long-range propagation of radio waves. After hearing about Marconi's transatlantic wireless test, Macdonald swiftly worked out a diffraction-oriented account of the phenomenon, which the Royal Society published in 1903.[14] To explain Marconi's results, Macdonald the Cambridge wrangler invoked the mathematical physics with a distinct "Tripos style"—approximating complex physical problems with simple geometry and equations, seeking closed-form solutions, and making explicit analogies to other areas of physics. Since his solution grounded much of the theoretical debate in the following decades, I outline some of its details.

First, Macdonald began with a simple model representing the supposedly relevant physical characteristics of radio-wave propagation. In the model, the earth was a sphere. The transmitting antenna was a vertically polarized point current source—the so-called Hertzian dipole. The dipole and the point of observation could be either above or on the ground. The region exterior to the sphere was free space with a uniform free-space dielectric constant ($\varepsilon_0 = [1/36\pi] \times 10^{-9}$ MKS unit), free-space permeability ($\mu_0 = 4\pi \times 10^{-7}$ MKS unit), and zero conductivity ($\sigma_0 = 0$). The region inside the sphere was a perfect conductor ($\sigma = \infty$). Macdonald's physical model (see figure 2.1) was simplistic and did not include the effects of the atmosphere. Yet it captured some degree of reality: the earth approximated a sphere, both land and sea conducted electricity well, and the vertical antenna tower was indeed so much shorter than the earth's radius that the point-source assumption was legitimate. This constituted the prototype model for all diffraction theorists after Macdonald too.[15]

Second, Macdonald translated the model into a mathematical expression.

13. Whittaker, "Macdonald" (1935), 553–55; Macdonald, *Electric Waves* (1902), v–ix.

14. Macdonald, "The bending of electric waves" (1903), 251–58.

15. Some of them worked on the case where the earth had a finite conductivity and a finite dielectric constant, but others continued to use the perfect-conductor case.

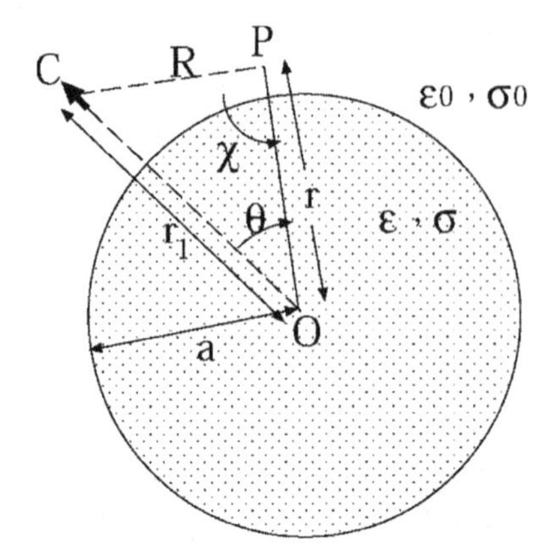


FIGURE 2.1. Macdonald's model of the spherical boundary condition for the diffraction theory. The sphere represents the earth, and the small arrow above the sphere the transmission antenna modeled as a vertically polarized Hertzian dipole.

To calculate the intensity of the electric and magnetic fields at any observation point on the earth, he formulated Maxwell's equations for the electromagnetic fields with a boundary condition deriving from the physical model in figure 2.1. He wrote down the wave equation for the azimuth component of the magnetic field's intensity ($\gamma \equiv H_\phi$) in a spherical coordinate system (the azimuth direction circulated around the axis that connected the earth's center and the dipole's location). This scalar wave equation was a Helmholtz-type equation:

$$(2.1)\quad \frac{\partial^2}{\partial r^2}(\rho\gamma)+\frac{1-\mu^2}{r^2}\frac{\partial^2}{\partial \mu^2}(\rho\gamma)+k^2\rho\gamma=0,$$

where $r$ was the distance between the point of observation and the earth's center, $\rho$ was the distance between the point of observation and the axis connecting the earth's center and the location of the current source, $k = 2\pi/\lambda$ was the wave number ($\lambda$ was the wavelength), $\mu = \cos\theta$, $i=\sqrt{-1}$, and $\theta$ was the angle of the wedge formed by the line connecting the current source's location and the earth's center and the line connecting the point of observation and the earth's center.

Third, he solved the mathematical problem. Mathematical physicists since the mid-nineteenth century had known how to express solutions of the Laplace-type equation (a generalization of the Helmholtz-type equation frequent in acoustics, electrostatics, and optics) in terms of "spherical harmonics"—products of Bessel functions and Legendre polynomials (another special function appearing in second-order differential equations with cylindrical or spherical symmetry).[16] Following Rayleigh's *Theory of Sound*, Macdonald employed a crucial technique: he expressed the solution of equation (2.1) as a sum of basis functions in which the $n$th basis function—the spherical harmonic of order $n$—was a combination of the Bessel and Hankel functions of order $n+½$ (Hankel and Bessel functions were different solutions to the same specific differential equations) and the Legendre polynomial of order $n$.[17] Then he determined the coefficient of each term in this sum from the boundary condition that the field intensity should be infinite at the Hertzian dipole (the nature of the point source) and that the tangential component of the electric field at the surface of the perfect conducting sphere should vanish (an outcome of Faraday's law). Macdonald could thereby express in series form the analytical solution to Maxwell's equations under the boundary condition in figure 2.1. The solution was the diffracted field[18]

(2.2)

$$\gamma = -\frac{ikr^{1/2}}{\rho}\sum_{n=1}^{\infty} g_n(r_1)\left[ J_{n+1/2}(kr) - \frac{(\partial/\partial a)[a^{1/2}J_{n+1/2}(ka)]}{(\partial/\partial a)[a^{1/2}K_{n+1/2}(ika)]}K_{n+1/2}(ikr)\right](1-\mu^2)\frac{\partial P_n(\mu)}{\partial \mu},$$

where $r_1$ was the distance between the current source's location and the earth's center, $a$ the earth's radius, $P_n$ the $n$th-order Legendre polynomial, $J_{n+1/2}$ the Bessel function of order $n+1/2$, $K_{n+1/2}(iz)$ proportional to $H_{n+1/2}^{(1)}(z)$—the first-kind Hankel function of order $n+1/2$—and $g_n$ a linear combination of $K_{n+1/2}$ and $K_{n+3/2}$, and $\mu = \cos\theta$. According to Macdonald, the terms in the sum consisting of the Bessel functions $J_{n+1/2}$ were the radiating field of the point current source in free space, and the terms consisting of the Hankel functions $K_{n+1/2}$ were the secondary field induced from the free-space field by the spherical conductor.

Equation (2.2) was the analytical solution to the wave equation describing radio waves propagating along the earth's surface. Most diffraction theorists until about 1920 agreed with all of Macdonald's steps up to this point. But the esoteric form of the solution did not allow for physical prediction or retrodiction, for it did not yet provide numerical results. To derive quantitative

16. Kline, *Mathematical Thought from Ancient to Modern Times* (1972), 500–700; Watson, *A Treatise of the Theory of Bessel Functions* (1952).

17. Rayleigh, *The Theory of Sound* (1945), chap. 17.

18. Macdonald, "The bending of electric waves" (1903), 253.

information from this formula, Macdonald could have embarked on direct numerical computations, possibly with the assistance of mechanical or human computers, as many astronomers and engineers did. Instead, he chose an alternative way that was more typical among mathematical physicists: he employed an approximation that converted the complicated solution in equation (2.2) into a much simpler and numerically tractable analytical form. This approximation turned out to be the source of controversy among diffraction theorists.

What Macdonald adopted was nothing but a smart trick of Rayleigh's. He noticed that his problem had exactly the same form as one that the master dealt with in *Theory of Sound*. Emulating him again, Macdonald proposed an asymptotic expansion of the Hankel functions in equation (2.2) with respect to $ka$ (reminder: $k = 2\pi/\lambda$ as wave number, $\lambda$ as wavelength, and $a$ as radius of the sphere). From the asymptotic properties of the Hankel functions, he established that when the wavelength was much shorter than the conductor's radius (i.e., $ka >> 1$), the field intensity in equation (2.2) obeyed a simple relation: the ratio of the electric field at the sphere's surface with separation angle $\theta$ (the angle between the oscillator P and the point of observation C as seen from the earth's center, O) to the electric field at $\theta = 0°$ was $1-\cos\chi$, $\chi$ being the angle between the dipole and the earth's center from the observer's viewpoint.[19]

This overly succinct result implied that the electric field that a Hertzian dipole produced did not vanish at any point on the earth except at the diametrically opposite point, $\theta = 180°$—*the earth never cast any shadow on the propagation of radio waves* , which thus could travel anywhere on the earth. Hence the magic of transatlantic wireless transmission: the field diffracted along the surface of the earth. Macdonald considered this conclusion a complement to Rayleigh's finding in acoustics.[20]

## *Rayleigh's Coup*

Unfortunately, Macdonald's approximation had a serious problem, which the master himself exposed. In May 1903, Rayleigh claimed that Macdonald's conclusion erred for two reasons—one physical and one mathematical. Physically, it was impossible to have a shadowless radio wave propagating around the earth, for the similar phenomenon does not occur in optics. The wavelengths

19. Ibid., 255.
20. Ibid., 257–58.

of radio waves (less than 50 km) have about the same ratio to the earth's radius as those of visible light to one inch. Because light shining on a small conducting ball with a radius of about one inch never creeps around the ball's surface to illuminate its rear shadow, it is equally inconceivable that radio waves can creep around the earth. Macdonald's diffraction effect seemed much higher than physical intuition would suggest.[21]

This, Rayleigh pointed out, was the outcome of a mathematical flaw: Macdonald's asymptotic approximation did not hold when the wavelength was much smaller than the radius of the sphere ($ka >> 1$). More specifically, his approximation was valid for all *finite* values of $n$, but the diffraction sum in equation (2.2) had *infinite* terms. From the properties of Bessel functions, the most important contribution of this sum came from the terms where the values of $n$ were approximately equal to $ka$. Thus when $ka$ reached infinity, the dominant terms were those with infinite $n$, at which point, Rayleigh conjectured, the ordinary asymptotic approximation of Hankel functions—the one that Macdonald used—did not hold. So Macdonald's scheme did not apply in the limiting case when the wavelength was much smaller than the dimension of the obstacle.[22] In other words, approximating equation (2.2) asymptotically was much trickier a mathematical issue than Macdonald had thought.

Rayleigh was not alone in pointing out this pitfall. Later in the same month, Poincaré also published a short paper on Macdonald's work. He too thought the asymptotic approximation faulty, because the Cambridge wrangler expanded the Hankel functions in equation (2.2) only to the leading order of $1/ka$, which was inadequate. Poincaré proposed an asymptotic expansion to higher orders.[23]

In the following decade, the mathematical problem of adequately approximating Macdonald's diffracted field in equation (2.2) enthralled mathematical physicists. Because the obstacle (the earth) was much larger than the radio wavelengths ($ka >> 1$), the technique of approximation in Rayleigh's scattering theory was not germane. Macdonald himself attempted expansion to the second order in 1903 but failed to obtain a clean analytical expression.[24] Mathematical physicists kept on proposing new methods of approximation to tackle the notorious Bessel functions. Practitioners at Cambridge dominated

21. Rayleigh, "On the bending of waves around a spherical obstacle" (1904), 40–41.

22. Ibid., 40–41.

23. Poincaré, "Sur la diffraction des ondes electriques" (1904), 42–52.

24. Macdonald, "The bending of electric waves round a conducting obstacle: amended result" (1904), 59–68.

many related discussions on approximating the diffracted field, but progress came from Poincaré.

### *Poincaré's Formula*

By the 1900s, Henri Poincaré had built up a reputation as one of the world's leading mathematicians. His findings on the three-body problem in astronomy and the related nonlinear dynamics foreshadowed the study of chaos that became popular decades later. His work on partial differential equations redefined the field. He was a pioneer of relativity before Einstein. His philosophy of science and mathematics were extremely influential. The American mathematician Oswald Veblen called him and David Hilbert the "mighty outlying summits" of modern mathematics after Gauss.[25]

Poincaré had studied at the École Polytechnique in Paris, which trained engineers for the armed forces and was the center of advanced mathematical education in France (as Cambridge was in Britain). As a military engineering school, it carried a tradition of attacking practical problems with theoretical, mathematical approaches. To Poincaré, such a mixture of abstract and concrete was a hallmark of Polytechnicians. During his professorship at the Université de Paris (Sorbonne) from 1881 until his death in 1912, he explored a variety of mathematical topics, most of them with physical or engineering implications.[26]

Poincaré's interest in the theory of Hertzian-wave propagation began with his work on partial differential equations. From the 1870s through the 1890s, he systematically explored the solvability, solutions, and boundary conditions of several differential equations—the Fourier equation of heat conduction, the Laplace equation, the sound-wave equation, and the telegraphic equation—important in acoustics, electrodynamics, electrostatics, and thermodynamics. The techniques and knowledge that he thereby developed helped him treat Maxwell's equations and propagation of electromagnetic waves.[27]

Poincaré turned to Marconi's transatlantic wireless experiment when he assumed a chair in electricity at the École Professionnelle Supérieure des

25. Veblen, "Jules Henri Poincaré" (1912), iii.

26. Gillispie, "Henri Poincaré" (1975), 51–52; Galison, *Einstein's Clocks and Poincaré's Maps* (2003), 48–49. The last volume of Poincaré's *Œuvres*, with the title *Physique mathématique*, contains papers on cathode rays, electromagnetic diffraction, elasticity, electrons, Hertzian waves, and telegraphy.

27. Webster, "Henri Poincaré as a mathematical physicist" (1913), 901–8.

Postes et des Télégraphes in 1902. He began to note Macdonald's research, presenting in the following year a critique that provided him with a blueprint for further work. He believed that Macdonald's physical model captured the physical reality but thought his method of approximation inadequate and sought an adequate and rigorous approximate solution to the mathematical problem that Macdonald formulated. To that end, he wrote nine articles on diffraction between 1909 and 1912, culminating with the ninety-one-page "Sur la diffraction des ondes Hertziennes" (1910).[28]

Since direct approximation of the terms in equation (2.2) did not work, Poincaré converted the infinite sum in Macdonald's closed-form solution into a definite *integral* and employed Cauchy's famous residue theorem in complex analysis to evaluate it. Like Macdonald, he expressed the analytical solution of the wave equation in terms of a series of spherical harmonic ratios consisting of Bessel and Hankel functions and Legendre polynomials. Each spherical harmonic ratio of order $n$ consisted of a ratio of a Hankel function with argument $kr_1$ to its derivative with argument $ka$—viz., $K_n(kr_1)/K'_n(ka)$. Then he showed that when the dipole oscillator was on the earth's surface, it was possible to convert the infinite sum into a sum of integrals relating to the poles of the spherical-harmonic ratios. (Note $x = x_0$ was a pole of a function $f(x)$ when $f(x_0)$ was infinite.)

From Cauchy's residue theorem, Poincaré expressed these integrals in terms of the poles of the spherical-harmonic ratios; the poles were the values of $\upsilon$ that satisfied $K'_\upsilon(ka) = 0$. He found that when the conducting earth was much larger than the wavelength, the term corresponding to the pole with the smallest imaginary part dominated the others. He also proved that the dominant pole's contribution to field intensity on the spherical surface was proportional to $\exp[-\kappa(ka)^{1/3}\theta]$, where $\theta$ was the angle of separation in figure 2.1 and $\kappa$ a constant. So the field intensity that a Hertzian dipole produced on a spherical conductor took the form of exponential decay with the angle of separation. Although Poincaré could have perhaps calculated the numerical value of $\kappa$ by pushing his approach further, he did not do so and did not specify the parameter.[29]

Poincaré's work involved esoteric theories of Bessel functions and complex analysis. Yet converting the series of basis functions into a sum of integrals and evaluating it through the minimum pole became a powerful technique for treating long-range propagation. More important, his conclusion

28. Poincaré, "Sur la diffraction des ondes Hertziennes" (1910), 169–259.

29. Ibid., 201.

had a straightforward physical meaning—the diffraction field on a conducting sphere *attenuated exponentially at a rate proportional to the −1/3 power of wavelength* (i.e., $\lambda^{-1/3}$); in other words, the diffracted field decreased exponentially in the form of $\exp[-\varsigma\lambda^{-1/3}\theta]$ ($\varsigma$ was another unspecified constant). This prediction was more consistent with physical intuition than Macdonald's, for the exponentially attenuating field guaranteed a large shadow area on the back of the earth. It also implied that the longer the wavelength, the less the attenuation, and the longer the propagation distance.

Nevertheless, Poincaré's conclusion lacked a crucial piece of information: it did not specify the numerical values of the decay rate (for $\varsigma$ was unknown) and the field's nonexponential amplitude factor. Without these values, the mathematical theory of diffraction could not produce quantitative results comparable to experimental data. The English physicist John William Nicholson noticed this shortcoming.

## *Nicholson's Numbers*

John William Nicholson was another Cambridge mathematician. After obtaining his master's degree at the University of Manchester, he entered Trinity College, Cambridge, and passed the Mathematical Tripos as twelfth wrangler in 1904. At Cambridge as student and then lecturer, he won the prestigious Smith's and Adam's prizes. Like Macdonald, he was a wrangler and pursued mathematical physics, but he focused on comparing theoretical with empirical, quantitative results. His best-known work, in the 1910s, on the solar corona and nebulae, presented a theory to explain the spectral lines of those astronomical entities the solar corona and nebulae, first in terms of the quantization of atomic angular momentum and then in terms of quantized vibrations of intra-atomic electrons in directions perpendicular to their orbits—a quantum atomic model before Niels Bohr's. Nicholson showed himself a master of numerical evaluations, as he went through the tedious computations of theoretically predicted values necessary for spectral data. He maintained this preoccupation with numbers in his work on surface diffraction, which contained intensive discussions on whether the predicted field attenuation was quantitatively consistent with wireless engineers' practical experience—a feature question that Macdonald and Poincaré neglected.[30]

30. Wilson, "John William Nicholson" (1956), 209–14; Arabatzis, *Representing Electrons* (2006), 122–23, 128–29.

Nicholson started his scientific career by investigating a set of problems that had preoccupied the Maxwellian Cambridge wranglers for several decades. From 1905 to 1910, he wrote on the diffraction of light around spherical obstacles, inductance and resistance between parallel wires, and scattering of sound waves by various objects. These explorations prepared him to take on the theory of transatlantic wireless that Macdonald had initiated. In his first publication on surface diffraction for long-range radio-wave propagation (February 1910), he reviewed and criticized Poincaré's work and claimed that he had been using a similar approach as early as 1908. His critique of the Frenchman was twofold: Poincaré did not correctly carry out the asymptotic approximation of Bessel functions, and his conversion of the infinite series into an integral was not rigorous.[31]

In 1910 and 1911, Nicholson published a series of papers, "On the bending of electric waves round a large sphere," in *Philosophical Magazine*. His approach resembled Poincaré's: converting the infinite series involving spherical harmonics into integrals and obtaining the approximate values of the integrals from the contribution of the dominant poles. Nicholson turned the infinite series into a sum of integrals corresponding to the poles of the spherical-harmonic ratios in equation (2.2), viz., the values of $\upsilon$ that made $K'_{\upsilon+1/2}(ka) = 0$. After examining the zero structure of $K'_{\upsilon+1/2}(ka)$, he concluded that when $ka$ was much larger than 1, the field intensity was approximately equal to the contribution from the zero of the derivative of the Hankel function—i.e., $K'_{\upsilon+1/2}(ka)$—with the smallest imaginary part, and the imaginary part of this zero had the form $\beta(ka)^{1/3}$. Such a zero appeared in the exponent of the attenuated field intensity. Thus the approximate field was proportional to $\exp[-\beta(ka)^{1/3}\theta]$.[32]

This conclusion was identical to Poincaré's: both showed that the diffracted field had an exponential attenuation with the transmitter-receiver separation proportional to the $-1/3$ power of wavelength. Nonetheless, Nicholson solved for the numerical values of the field's amplitude factor and the attenuation coefficient $\beta$, for which he obtained the value 0.696. So he could write down explicitly the ratio of the diffracted field intensity at any point to the field intensity of the wave propagating from the source to the same point of observation when the conducting sphere was absent. Having this numerical information, Nicholson constructed tables to provide quantita-

31. John W. Nicholson, "On the bending of electric waves round the earth" (1910), 276–78.
32. Nicholson, "Bending II" (1910), 166–72.

TABLE IV.

| $\theta$. | $\frac{\text{Amp. at } \theta^\circ}{\text{Amp. at } 1^\circ}$ | $\frac{\text{Energy at } \theta^\circ}{\text{Energy at } 1^\circ}$ | Terr. miles. | $\theta$. | $\frac{\text{Amp. at } \theta^\circ}{\text{Amp. at } 1^\circ}$ | $\frac{\text{Energy at } \theta^\circ}{\text{Energy at } 1^\circ}$ | Terr. miles. |
|---|---|---|---|---|---|---|---|
| 1° | 1 | 1 | 69 | 16° | $\cdot 0^{3}96$ | $\cdot 0^{6}93$ | 1104 |
| 2° | ·812 | ·659 | 138 | 17° | $\cdot 0^{3}57$ | $\cdot 0^{6}33$ | 1173 |
| 3° | ·571 | ·326 | 207 | 18° | $\cdot 0^{3}34$ | $\cdot 0^{6}11$ | 1242 |
| 4° | ·378 | ·143 | 276 | 19° | $\cdot 0^{3}20$ | $\cdot 0^{7}39$ | 1311 |
| 5° | ·243 | ·059 | 345 | 20° | $\cdot 0^{3}12$ | $\cdot 0^{7}14$ | 1380 |
| 6° | ·153 | ·023 | 414 | 21° | $\cdot 0^{4}69$ | $\cdot 0^{8}47$ | 1449 |
| 7° | ·095 | $\cdot 0^{2}89$ | 483 | 22° | $\cdot 0^{4}40$ | $\cdot 0^{8}16$ | 1518 |
| 8° | ·058 | $\cdot 0^{2}34$ | 552 | 23° | $\cdot 0^{4}24$ | $\cdot 0^{9}56$ | 1587 |
| 9° | ·035 | $\cdot 0^{2}12$ | 621 | 24° | $\cdot 0^{4}14$ | $\cdot 0^{9}19$ | 1656 |
| 10° | ·021 | $\cdot 0^{3}46$ | 690 | 25° | $\cdot 0^{5}81$ | $\cdot 0^{10}66$ | 1725 |
| 11° | ·013 | $\cdot 0^{3}16$ | 759 | 26° | $\cdot 0^{5}47$ | $\cdot 0^{10}22$ | 1794 |
| 12° | $\cdot 0^{2}77$ | $\cdot 0^{4}59$ | 828 | 27° | $\cdot 0^{5}27$ | $\cdot 0^{11}74$ | 1863 |
| 13° | $\cdot 0^{2}46$ | $\cdot 0^{4}21$ | 897 | 28° | $\cdot 0^{5}16$ | $\cdot 0^{11}26$ | 1932 |
| 14° | $\cdot 0^{2}27$ | $\cdot 0^{5}75$ | 966 | 29° | $\cdot 0^{6}94$ | $\cdot 0^{12}89$ | 2001 |
| 15° | $\cdot 0^{2}16$ | $\cdot 0^{5}26$ | 1035 | 30° | $\cdot 0^{6}55$ | $\cdot 0^{12}30$ | 2070 |

TABLE V.

| $\theta$. | $\frac{\text{Amp. at } \theta^\circ}{\text{Amp. at } 1^\circ}$ | $\frac{\text{Energy at } \theta^\circ}{\text{Energy at } 1^\circ}$ | Terr. miles. | $\theta$. | $\frac{\text{Amp. at } \theta^\circ}{\text{Amp. at } 1^\circ}$ | $\frac{\text{Energy at } \theta^\circ}{\text{Energ y at } 1}$ | Terr. miles. |
|---|---|---|---|---|---|---|---|
| 35° | $\cdot 0^{7}37$ | $\cdot 0^{14}13$ | 2415 | 65° | $\cdot 0^{14}27$ | $\cdot 0^{29}74$ | 4485 |
| 40° | $\cdot 0^{8}24$ | $\cdot 0^{17}59$ | 2760 | 70° | $\cdot 0^{15}17$ | $\cdot 0^{31}30$ | 4830 |
| 45° | $\cdot 0^{9}16$ | $\cdot 0^{19}25$ | 3105 | 75° | $\cdot 0^{16}11$ | $\cdot 0^{33}12$ | 5175 |
| 50° | $\cdot 0^{10}10$ | $\cdot 0^{21}11$ | 3450 | 80° | $\cdot 0^{18}69$ | $\cdot 0^{36}47$ | 5520 |
| 55° | $\cdot 0^{12}66$ | $\cdot 0^{21}44$ | 3795 | 85° | $\cdot 0^{19}43$ | $\cdot 0^{38}19$ | 5865 |
| 60° | $\cdot 0^{13}43$ | $\cdot 0^{26}18$ | 4140 | 90° | $\cdot 0^{20}27$ | $\cdot 0^{41}73$ | 6210 |

FIGURE 2.2. Nicholson's tables of predicted radio-wave intensity diffracting around the earth. Nicholson, "Bending III" (1911), tables IV–V. Superscript numbers represent the number of leading zeros after the decimal points; e.g., $0.0^{2}77 = 0.0077 = 7.7\times10^{-3}$.

tive predictions of diffracting radio waves around the earth (figure 2.2). With Nicholson's work at hand, wireless experimenters could compare their measurements directly with the numerical values of diffracting field intensity from theoretical predictions.[33]

No quantitative experimental data on radio-wave propagation was available when Nicholson made predictions of field intensity. Yet that did not prevent him from making a surprising statement: "Diffraction must be a relatively insignificant agency in the success of experiments such as those of Marconi."[34] His confidence came from a dramatic discrepancy between the numerical scale of the diffraction theory and that of Marconi's observation. The exponential decay with the rate $0.696(ka)^{1/3}$ made the diffracting field diminish much faster than it should have. Were Nicholson's predictions valid, the Italian inventor on the Newfoundland coast would not have received any audible signal from England.

Nicholson's result thus appeared paradoxical. On the one hand, he had committed himself to research on the diffraction theory of radio-wave propagation, and he seemed to solve its major mathematical problem with sophisticated analytical techniques. On the other hand, his solution denied any significant role to diffraction in long-distance propagation of radio waves. Although Nicholson continued to work on the diffraction theory for a few years, he no longer believed that it could explain radio-wave propagation.

Yet his conclusion did not destroy the diffraction theory for two reasons. First, his mathematical solution to Macdonald's shortfall was not the final one: other mathematicians would develop alternative methods with different numerical results. Second, the British and French diffraction theorists all assumed that one could model the earth as a perfect conductor and that the curving shape of the perfect conductor guided the diffracted field around the earth. But some scientists and engineers working on the diffraction theory challenged this assumption. To learn why and how, we must turn from England and France to Germany.

## SURFACE DIFFRACTION THEORY IN GERMANY

As the Cambridge wranglers and the Parisian Polytechnician developed theories of surface diffraction, a group of Germans was studying the problem of radio-wave propagation over the ground. Unlike the Britons and French,

33. Nicholson, "Bending III" (1911), 65–68.

34. Ibid., 67–68.

who focused on the effect of the earth's shape, the German theorists stressed the relevance of *finite ground conductivity* in observed wireless phenomena. They contended that the "surface waves" would rise above the ground and creep along its surface because of finite ground conductivity. The surface wave's energy would concentrate around the ground's surface, and the finite ground conductivity could modify the wave's polarization. The German diffraction theorists between 1900 and 1920 were Jonathan Zenneck, an electrical engineer teaching at Braunschweig, and the renowned theoretical physicist Arnold Sommerfeld and his protégés at the University of Munich.

### *Antenna Directivity and Zenneck's Surface Waves*

If spherical diffraction was a product of Cambridge's mathematical training and the Polytechnique's theory-oriented engineering pedagogy, then the surface-wave theory was an outcome of Germany's reformed physics education, which stressed both theory and useful arts. Jonathan Adolf Wilhelm Zenneck was a product of this approach. He wrote an enormously influential textbook on wireless telegraphy[35] and helped to set up the Institute of Radio Engineers while in wartime detention in the United States. Between the world wars, he built an institute of radio physics in Munich that played a crucial part in ionospheric research. After World War II, he directed the Deutsches Museum in Munich and worked to preserve the history of engineering and science in modern Germany.

Zenneck entered the University of Tübingen in 1889 to pursue a doctoral degree; there he became an assistant to physicist Konrad Ferdinand Braun, a radio pioneer and inventor of the oscilloscope. During his years with Braun, first in Tübingen and then in Strasbourg (which Germany annexed after the Franco-Prussian War) from 1892 to 1906, their research group was competing fiercely with Marconi's team in wireless telegraphy. In 1899, Braun launched tests on long-range wireless telegraphy in Cuxhaven, a port near Hamburg on the North Sea, and this project raised Zenneck's curiosity about the theoretical foundations for radio-wave propagation.[36]

35. Zenneck, *Wireless Telegraphy* (1915).

36. For Zenneck's biographical data, see Kurylo and Susskind, *Braun* (1981), 73–74, 130–73; Von Oettingen, *Poggendorff's Biographisch-Literarisches Handwörterbuch* (1904), 1408; and Schmucker, "Jonathan Zenneck" (1999), 150–200.

A wireless experimenter himself, Zenneck studied wave propagation not for the sake of abstract representations of generic facts or to work out neat mathematical solutions. Rather, he sought to understand novel effects that he knew or even observed at first hand in experiments. One such novelty was of course the feasibility of long-range radio-wave propagation beneath the horizon, which he (like Marconi) had witnessed. But the so-called directional antenna also intrigued him. His own theory of surface diffraction thus grew out of his experience with both long-distance propagation and antenna directivity.

Before the mid-1900s, most wireless antennae followed Marconi's design of a grounded and erected aerial in his famous "7777" patent: a long, vertical wire conductor connected at its lower end to the spark-gap generator. This vertical-tower antenna was nondirectional—when it transmitted an electromagnetic wave, the amplitude of the current at a receiver was independent of the direction of its location with respect to the antenna, if the receiver was also on the ground. Similarly, when it received waves from a fixed transmitter on the ground, the arriving current was independent of the receiving antenna's orientation vis-à-vis the transmitter.

Despite the numerous advantages of Marconi's isotropic antenna, however, it was sometimes desirable—especially for fixed point-to-point communications—to have a directional antenna, in which the radio-wave intensity from or into the antenna was prominent only within a narrow range of angular directions. Marconi was well aware of the directional antenna's importance and had experimented with ways to achieve it at his Poldhu laboratory. In 1906, he reported the empirical discovery of antenna directivity. He had done so through means that seemed too simple to be true: by laying down the aerial on the ground to make it horizontal (figure 2.3). When the antenna aerial was horizontal, its radiation pattern became nonuniform, with one maximum and two minima. The intensity reached maximum when the transmitter-receiver

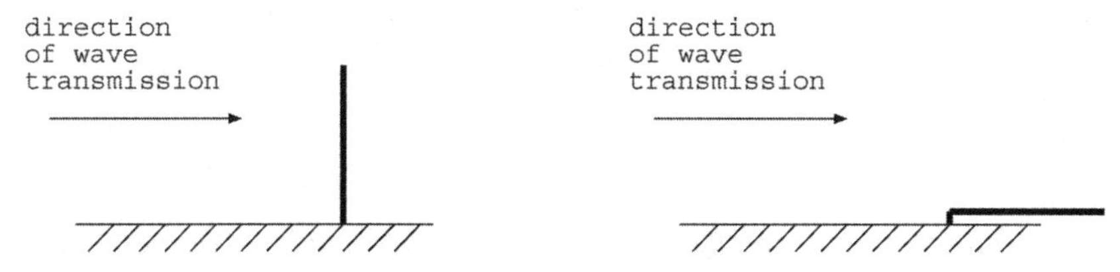


FIGURE 2.3. Nondirectional vertical antenna (left panel) and directional horizontal antenna (right panel). The thick lines represent the aerials.

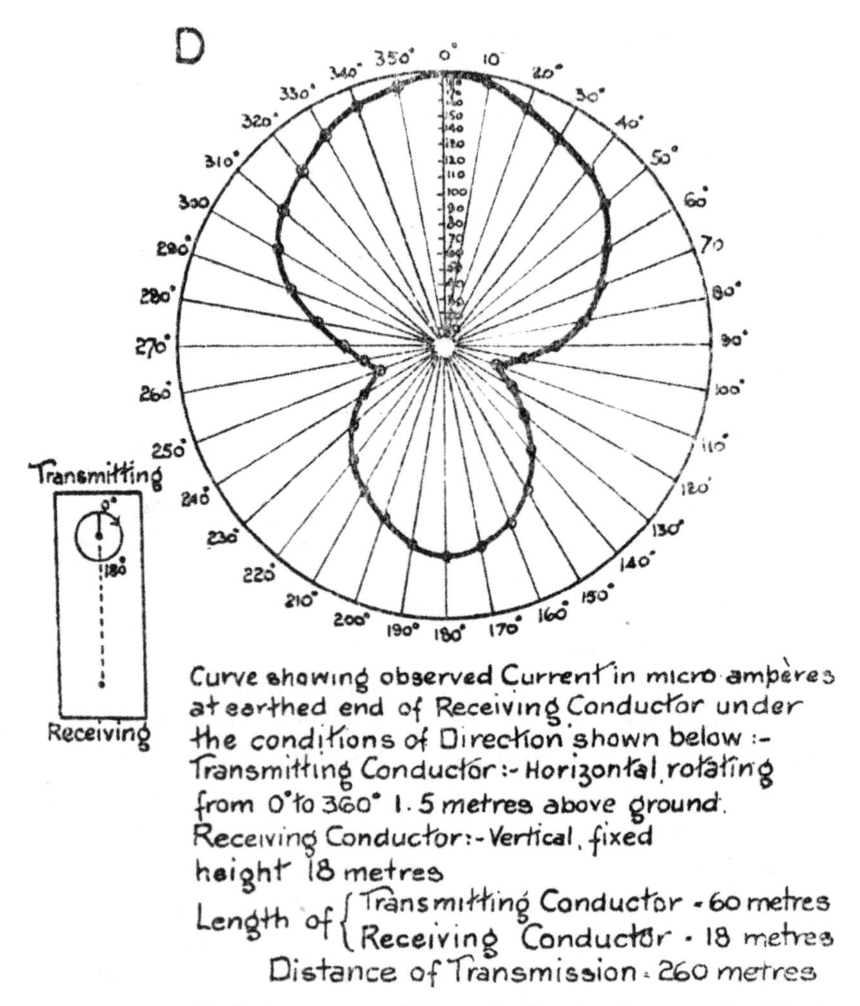


FIGURE 2.4. Radiation patterns of Marconi's directional antenna. Panel D: the transmitter's antenna rotates from 0° to 360°; Panel B: the receiver's antenna rotates from 0° to 360°. Marconi, "On methods" (1906), figures 2 and 4.

direction was 0° from the aerial direction of the transmitter or receiver antenna. Typically, the intensity was minimal when the directional angle was about 110° or 250°. The minimum loci and the shape of the radiation pattern varied with antenna parameters, such as aerial length and height (figure 2.4).[37] In the same year, Braun also published his discovery on directed

37. Marconi, "On methods" (1906), 413–21.

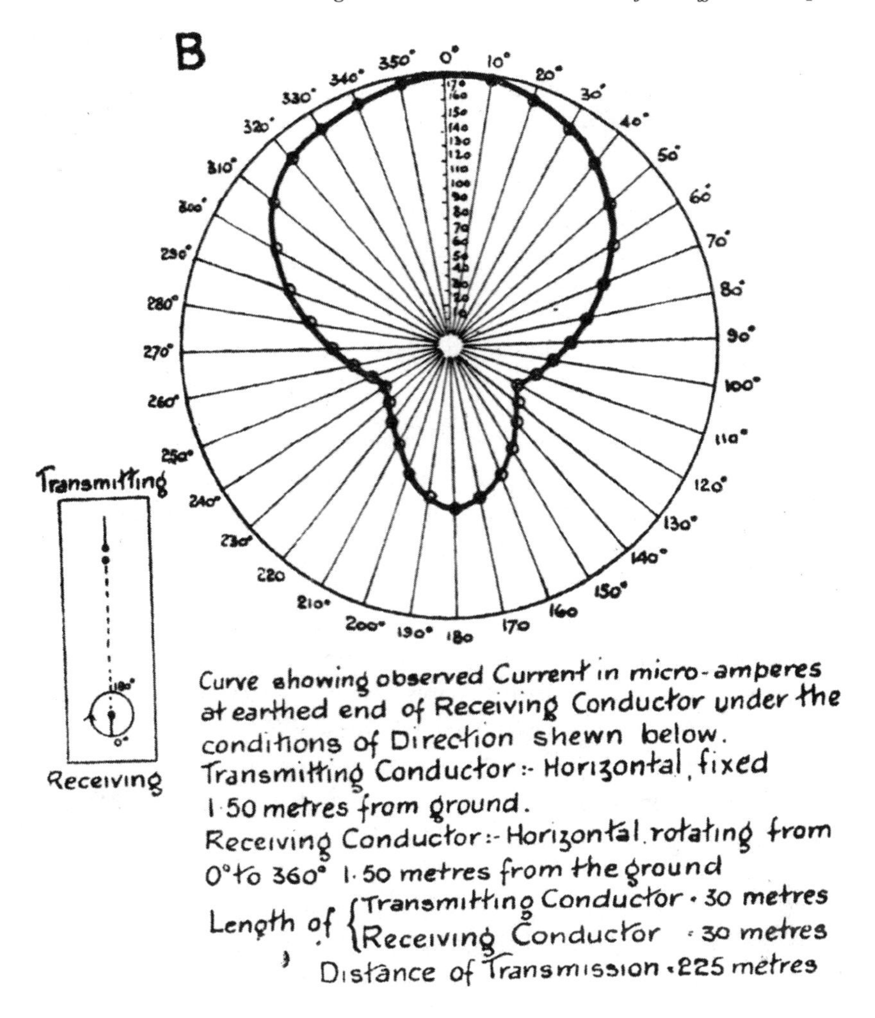


wireless telegraphy as a result of his Cuxhaven and Strasbourg experiments, which Zenneck helped to perform.[38]

In 1906, a year after leaving Strasbourg, Zenneck obtained a professorship in physics at the Braunschweig Technical University. He prepared for new

38. Braun, "On directed wireless telegraphy" (1906), 222–24, 244–48. See also Kurylo and Susskind, *Braun* (1981), 134, 143, 170.

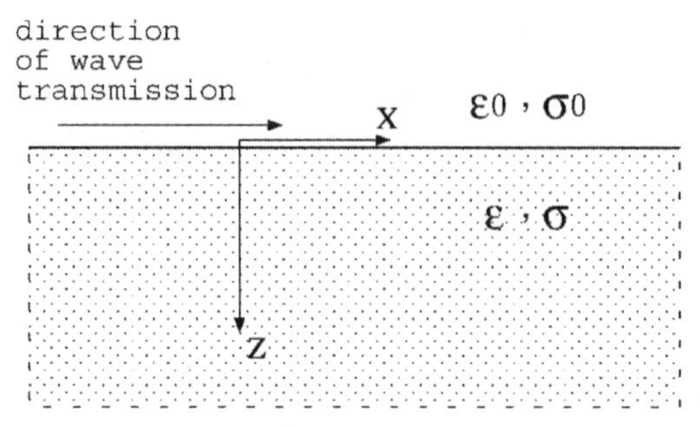


FIGURE 2.5. Zenneck's boundary condition.

courses, set up a laboratory, supervised students, wrote a textbook on wireless telegraphy, and explored photographic techniques for the oscilloscopes (the Braunian tubes) that his adviser had invented. Yet, busy as he was, he could not forget his experience with radio-wave propagation on the North Sea:[39]

> After the days at Cuxhaven, I could not shed the idea that the whole picture at the time about propagation of wireless telegraphic waves—that the waves should travel rectilinearly like light waves—was not correct. Since the transmitting antenna was connected to the [electrically] conducting ground, I believed the waves produced in this manner must resemble more the waves generated by a dipole and propagating along a Lecher conductor [a pair of parallel wires] connected to the dipole than the rectilinear waves radiating in free space.

Another factor that helped convince Zenneck that the optical model did not fit: while the free-space propagation theory predicted longer reach for short wavelengths, his experiments indicated the opposite for wireless telegraphy. To cope with all these issues, he worked on his own theory.

Zenneck's theoretical work resulted in a 1907 article in *Annalen der Physik*. Unlike his British and French counterparts, who assumed the boundary condition of a spherical and infinitely conducting earth, he hypothesized a flat and finitely conducting earth, where radio waves propagated along an infinite planar interface between the air and an imperfect conductor (figure 2.5). The novelty of his approach lay not only in the geometry, however; he did not use

39. Zenneck, *Erinnerungen* (1961), 162 (my translation). I thank the library of the Max Planck Institute for the History of Science in Berlin for providing me with a photocopy of relevant pages of the book.

any information about the dipole oscillator to explain the overall field that the source generated and that the boundary condition shaped. Instead, he *supposed* a particular form of electric and magnetic field and confirmed that it satisfied Maxwell's equations and his boundary condition. This form was a plane wave with field components containing a factor of $\exp[i(\omega t+sx)]$, where $\omega = 2\pi f$ was the angular frequency and $S =$ the wave number along the horizontal direction. Plugging this expression into Maxwell's equations and the boundary condition, he found that the field quantities above the surface were proportional to $\exp[i(\omega t+sx-r_0 z)]$ and those below it to $\exp[i(\omega t+sx+r_1 z)]$, where $r_0 = (k_0^2-s^2)^{1/2}$, $r_1 = (k_1^2-s^2)^{1/2}$, and $k_0$ and $k_1$ denoted the wave numbers in the air and on the ground, respectively. The values of $s$, $r_1$, and $r_0$, all of which he obtained by solving simple algebraic equations, were products of the dielectric constants and conductivities of air and ground. They were complex numbers.

The physical implications of Zenneck's field solution were extraordinary. First, unlike optical plane waves in free space, his wave not only propagated but also *attenuated* along both $x$ (horizontal) and $z$ (vertical) directions. In addition, the polarization of his wave was a function of the ground conductivity and the dielectric constant. Unlike free-space plane waves, his wave had a fixed polarization that could not be chosen freely.

Second, when ground conductivity was low (in the scale of earth, stone, or sand conductivity), the polarizing direction of his wave inclined along the direction from which the wave came. This, Zenneck pointed out, fitted Marconi's experimental results on directional antennae, where a receiving antenna received maximum power when the vertical aerial inclined along the line of sight between transmitter and receiver (see figures 2.3 and 2.4). According to Zenneck's theory, the finite ground conductivity inclined the propagating wave's direction of polarization towards the direction of propagation; and the antenna most efficiently converted the field into an oscillating current when it aligned with the polarizing direction of the field (figure 2.6).

Third, Zenneck's wave along the direction of propagation attenuated most rapidly at finite ground conductivity. When ground conductivity was either zero or infinite, the attenuation was zero. Thus either a perfect conducting ground or a ground with high resistance (such as a dielectric material) could support long-distance propagation. In addition, the attenuation decreased with wavelength and was much stronger along the vertical than along the propagating direction. The field intensity decreased to $1/e$ of the original value at $z = 0$ within a few centimeters but to $1/e$ of the original value at $x = 0$ in a few kilometers ($e \approx 2.71828$ is the base of the natural logarithm). Thus most

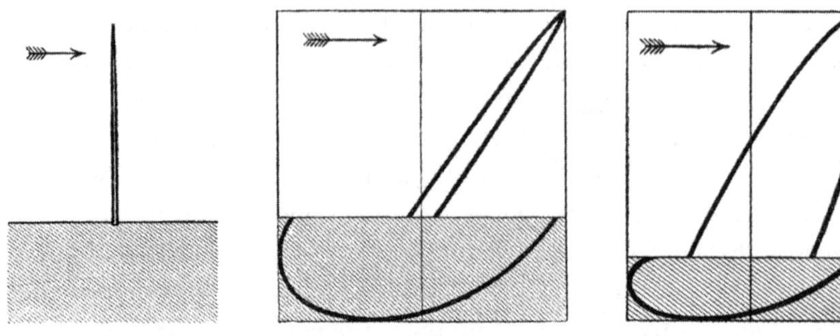

FIGURE 2.6. The polarizations of Zenneck's waves. The electric field's direction starts from the intersection of the vertical line and the ground and ends at points on the curves. The three figures represent three conditions: left: substantial ground conductivity; middle: small ground conductivity and dielectric constant; right: ground conductivity neither large nor small, and ground dielectric constant small. The arrows denote the direction of wave propagation. Zenneck, "Fortpflanzung ebener elektromagnetischer Wellen" (1907), figures 4-6.

energy concentrated near the air-ground boundary. In this sense, Zenneck's wave functioned as a surface wave.

Zenneck's work provided a novel insight. Unlike the British and French theorists who held that the shape of the earth itself enabled long-distance wave propagation, he suggested that ground resistance also played a crucial role. In addition to explaining long-range transmission, moreover, his model could also shed light on antenna directivity. Nevertheless, his paper of 1907 did not offer a complete theory. An obvious shortcoming—especially for beyond-the-horizon propagation, on which he was so keen—was that he did not model the curvature of the earth's surface. A more fundamental problem concerned the uniqueness of his expression. Zenneck's plane wave was only one *possible* solution to Maxwell's equations for the given boundary conditions, not necessarily *the* solution describing physical reality. In contrast to the British and French diffraction theories, he did not specify the source that generated his wave. He did not know how to generate this kind of wave and, worse, whether it was possible to generate this wave at all!

This issue arose because Zenneck did not give a rigorous solution to Maxwell's equations for the given boundary *and* radiation-source conditions. He was well aware of both inadequacies, but he could do little to alleviate them. As he confessed in his memoirs, "My assumptions of a flat ground surface as well as a plane electromagnetic wave were based not on a factual but on a personal reason: without such an assumption, the task [of solving the problem] would be too difficult for me, given my [limited] mathematical knowledge."[40] A more mathematical compatriot—Arnold Sommerfeld—took on the task.

### *Sommerfeld's Refinement of Surface Waves*

Arnold Sommerfeld personified German leadership in theoretical physics at the turn of the century. While Albert Einstein, Werner Heisenberg, and Max Planck were articulating the revolutionary quantum concept in physics, Sommerfeld refined the "old" quantum theory, perfecting Planck and Einstein's ideas in spectroscopic calculations before Heisenberg introduced his new theory. Like the Cambridge wranglers and Poincaré, Sommerfeld epitomized the complex relationship between theory and practice. Yet, whereas the British and French mathematicians had only indirect contact with empirical investigators, he was closely in touch with experimenters and engineers, as was common among theoretical physicists in Germany.

A native of Königsberg (then in East Prussia), Arnold Johannes Wilhelm Sommerfeld entered the local university in 1886, majoring in mathematics. Between 1893 and 1897, he worked at the Mineralogical Institute in Göttingen as an assistant to the mathematician Felix Klein, following his employer's research program to apply advanced real and complex analysis of differential equations to various physical and engineering problems. One such problem involved optical diffraction from a straight edge. In 1895, Sommerfeld presented an exact solution to a diffraction problem in the form of an integral on the complex plane. This novel approach—converting the solution to a differential equation with a proper boundary condition into a closed-form complex integral suitable for numerical evaluation—would become a hallmark of his work.[41]

Sommerfeld learned about Hertzian waves in the early 1890s.[42] His first research paper on this topic, in 1899, examined the propagation of electromagnetic waves along a conducting wire. He demonstrated that as the current flowed in a wire, the Hertzian wave that the current produced also propagated along the wire. Once scientists could grasp the electrical phenomena in terms of waves in the ether, wired and wireless were not essentially different to them. Since they could understand the energy transfer deriving from the flow of an electric current in a wire as the propagation of an ethereal wave along the wire, they could readily perceive a wave propagating above the ground in terms of

40. Zenneck, *Erinnerungen* (1961), 163 (my translation).

41. Gillispie, *Dictionary of Scientific Biography* (1975), 526–29.

42. Sommerfeld, "Autobiographische Skizze" (1968), 674.

the flow of energy guided by the ground's boundary condition.[43] Sommerfeld thus opened up a conceptual ground for his later treatment of wireless waves above the earth.

In 1906, the young man from Königsberg took the chair at the institute of theoretical physics at the University of Munich, replacing Ludwig Boltzmann. Munich was still a minor player in German theoretical physics, no real challenge to Berlin, Göttingen, and Königsberg, which had been centers since the days of Gauss, Helmholtz, Kirchhoff, Neumann, and Weber. Sommerfeld, however, soon made Munich a powerhouse of theoretical physics, principally by combining research and teaching: he built a "theoretical laboratory" and launched an intellectual tradition of physics study. He emulated experimental physicists by organizing young graduate students and postdoctoral assistants to conduct team research in theoretical physics—the Sommerfeldschule (the Sommerfeld school). Between 1908 and 1914, he supervised sixteen doctoral dissertations on crystal optics, hydrodynamics, quantum theory, relativity, rigid-body mechanics, wired electromagnetic waves, and wireless waves.[44]

The Sommerfeld school specialized in dazzlingly sophisticated computations, displayed an instrumental attitude towards theory, and closely followed empirical results. As his peer Max Born commented, Sommerfeld's gift lay not so much in "the divination of new fundamental principles . . . or the daring combination of two different fields of phenomena into a higher unit," but rather in conceptual clarity and in the "logical and mathematical penetration of established or problematic theories and the derivation of consequences which might lead to their confirmation or rejection."[45] Recently, Suman Seth has pointed out that Sommerfeld coined the phrase "die Technik der Quanten" (the craft of the quantum), which hinted at the engineering-like character of his theoretical research.[46] Despite its theoretical focus, the Sommerfeld school also took on practical work such as the gyroscope.[47] Clearly Sommerfeld and his disciples learned a great deal from engineers and their methods.

43. Sommerfeld, "Fortpflanzung elektrodynamischer Wellen längs eines Drahtes" (1899), 233–90.

44. Eckert and Märker, *Sommerfeld* (2000), 278–79; Jungnickel and McCormmach, *Mastery of Nature*: (1986), vol. 2, 281–84.

45. Jungnickel and McCormmach, *Mastery of Nature* (1986), vol. 2, 284.

46. Seth, "Crafting the quantum" (2008), 335–48, and *Crafting the Quantum* (2010), 6; Sibum, "Science and the changing sense of reality circa 1900" (2008), 295–97.

47. Werner Heisenberg, "Vorwort für die Sommerfeld-Gesamtausgabe," in *Sommerfeld: Gesammelte Schriften* (1968), vol. 1, i–v.

Sommerfeld had been in touch with wireless experimenters as early as the 1890s; he wrote the paper on wire-propagating waves with them in mind. In 1899, he visited Braun's laboratory in Strasbourg, where he met Zenneck, who became a lifelong friend.[48] After reading Zenneck's 1907 work, he figured out a way to solve the puzzle the author posed—how the surface wave emerged or whether it existed at all—using his own mathematical techniques introduced in the mid-1890s on optical diffraction and wired waves. In 1909, he showed that Zenneck's "surface wave" was the product of a vertically polarized Hertzian dipole oscillator above the conducting flat surface.[49]

Sommerfeld's problem shared the same boundary condition as Zenneck's: an infinite flat interface separating the conducting material below and the air above. But Sommerfeld specified the form of the radiating source—a vertically polarized electric Hertzian dipole oscillator on the ground's surface. To solve the problem with such a radiating source, he expressed the electric and magnetic field in terms of a Hertzian potential function (the electric and magnetic fields are the spatial derivatives of the Hertzian potential). Like Macdonald, he expanded the Hertzian potential into a set of basis functions more convenient for mathematical manipulations. Yet his expansion was not a discrete sum of spherical harmonics with multiple half-integer orders. Following the mathematical tools he had developed in the 1890s, he employed rather *an integral expansion involving* only the Bessel function of order 0. He noticed that any cylindrical wave in the form of $J_0(qr)\exp[(q^2-k^2)^{1/2}z]$ solved the wave equation, where $q$ was a free parameter and $J_0$ the 0th-order Bessel function. Thus his solution of the Hertzian potential above and below the flat surface took the form of an integral expansion of cylindrical waves over $q$:[50]

(2.3)

$$\Pi_p = \int_{-\infty}^{\infty} \frac{(k_1^2/\mu_1 + k_2^2/\mu_2)}{2\left[(k_1^2/\mu_1)\sqrt{q^2-k_2^2} + (k_2^2/\mu_2)\sqrt{q^2-k_1^2}\right]} H_0(qr)\exp(-\sqrt{q^2-k_p^2}\,z)\,q\,dq,$$

48. Decades later, Zenneck recalled that one day a young man appeared to him and introduced himself as Sommerfeld. The young man invited Zenneck for a dinner talk with him on his research. Sommerfeld had especial interest in the technology of wireless telegraphy that Braun's laboratory had recently been developing. Eckert and Märker, *Sommerfeld* (2000), 285, citing Zenneck, "Persönlische Erinnerungen an Arnold Sommerfeld," a lecture note at the University of Munich, 30 Nov. 1951.

49. Sommerfeld, "Ausbreitung der Wellen in der drahtlosen Telegraphie" (1909), 665–736.

50. Ibid., 687.

where $p = 1, 2$; $\Pi_p$, $k_p$, and $\mu_p$ were the Hertzian potential, the wave number, and the permeability above (when $p = 1$) or below (when $p = 2$) the ground, respectively; and $H_0(x)$ was the 0th-order Hankel function of the first kind.

To evaluate equation (2.3), Sommerfeld used Cauchy's residue theorem to convert the integral over the entire real axis into a contour integral circumscribing the entire upper half of the complex plane, which equaled the sum of the residues from the integrand's poles. This computation was tricky because of a strange pole structure. On the upper half of the complex plane, the integrand in equation (2.3) had two singularities—the branch points $q = k_1$ and $q = k_2$ associated with the multivalue functions $(q^2-k_p^2)^{1/2}$. To specify single values for such functions at a given $q$, it was necessary to define the branch cuts—curves emanating from the branch points—and to specify that any path on the complex plane across a branch cut corresponded to a sudden change of the sign of the functional value. (In mathematical terms, this abrupt sign change, which resulted from crossing a branch cut, formed a jump from one Riemann sheet to another.) One could choose branch cuts freely, as long as one consistently specified single values for multivalue functions. Sommerfeld chose as branch cuts for $(q^2-k_p^2)^{1/2}$ the hyperbolae passing through the branch points $q = k_p$ so that the real parts of $(q^2-k_p^2)^{1/2}$ were zero along these hyperbolae. Under this specification, the integrand had one simple pole with nonnegative real and imaginary parts.

From Cauchy's residue theorem, the integral was thus the sum of three terms—the contributions from the two branch points and the one from the simple pole (figure 2.7). The simple-pole contribution turned out to have the same form as Zenneck's surface wave, whereas the terms for the branch points attenuated much faster with horizontal distance than the surface-wave term did. Only the surface wave was nontrivial at long distances. In short, Sommerfeld demonstrated that Zenneck's surface wave was the asymptotic solution of the diffracted field that emerged from a vertically polarized Hertzian dipole sitting just above the flat boundary surface.[51]

Sommerfeld's work in 1909 highlighted the differences between the Ger-

51. In 1919, however, the German mathematician Hermann Weyl solved Sommerfeld's diffraction problem with a different approach, and the new solution did not contain Zenneck's surface wave. Sommerfeld revised his early work and confirmed Weyl's result in 1926. The mathematical consistency of Sommerfeld's 1909 theory has been a point of contention among physicists and engineers even up to this day. For instance, K. A. Norton at the U.S. National Bureau of Standards claimed in 1935 that Sommerfeld incorrectly reversed the sign of a term in his mathematical expressions, which led to an erroneous solution. This belief in the sign error of Sommerfeld's 1909 work was popular in the following decades, but was later proven to be a

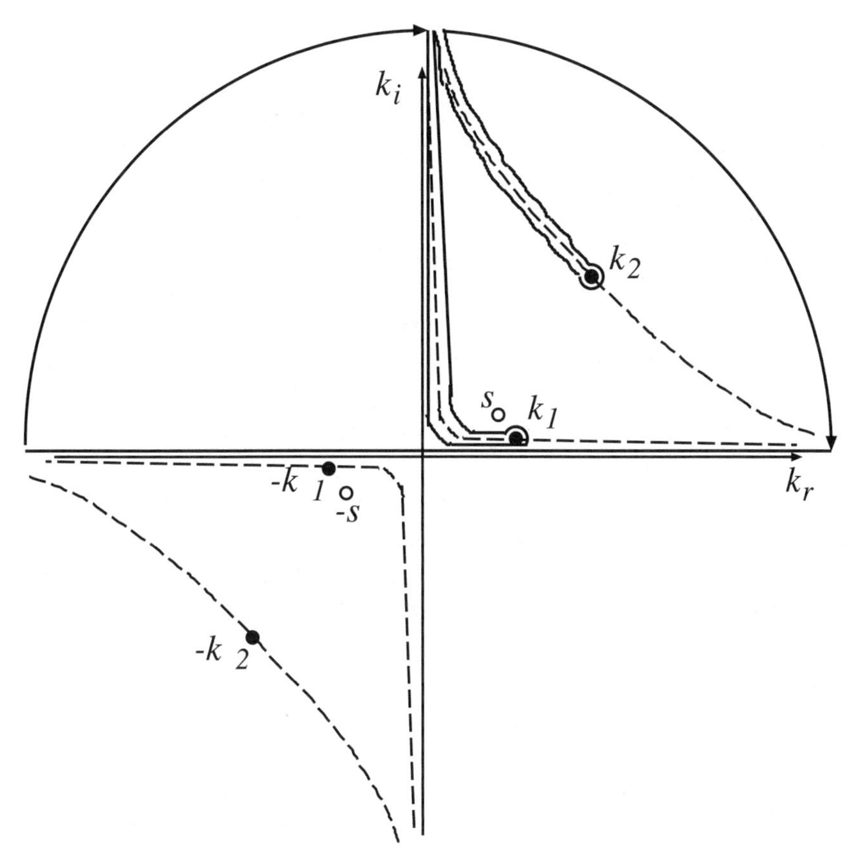


FIGURE 2.7. Sommerfeld's path of integral, branch points, branch cuts, and simple poles.

man and the British-French diffraction theories. The Germans suggested that not only the geometric shape of the boundary surface, but also the resistance of the ground, affected wave-propagating distance. In a strict sense, Zenneck and Sommerfeld's problem was different from Macdonald, Nicholson, and Poincaré's. While the Germans examined a flat, finitely conducting boundary condition attempting to understand the surface waves that had more to do with antenna directivity, the Britons and the French focused on a spherical, infinitely conducting boundary condition to model specifically transatlantic wave propagation. But this separation was only temporary, as the theoretical

false accusation. For one of the most recent technical discussions on Sommerfeld's 1909 theory, see Collin, "Hertzian dipole radiating over a lossy earth or sea" (2004), 64–79.

physicists at Munich quickly extended the notion of surface waves into the spherical boundary condition.

Moreover, Sommerfeld's approach to the diffraction problems created a unique tradition of practice. Expanding the field with respect to an integral of cylindrical waves differed significantly from the common British technique of expanding the field over a sum of discrete-order spherical harmonics. Sommerfeld saw the potential of this approach and asked his students at Munich to pursue it as a dissertation topic. As we see in chapter 4, Herman William March prepared a thesis in 1911 extending Sommerfeld's integral approach to the diffracted waves along a spherical conductor. Witold von Rybczyński wrote on a similar topic in 1913. The Sommerfeld school put a finger on the problems of long-distance radio-wave propagation.

From 1901 to 1909, theoretical physicists and mathematicians in Britain, France, and Germany proposed various versions of a surface diffraction theory to describe radio-wave propagation above the earth's surface. Following a Duhemian doctrine, they aimed to develop a formal representation of two empirical facts that Marconi and other wireless inventors had identified: transatlantic transmission and antenna directivity. The core of these theorists' practice consisted of sophisticated mathematical manipulations of complex analysis and special functions for solving differential equations. They prized theory not for the breadth of empirical information it could account for, but for its logical consistency and accessibility in form: a theory should be mathematically rigorous and compatible with physical intuition. It had to be presentable as a closed analytical formula in which every term had a physical meaning, rather than as a numerical table, even if achieving that outcome required approximation.

The surface diffraction theory of the 1900s was thus a quantitative representation of radio-wave propagation. Ironically, there were few, perhaps no, numerical data against which to compare the theory. Although the theory emerged to explicate empirical observations in wireless transmission (especially Marconi's transatlantic wireless), a systematic experimental study of radio-wave propagation was lacking throughout most of the decade. Where was the quantitative evidence for long-range propagation? It turned out that such evidence came not from European academicians, but from the armed forces of a rising military power across the Atlantic.

CHAPTER 3

# The U.S. Navy and the Austin-Cohen Formula

Although Marconi's transatlantic test of 1901 inspired mathematical physicists to work on long-range propagation of radio waves, it offered no quantitative data to support any of the candidate theories. In 1909, a group of researchers from the U.S. Naval Wireless Telegraphic Laboratory in Washington performed a propagation experiment in the Atlantic Ocean, producing the first set of such data. They measured the intensity of radio waves at different frequencies and various distances from the transmitter. They synthesized the results into the so-called Austin-Cohen formula—after the chief experimenter, Louis Austin, and his aide, Louis Cohen. The Austin-Cohen formula soon became the most useful quantitative empirical law for radio-wave propagation. In the 1910s, it dominated thinking on wireless spectrum usage and shaped the theoretical debate over the nature of long-range propagation.

The story of the Austin-Cohen formula marked the beginning of the military's close involvement in radio-wave propagation studies and ionospheric science. Throughout the first half of the twentieth century, navies and armies in North America and Europe played a major part in this research area. Unlike academic theorists, military investigators were experimental physicists and engineers seeking the most practical applications of radio science, dirtied their hands playing with machines, and had close contact with line officers on the front lines. The military had a huge advantage in long-distance propagation experiments: it could afford large-scale testing much more readily than a university or corporate laboratory. Even though Marconi's Poldhu center and Braun's physics department at Strasbourg could recruit talented researchers, build high-power stations, and arrange expeditions, they could not send cruisers to midocean, coordinate teams far away, and manage the logistics. The U.S. Navy, for instance, was the only producer of quantitative empirical data on propagation in the 1910s. Not until the early 1920s did engineers at the British Marconi Company perform another set of measurements to confirm the Austin-Cohen formula.[1]

The propagation research had a meaning to the military distinct from that of other field-scientific inquiries, however. While armed forces sponsored research in geodesy, geology, meteorology, navigation, and oceanography to obtain information about the environment for planning their operations in general, they studied radio-wave propagation to appropriate a particular technology—wireless communications. Their propagation studies inevitably related to the functionality of specific equipment. Unlike the other field sciences that used instruments to measure natural phenomena *independent* of them, in propagation research the measured phenomena (wave transmission) emerged from the instruments of interest (radio), and the goal was to gauge their performance.

The U.S. Navy's measuring campaign in 1909 epitomized the military's early empirical investigations on radio-wave propagation. The navy planned and executed it centrally. The work involved large geographical scales with varying distances. The setting consisted of a fixed wave generator (in this case, a transmitting station on the coast) and a few mobile measuring platforms (in this case, two warships). And above all, the original aim was not to investigate scientifically long-distance propagation of radio waves, but to test the equipment of the navy's new high-power wireless-telegraph station in Arlington, Virginia.

## A NAVAL WIRELESS LABORATORY

About 1900, the U.S. Navy undertook actions to modernize its fleets, as a result of post–Civil War industrialization and as a response to the growing naval arms race in Europe that American strategist Alfred Mahan had advocated in his *Influence of Sea Power upon History*.[2] The rapid growth of wireless telegraphy in the 1890s convinced a few top-rank officers to incorporate this novel technology into their "New Navy." Many observers had noted the military

1. Round et al., "Report on measurements" (1925), 933–34.

2. Mahan, *The Influence of Sea Power upon History* (1903).

potential of radio since its invention. Marconi had sold the technology to the British and Italian navies in the early 1890s. The U.S. Navy introduced radio in 1899, when its Bureau of Equipment invited Marconi's company to install wireless-telegraph sets on two warships for onboard testing. Over the next decade the navy continued to procure radio transmitters and receivers from wireless companies.

The bureau demonstrated its commitment to wireless by establishing a Radio Division, planning eventually to replace the existing flag-and-light arrangement with radio as a standard fleet communications link. But the plan failed, principally because of cultural gaps and organizational inertia. Experienced seamen in combat units rejected the new gadget because it had no performance record; engineering officers in the conservative military culture often clashed with wireless inventors who preferred flexibility and novelty; and the navy's decentralized, bureaucratic structure made it difficult to overcome these barriers, for the navy's eight bureaus did not have sufficient administrative power to coordinate their joint responsibilities.[3]

In addition, wireless technology was relatively unreliable. Up to the mid-1910s, mainstream transmitter technology—the spark-gap discharger—suffered from its reliance on highly damped waveforms, which produced broadband signals. And mainstream receiver technology—the coherer—was not able faithfully to follow the continuous variation of a signal and was moreover unstable. New continuous-wave transmitter technologies and new sensitive receiver technologies were under development. But their operational characteristics were largely unknown. In order to incorporate radio successfully, the navy required reliable data for the performance of the new instruments under real-world conditions. It could not otherwise develop standard operational and maintenance procedures or produce systematic and rational schemes for selecting appropriate wireless devices. The navy needed a research laboratory to test equipment, measure wireless devices, and evaluate new technologies.

The head of the navy's Radio Division, Cleland Davis, wished to set up such a facility. The director of the U.S. National Bureau of Standards (NBS), Samuel Wesley Stratton, was keen to help. Davis and Stratton decided that the lab should fit within the NBS, in order to harness the people and facilities already active in electrical research. NBS accordingly set up the Naval

3. Douglas, "Technological innovation" (1985), 117–73; Howeth, *History of Communications-Electronics* (1963), chaps. 12 and 13.

Wireless Telegraphic Laboratory in the spring of 1908 at its headquarters in Washington, DC.[4]

The Department of the Treasury had established the National Bureau of Standards in 1901. Following the models of Britain's National Physical Laboratory and Germany's Physikalisch-Technische Reichsanstalt, NBS aimed to develop new standards for various physical quantities. Fabrication of precisely calibrated measuring instruments and development of accurate experimental methods for measuring electrical quantities were among its early activities. Its interest in electrical measurement overlapped with the agenda of the Naval Bureau of Equipment for testing wireless instruments. Their cooperation was hardly coincidental.[5]

The first head of the Naval Wireless Telegraphic Laboratory was Louis Winslow Austin, a physicist then working at NBS. Austin belonged to the post–Civil War generation of American scientists who pursued advanced education in Germany and adopted the German model in building a native scientific research infrastructure. A native of Maine, Austin went to the University of Strasbourg to study physics, particularly techniques of exact measurement. After receiving a doctoral degree in 1893, he taught at the University of Wisconsin, where he was among the first to introduce the German experimental physics curriculum into the American college system. In 1902, he returned to Germany to work at the Reichsanstalt, heading back to the United States two years later to work at NBS. When the plan to establish the Naval Wireless Telegraphic Laboratory became final in 1908, Austin officially transferred to the navy to head the facility.[6]

The initial setup was provisional—the lab was to conduct experiments of tabletop scale at the bureau, while field measurements and operational tests were supposed to be conducted at naval wireless stations. Throughout its first year the laboratory had only two formal employees—Austin and his regular assistant, George H. Clark, an MIT graduate and engineer at Stone Telephone & Telegraph Company.[7] It occasionally hired technicians to help with experiments. Though formally a naval organization, it appeared to be a temporary

4. Austin, "Naval Radio-Telegraphic Laboratory" (1912), 122–41. The organization became the Naval Radio-Telegraphic Laboratory in 1912.

5. Cochrane, *Measures for Progress* (1966), chap. 2. NBS was a crucial participant in radio ionospheric research in the first half of the twentieth century.

6. For a résumé of Louis Austin before 1922, see Weinmeister, *Poggendorff's Biographisch-Literarisches Handwörterbuch* (1922), 42–43.

7. Howeth, *History of Communications-Electronics* (1963), 172–73.

research unit within NBS. This situation changed markedly in 1909, when the navy asked it to perform its experiment with long-distance propagation.

## THE EXPERIMENT

The idea of long-distance wireless transmission had circulated among interested members of the U.S. Navy for some time. Naval officers were aware that building a wired worldwide communications network would be prohibitively difficult and expensive, because British firms controlled most submarine telegraph cables. On the other hand, key to a long-range wireless communications network was a set of geographically dispersed high-power transmitting stations.

In 1908, the navy decided to install the first such outlet in Arlington, Virginia, near Washington, DC, and it publicized a contract that laid out the technical requirements, calling for proposals. The contract stated that the transmitter should be capable of transmitting at all times to a radius of 3,000 miles in any direction from Washington and that shipboard transmitters should have a range of 1,000 miles.[8] The contract went in early 1909 to the National Electric Signaling Company (NESCO), a firm that Canadian-American inventor Reginald Fessenden had set up. The transmitter was of the same type as the 100 kw synchronous rotary-spark discharger that Fessenden designed and then installed at NESCO's experimental wireless station in Brant Rock, Massachusetts.

The award to NESCO was not without controversy, however. Several wireless corporations challenged the bidding process. They asserted that the navy had set technical specifications with NESCO in mind, that this company was not the lowest bidder, and that wireless companies such as Marconi and Telefunken had more experience.[9] Worse, both the navy and NESCO already knew from operational records that NESCO's 100 kw machine was not in fact able to match the contract's long-distance specification. No engineering

8. Bureau of Steam Engineering, *Memorandum for Chief of Bureau*, 1 Feb. 1911, 3, Bureau of Ships Records (RG 19), Radio Division, job no. 445-1, 1908–39, E 1084, box 9, U.S. National Archives I, Washington, DC.

9. Ibid., 2–9. Aitken suggested that NESCO built better relations with the U.S. Navy when the company hired Col. John Firth. Firth was on excellent terms with the leaders of the Bureau of Equipment. "It was largely," asserted Aitken, "if not wholly through his effort that NESCO won the important 1909 contract for the Arlington station." See Aitken, *The Continuous Wave* (1976), 82.

company had ever developed a successful commercial wireless system with an operational range greater than 1,000 miles.

Nonetheless, the navy stuck with NESCO, for it preferred an American over a British or German company and believed strongly that the Fessenden rotary-spark discharger was the best that a U.S. company could offer. To quell complaints and to legitimate its choice, the navy wrote an addendum to the contract that called for further technical tests and measurements of the existing system in Brant Rock while construction at Arlington went ahead.[10] The navy asked the Naval Wireless Telegraphic Laboratory to test and measure the 100 kw rotary-spark discharger at Brant Rock.

To embark on this project, the laboratory did tabletop measurements of the electrical characteristics of the devices and tested their communications qualities under field conditions. It divided the task so that NBS and the Brant Rock station could perform the measurements and field tests simultaneously. During late summer and autumn 1909, lab technicians conducted preliminary indoor measurements on the wireless sets at Brant Rock. Meanwhile, several long-range field tests took place between Brant Rock and two U.S. Navy cruisers, the USS *Birmingham* and *Salem*, which the navy dispatched into the Atlantic. During these voyages, experimenters tried different arrangements of the wireless sets on board. The long-range field tests continued in winter 1909 and spring and summer 1910. Short-range tests between NBS in Washington and Brant Rock followed in late 1910.[11]

Invaluable results came from field tests in July 1910, out of which Austin synthesized an empirical formula that governed the relationship between radiation intensity and a number of physical parameters, including antenna height, distance, and wavelength. Yet the tests were aiming merely to assess the signal strength at various distances of a *particular* wireless set—the Fessenden rotary-spark transmitter—rather than producing an empirical law for propagation of electromagnetic waves. Other engineers had tried to establish a quantitative relationship between incoming antenna current and transmitter-receiver distance. In 1904, William Duddell and his assistant J. E. Taylor tested wireless telegraphs for the British Post Office at Bushy Park, Teddington, to measure intensity of incoming current to determine its dependence on the arrangements of transmitting and receiving apparatuses and

10. Ibid., 88.

11. For a description of the facility's tests and measurements in 1909 and 1910, see Austin, "Naval Radio-Telegraphic Laboratory" (1912), 125, 147–53, and "Some quantitative experiments" (1911), 315–63.

their distance. Within about 6,000 feet (2.7 km), the numerical product of measured current and distance was approximately a constant—a finding consistent with Hertz's inverse law. Later the Britons tried the experiment between a land station and the steamship *Monarch* in the Irish Sea and obtained the same relationship. From 1902 to 1904 Lieutenant Camille Tissot of the French navy performed similar tests and found the same inverse relationship for distances up to 40 km.[12]

Nevertheless, no one before the U.S. Navy in 1910 had worked on distances as far as 1,000 miles. The naval project began with preliminary short-range tests in early July, with both the *Birmingham* and the *Salem* in harbor at Provincetown, Massachusetts. The *Birmingham* set sail on 14 July, traveling south until it was 1,200 miles from Brant Rock. The *Salem* left the next day and sailed to a point 450 miles from Brant Rock (figure 3.1). Throughout their voyages, wireless sets in Brant Rock and on the two vessels were regularly transmitting signals to and receiving signals from one another. The tests began on 14 July and ended on 22 July. Every day the three parties exchanged messages two to four times, with one or two of these at night. Technicians working for the Naval Wireless Telegraphic Laboratory and NESCO's engineers executed the entire process, including instrument calibration and maintenance, on-site measurements, and data analysis.[13]

Austin's report documented the instruments' performance. The transmitters were Fessenden synchronous rotary-spark dischargers, with 100 kw of power at the Brant Rock station and 2 kw each on the cruisers. He had chosen two wavelengths—1,000 m (300 kilohertz, or kHz, denoting 300,000 cycles per second) and 3,750 m (80 kHz). Sensitive detectors were crucial, and the lab chose a crystal rectifier with a galvanometer or a shunted-telephone circuit with an electrolytic detector called a "barretter" (figure 3.2).[14]

12. For the British experiments, see Duddell and Taylor, "Wireless telegraphy measurements" (1905), 258–61, 299–302, 349–51. Austin mentioned their test in the Irish Sea in "Some quantitative experiments" (1911), 315. For the French experiment, see Tissot, "Note on the use of the bolometer" (1906), 848–49.

13. Austin, "Some quantitative experiments" (1911), 320–30.

14. For a description of the instrumental conditions, see ibid., 315–18. The circuit worked in the following way. In figure 3.2, the antenna current at the top of the aerial could be directed along either path A connecting the rectifier-galvanometer device or path B connecting the shunted-telephone device. The symbol D on the left-hand side represented a crystal rectifier. The alternating antenna current in path A was coupled to the circuit on the left through an inductor and a variable capacitor. The peaks of the alternating current were captured by the rectifier D and the capacitor across the galvanometer G. Thus the deflection of G indicated the

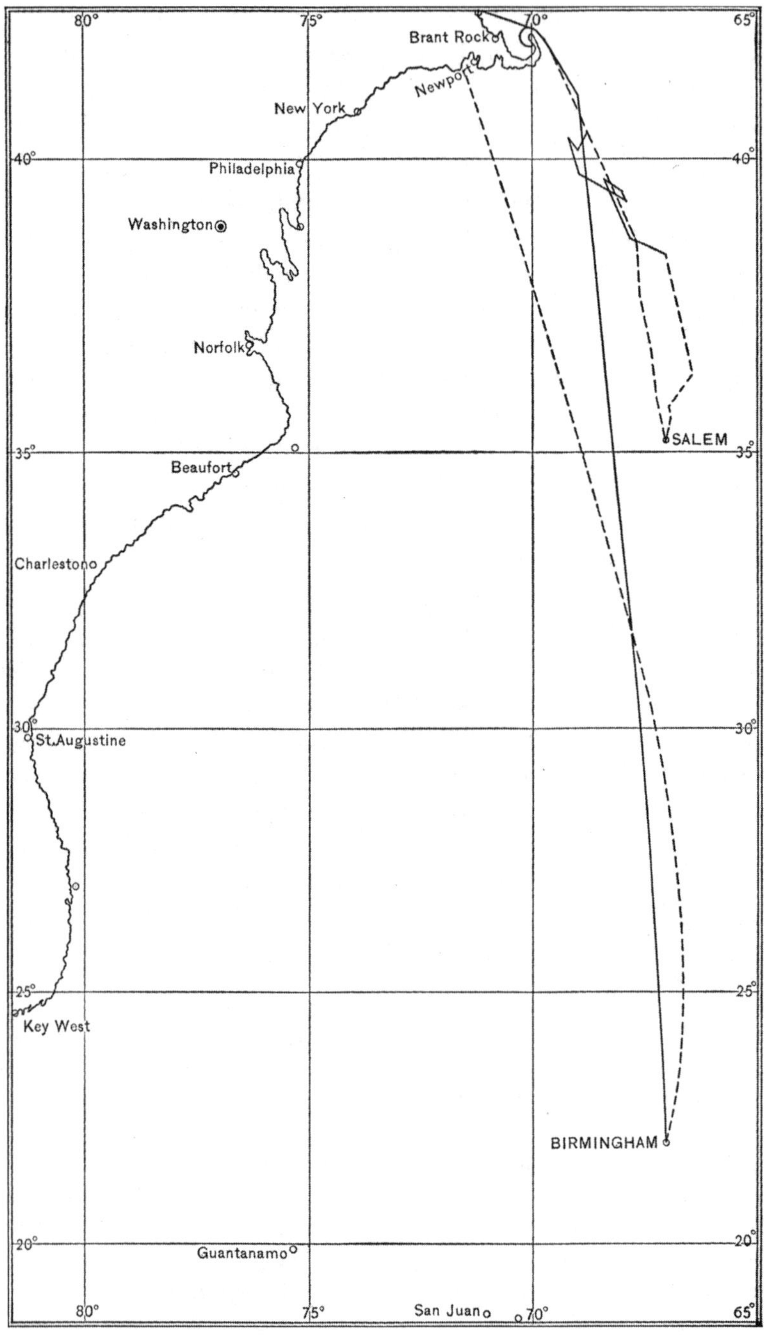


FIGURE 3.1. Routes of the *Salem* and the *Birmingham* in July 1910. Austin, "Some quantitative experiments" (1911), figure 2.

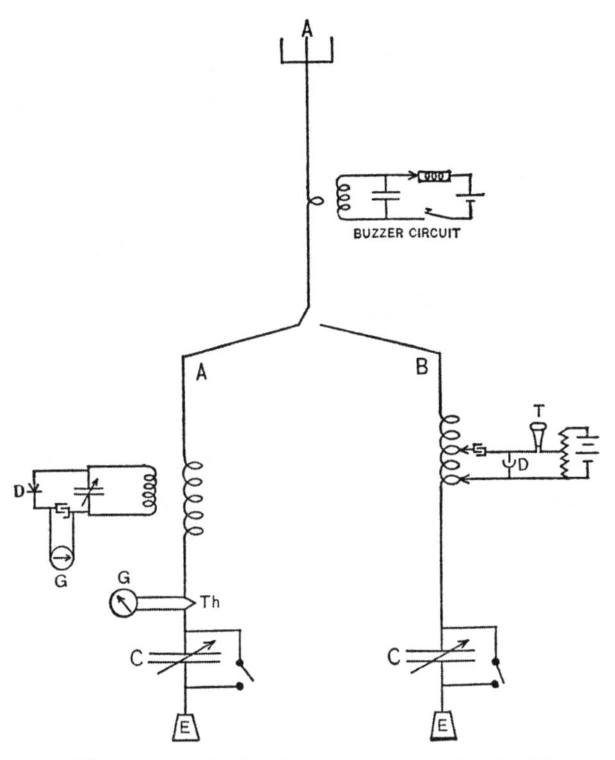


FIGURE 3.2. The detector in Austin's experiments. Austin, "Some quantitative experiments" (1911), figure 1.

The nine-day test produced five tables of signals: *Birmingham*–Brant Rock, *Birmingham–Salem* (wavelength 1,000 meters), Brant Rock–*Birmingham*, *Salem–Birmingham*, and *Salem*–Brant Rock (wavelengths 1,000 and 3,750 meters). During the voyages receiving technicians noticed an unusual phenomenon: signals at night were significantly more erratic than those during the day. The night level was usually stronger but fluctuated more and experienced greater disturbances.

## THE FORMULA

Data analysis by Austin and his colleagues produced the most influential result of the field tests. They straightaway scoured the data for a simple mathematical

amplitude of the alternating current. D on the right side was a NESCO barretter, including a cathode plate and an anode needle dipped into an electrolytic liquid. Normally the two oxide-covered electrodes remained insulated from each other.

formula that could serve as a useful approximation for the measured values. This required theoretical considerations for several reasons. First, the nighttime measurements fluctuated so much that Austin decided to use daytime data only, which he tried to fit by using the Duddell-Taylor-Tissot formula, according to which the receiving antenna's current is inversely proportional to distance from the transmitter. This inverse law did nicely fit the data up to 200 miles, but it also gave values that were much too high at greater distances. Austin argued that the discrepancy occurred because the atmosphere absorbed additional energy. To accommodate the effect, he assumed (in conformity with simple absorption laws elsewhere in physics) that the atmosphere produced an exponential decay of the form $\exp(-Ad)/d$, where $d$ denoted distance and $A$ the decay coefficient. This worked reasonably well.

Louis Cohen, a NESCO engineer assisting Austin's experiment, found that a fixed $A$ could approximately match data from different stations with a given wavelength. There was a clear difference in the decay coefficient at different wavelengths, but one could accommodate this by making the coefficient proportional to the reciprocal of the square root of wavelength, namely $A = \Sigma/\lambda^{1/2}$. Thus the antenna current was proportional to $\exp(-\Sigma d/\lambda^{1/2})/d$. Austin and Cohen found that $\Sigma$ would equal 0.0015 when one expressed distance $d$ and wavelength $\lambda$ in kilometers.[15]

---

When an electromagnetic impulse hit the electrodes, it induced an electrolytic reaction that made the liquid between the cathode and the anode conductive. Because the barretter's resistance varied with the intensity of the excitation, it could serve as a detector. The antenna current through path B was coupled into the circuit on the right through a variable inductor. In this circuit, a battery was connected with a variable resistor partially shunted with an earphone T and the barretter D. This arrangement indirectly measured the antenna current by measuring the barretter's resistance, which fell under electromagnetic excitation. Then a part of the current from the battery passed through the path of T and D. So the operator heard a sound from the earphone.

In order to measure D's resistance, the operator adjusted the variable resistor to render the sound barely recognizable. At that sound level the current on the T path remained a constant, and thus the barretter had a simple and fixed relationship with the resistance of the variable resistor. So the measure of the barretter's resistance came from the resistance reading of the variable resistor.

The two methods had different advantages. The crystal rectifier was a stable sensor, though not as sensitive as the electrolytic barretter. The barretter was extremely sensitive, but the shunted-telephone method was not stable, because it relied on the operator's ears to recognize the "barely audible" level, which ambient noise from shipboard sounds could seriously disturb.

15. Ibid., 326–27.

Measuring a well-defined physical quantity (antenna current) and representing the data through a simple mathematical formula were not the only possible choices for an engineering experiment. Austin and his colleagues could have measured a quantity that had less to do with the physical characteristics of transmitters and receivers than with the efficiency of information exchange, such as the percentage ratio of correctly received binary messages for a given signal rate, like what many wireless practitioners chose to do.[16] They could also have tabulated numbers or plotted diagrams, as many engineers and inventors did,[17] instead of synthesizing a formula. Their manner of presenting their results reflected both the tradition of German experimental physics in which Austin had trained and NBS's scientific culture. Their choice, in other words, did not derive solely from desiderata vis-à-vis equipment testing, but it helped later to convert Austin and Cohen's empirical results into a scientific law.

In addition to the effect of distance, the laboratory also probed four other physical parameters: transmitting-antenna current, height of transmitting antenna, height of receiving antenna, and wavelength. To do so, Austin and his colleagues performed control-variable experiments by varying one physical parameter while holding fixed all the others in order to tease out simple dependencies in the resulting data. These experiments took place after the sea voyage in mid-July 1910 and at comparatively short ranges. Examinations of transmitted and received antenna currents took place between Brant Rock and a wireless station at NBS in Washington, DC, 380 miles away. The results indicated that the receiving antenna's current was proportional to the products of transmitting- and receiving-antenna heights, the transmitting antenna's current, and the reciprocal of wavelength (in addition to the exponential decay factor). Combining these relations with the previous long-distance transmission formula, Austin produced what rapidly became known as the Austin-Cohen formula:[18]

$$(3.1) \quad I_r = \mathbf{4.25}\frac{I_s h_1 h_2}{d\lambda}\exp\left[-\frac{\alpha d}{\sqrt{\lambda}}\right],$$

where $I_r$ was received current through an antenna with an impedance of 25 ohm, $I_s$ was the transmitting antenna's current, $h_1$ and $h_2$ were the heights

16. For example, see Jackson, "On the phenomena" (1902), 254–72.

17. For example, see Marconi, "On methods" (1906), 413–21.

18. Austin, "Some quantitative experiments" (1911), 330–39.

of transmitting and receiving antennae, respectively, $\lambda$ was wavelength, $d$ was distance, and $\alpha = 0.0015$. All lengths were in kilometers and current in ampères.

Austin and Cohen did not at first intend the formula to be a general law under a broad range of circumstances. According to Austin, it was rather "an equation which will cover the normal day received current over salt water through 25 ohms for two stations with flat-top antennas of any height, with any value of sending current and any wave length, provided the sending station is so coupled as to give but one wave length."[19] By the time of the experiments, they had tested it with only one type of transmitter—the Fessenden rotary-spark discharger.

How well did measurement support the formula? Austin plotted the data, the formula $I_r \propto \exp(-\alpha d/\lambda^{1/2})/d$, and the inverse-distance law, in which $I_r \propto 1/d$ (figure 3.3 and figure 3.4 are examples). The comparison indicated that the exponential relation in the formula fitted the measured data significantly better than did the Duddell-Taylor-Tissot inverse law. At distances comparable to a thousand miles, the inverse law was four to twenty times higher than the measured data. The Austin-Cohen formula, in considerable contrast, was within 10 percent of measurement for distances less than 300 miles and for the most part within 25 percent for distances below 1,000 miles. In addition, Austin compared his formula with the results from other experiments that took place in 1909 and 1910.[20] Some agreed well, but experiments before July 1910 did not. Austin attributed this discrepancy to atmospheric disturbances and human errors deriving from the shunted-telephone method when the ship was rocking heavily.[21]

19. Ibid., 340–41.

20. The experiments included the voyages of the *Birmingham* and the *Salem* in December 1909, March 1910, and May 1910 and a test between two torpedo boats, the *Stringham* and the *Bailey*, in Chesapeake Bay during November 1910, which involved a short wavelength (300 meters) and lower antenna heights (about 40 feet).

21. The shunted-telephone method relied on operators to decide the level of bare audibility. In a noisy environment (e.g., on a rolling ship), measurement error could be significant. For instance, when conducting a long-distance radio experiment in March 1913, the commanding officer of the *Salem*, E. T. Pollock, reported to the secretary of the navy that the rolling of the ship due to severe weather posed serious problems for signal detection. E. T. Pollock to the Secretary of the Navy (letter), 5 March 1913, Bureau of Ships Records (RG 19), E 988, 841(26), box 1926. In addition, because the barretter contained an electrolytic solution, heavily rolling seas sometimes caused the liquid to splash around, and the detectors would then fail.

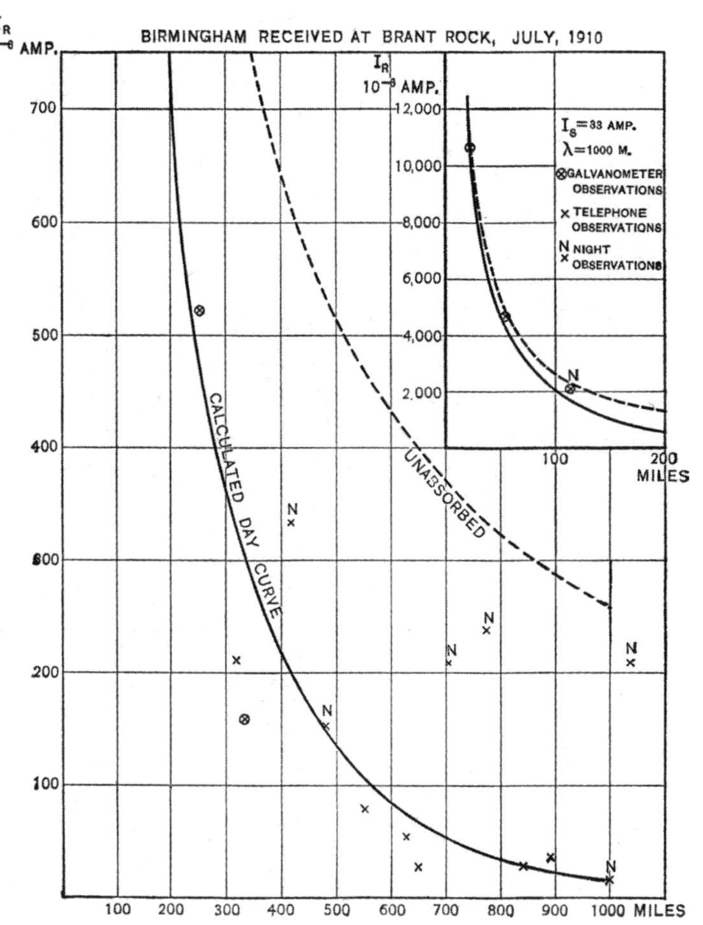


FIGURE 3.3. An example of Austin's experimental results in July 1910. The signal traveled from the *Birmingham* to Brant Rock. The wavelength was 1000 m. The vertical axis represents incoming-antenna current, and the horizontal axis distance. The solid curve was calculated from the Austin-Cohen formula, and the dashed curve from the inverse law. The points with the legend "N" represent readings at night. Austin, "Some quantitative experiments" (1911), figure 3.

The Austin-Cohen formula's initial status as an approximate representation of a specific transmitter's performance did not restrict its later use. Austin published his data and conclusions, making them available not only to the navy but also to engineers and physicists in general. The empirical formula soon reached wave-propagation theorists through this publication and

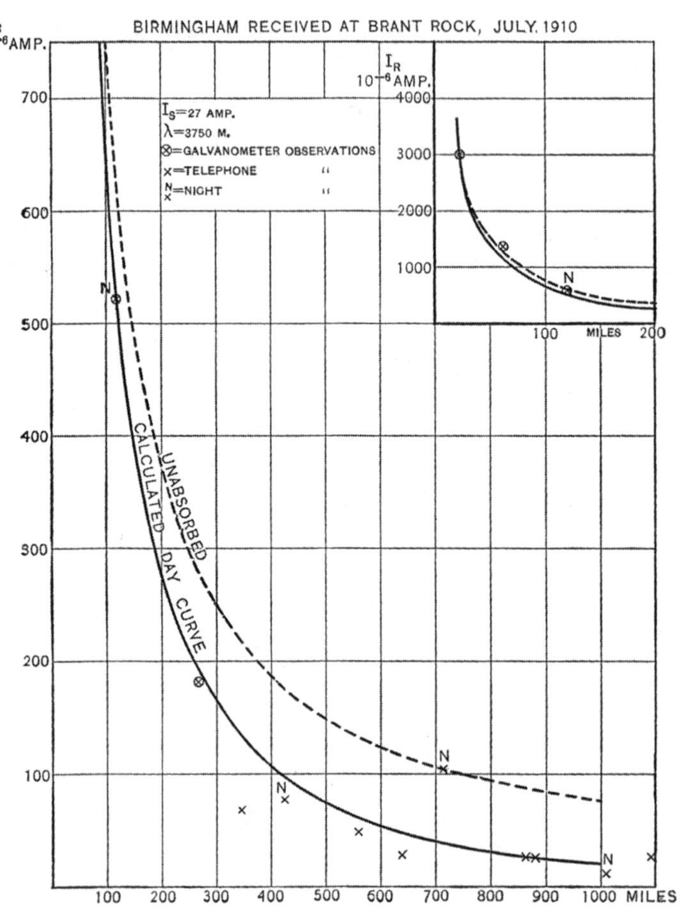


FIGURE 3.4. Another example of Austin's experimental results in July 1910. The wavelength was 3750 meters, and the other parameters were as in figure 3.3. Austin, "Some quantitative experiments" (1911), figure 4.

probably via Austin's personal connections with Zenneck and Sommerfeld.[22] In direct consequence, the formula became the most important, and perhaps the only quantitative, empirical basis for all theories of long-distance

22. Braun taught at the University of Strasbourg when Austin attended it. Zenneck was Braun's student, and Sommerfeld was familiar with Braun and Zenneck. Austin's German connection developed during these years. In the 1900s and 1910s, Austin occasionally wrote postcards and letters to Zenneck and Sommerfeld, some of which are in the Jonathan Zenneck Papers (NL053) and the Arnold Sommerfeld Papers (NL056) in the Deutsches Museum Archives in Munich.

radio-wave propagation throughout the 1910s. Before 1911, scientists worried whether their theories were consistent with physical intuition or wireless practitioners' general qualitative knowledge. After 1911, they worried whether their predicted numerical results fitted Austin's quantitative data or the Austin-Cohen formula. That altered the epistemic situation significantly.

CHAPTER 4

# Synthesis with Atmospheric Reflection

If the success of Marconi's transatlantic test opened up the possibility of propagating radio waves over the long range, the Austin-Cohen formula offered substantial quantitative evidence for the phenomenon. With the formula, mathematical physicists could now check the validity of various surface-diffraction theories and perhaps resolve the debate over them. This was not to be the case, however. Introduction of the formula did not favor one theory over others, for the empirical regularity agreed with none of them. The source of discrepancy lay in the form of wavelength $\lambda$ (or, equivalently, frequency, as frequency was supposedly propagation speed over wavelength) dependence at the exponential decay factor: while Austin-Cohen gave an exponential decay proportional to $\lambda^{-1/2}$, all versions of surface diffraction yielded a factor of $\lambda^{-1/3}$.

Surface-diffraction theorists responded to this situation in various ways. For Sommerfeld's protégé Witold von Rybczyński, the problem lay in Austin and Cohen's incorrect synthesis of measured results, not in the theory. For English physicist A. E. H. Love, the problem consisted of the "device law" the experimenters used to translate detector output into antenna current. Both men denied that the surface-diffraction theory had failed and instead questioned the efficacy of the empirical formula. But such contentions did not hold long, as Austin managed to reaffirm his formula from more propagation measurements that he performed for the U.S. Navy when it tested wireless equipment in 1913. During the 1910s, most practitioners of wireless technology and many mathematical physicists (such as Sommerfeld) ceased to believe in surface diffraction as a valid explanation for long-range propagation. The discrepancy between versions of the theory still existed, and it needed resolution. But to the majority of researchers, this was an issue of mathematical consistency regarding analytical approximations; such an issue lacked empirical significance.

As surface diffraction was losing credibility, another theory entered the scene. The "atmospheric reflection" theory stated that radio waves overcame the earth's curvature to reach long distances by bouncing back and forth between the earth and a reflective layer—a huge conducting shell—in the upper atmosphere. The theory contrasted starkly with surface diffraction in several important aspects. Maverick English Maxwellian Oliver Heaviside and Harvard engineering professor Arthur Kennelly first proposed it in 1902, and it gained wider support and attention among wireless explorers, electrical engineers, and experimental physicists (later "radio scientists" and "radio engineers"), rather than among mathematical physicists. Furthermore, it began as an explanatory model for a variety of wireless phenomena. Unlike surface diffraction, which dealt only with formal representation of long-distance propagation along the earth's curvature, atmospheric reflection accounted for diurnal and seasonal variations of signal transmission and the existence of static noise—issues that caught the attention of the engineering and experimenter communities, as they experienced such effects quite often in their daily practice.

The broader epistemic net of atmospheric reflection, however, came with a price. An immediate problem concerned the need to account for the new entity. Whereas the boundary condition of surface diffraction (the earth) was self-evident, atmospheric reflection contained an extra component: the wave-reflecting upper layer of the atmosphere. Why is there such a layer? What is its constitution? And why can it deflect radio waves? By proposing a novel object to explain some propagation effects, theorists opened a Pandora's box that raised more questions about the object's nature. In 1914, British wireless experimenter William Eccles made the first move to tackle these issues as he developed an argument for the bending of radio waves in the upper atmosphere. According to him, sunlight induced ions and electrons in the upper layer (which Kennelly and others had already surmised), and the resulting change in refractive index forced radio waves downward.

Eccles's model was the earliest quantitative formal theory of atmospheric reflection. Yet until about 1920, the theory remained a qualitative hypothesis for explaining phenomena. It never reached the degree of quantification and mathematization of surface diffraction. Even Eccles gave only the formulation of the refractive index, not the intensity of the propagating wave. While surface diffraction produced quantitative predictions inconsistent with the Austin-Cohen formula, atmospheric reflection could not generate any quantitative predictions comparable to Austin-Cohen.

Finally another Cambridge wrangler, George Watson, synthesized surface diffraction and atmospheric reflection. In 1918, he attacked the longstanding mathematical puzzle of surface diffraction regarding the correct analytical approximation of the solution. He developed a new technique for manipulating complex integrals to alleviate the problem of convergence in most of the previous diffraction theories, which helped settle the debate over the theories' logical consistency. More important, however, Watson's novel approach allowed him to solve the problems of electromagnetic-wave propagation with boundary conditions other than those the surface-diffraction theory had stipulated. Kennelly and Heaviside's model of atmospheric reflection—with a conducting spherical earth and a concentric conducting layer—constituted another boundary condition whose wave-propagation behavior was tractable in the light of the new mathematical technique.

Watson did just that. In 1919, he extended his approach to treat the atmospheric-reflection boundary condition. Without surprise, he found that the solution exhibited an exponential decay factor proportional to $\lambda^{-1/2}$, a result that Austin-Cohen predicted. By integrating the mathematical prowess of surface-diffraction theorists and the physical insight of atmospheric-reflection theorists, therefore, Watson was able to fit the propagation theory with the empirical law for the first time.

## SURFACE DIFFRACTION AFTER AUSTIN-COHEN

The surface-diffraction theorists in Munich and Cambridge had difficult days after 1911. On the one hand, they had not fully bridged the conceptual distinction between geometry-directed deflection and the surface wave, and analytical approximations still yielded mutually exclusive predictions whose sources of discrepancy were hard to pin down. On the other hand, none of these predictions demonstrated the frequency dependence of the Austin-Cohen formula. The 1910s marked efforts by British and German mathematical physicists to overcome such difficulties. On the other hand, the opportunity for fitting data into their theoretical shoes decreased as a new naval experiment strengthened the credibility of the empirical formula.

## *The German Story*

In the early 1910s, the Munich diffraction theorists broadened the boundary condition for surface waves from a flat earth to a spherical one. In 1909, Sommerfeld solved the electromagnetic field induced by a vertical Hertzian dipole on top of an infinite planar conductor. This result encouraged him to pursue with his students several other diffraction problems using the same analytical techniques. One of the major problems was the field generated by the same source on top of a large conducting sphere. He assigned this problem to his American doctoral student Hermann March.

Born in Ocheyedan, Iowa, Hermann William March studied physics at the University of Michigan and later at the University of Wisconsin. After teaching mathematics in Wisconsin for years, he went to Munich to pursue a PhD under Sommerfeld.[1] At his supervisor's request, he wrote a doctoral dissertation on radio-wave diffraction along the earth's surface in 1911 and published it in 1912.[2]

This thesis was a direct extension of Sommerfeld's work on surface waves in 1909. March expressed the dipole-generated field above a large conducting sphere (see figure 2.1)—the same boundary condition on which the Britons and Poincaré had focused—as the spatial derivative of a Hertzian potential. He also represented the general solution of the wave equation under this condition as an expansion in spherical harmonics. The British theorists had taken such an expansion as a *sum* over a *discrete* index. In contrast, March expanded the solution as an *integral* over a *continuous* index, as his adviser had done with flat-surface diffraction. Then he matched the integral with the spherical boundary condition to determine the functional form of the integrand, which turned out to be inversely proportional to the derivative of the Hankel functions with argument *ka* and order equaling the continuous index of the integral:

$$\Pi=-\frac{2}{ka^2}\int_0^\infty \alpha P_{\alpha-1/2}(\cos\theta)\frac{\varsigma_{\alpha-1/2}(ka)}{\varsigma'_{\alpha-1/2}(ka)}d\alpha \tag{4.1}$$

where $\varsigma_{\alpha-1/2}(ka)\equiv(\pi ka/2)^{1/2}H_\alpha^{(2)}(ka)$ and $H_\alpha^{(2)}(x)$ was a Hankel function of the second kind. To evaluate the integral, March approximated asymptotically

1. Stobbe, *Poggendorff's Handwörterbuch*, 6 (1931), 1643.

2. Hermann William March, "Über die Ausbreitung" (1912), 29–50; March submitted the manuscript on 21 October 1911.

the Hankel function for large *ka* in the same way as Macdonald had done. By so doing, he could calculate the integral analytically. In March's final result, the Hertzian potential was proportional to $\exp[-ika\theta]/(\theta\sin\theta)^{1/2}$ for large *ka* ($\theta$ was the angle of transmitter-receiver separation with respect to the earth's center). March's solution did not have any exponential decay with respect to $\theta$ (cf., $\exp[-\beta(ka)^{1/3}\theta]$); the exponential $\exp[-ika\theta]$ was a sinusoidal function of $\theta$.

Although March's integral approach was fresh, his result had the defect of decaying much more slowly than both the British theorists had predicted and the wireless practitioners had observed—his field intensity decreased only with $(\theta\sin\theta)^{1/2}$, not exponentially with $\theta$. This was the outcome of the same mathematical problem that made Macdonald's original approach problematic: when integrating the functions containing the derivatives of Hankel functions for large *ka*, one cannot perform the asymptotic expansion in the usual way for Hankel functions with orders comparable to *ka*.

This question concerned Poincaré when he read March's paper. He pointed out the possible source of error to Sommerfeld[3] and in March 1912 in a note to *Comptes rendus*, the journal of France's Acadèmie des Sciences,[4] which also commented that March's predictions considerably disagreed with Austin's recent experimental data.

Sommerfeld put another student, Witold von Rybczyński from Poland, on the problem that March left. In March 1913, Rybczyński published his results for improving the approximation to March's integral in equation (4.1).[5] Following Poincaré, he took the dominant contribution of the integral from the pole of the factor $\zeta_{\alpha-1/2}(ka)/\zeta'_{\alpha-1/2}(ka)$ in the integrand with the smallest imaginary part. He used Poincaré's calculation to obtain the numerical value of this dominant pole. Then he replaced March's asymptotic approximation of $\zeta_{\alpha-1/2}(ka)/\zeta'_{\alpha-1/2}(ka)$ with another function with a similar form. But the value of Rybczyński's function at the projected point of the dominant pole on the real axis equaled that of Poincaré's integrand at the same point. By doing this ad hoc manipulation, Rybczyński retained the virtues of both March and Poincaré's approximations: the new integrand was more precise than March's in the region that gave the dominant contribution, and much easier to inte-

3. Poincaré to Sommerfeld (letter), 1 Jan 1912, Handwriting Collection (HS 1977–28), A266, Deutsches Museum Archives, also available at http://sommerfeld.userweb.mwn.de/KurzFass/01381.html (last accessed on 3 December 2012).

4. Poincaré, "Sur la diffraction des ondes Hertziennes" (1912), 795–97.

5. Von Rybczyński, "Über die Ausbreitung" (1913), 191–208.

grate analytically than Poincaré's. The new approximation led Rybczyński to obtain a field intensity proportional to $\exp[-0.33(ka)^{1/3}\theta]/(\theta\sin\theta)^{1/2}$.

Rybczyński showed more persuasively than March that the diffraction theory could produce an exponentially damped wave along a large spherical conductor. But his factor $\exp[-0.33(ka)^{1/3}\theta]$ differed from Nicholson's factor $\exp[-0.7(ka)^{1/3}\theta]$, and both disagreed with Austin-Cohen in their wavelength dependence. The diffraction theories stuck with decay rates inversely proportional to $\lambda^{-1/3}$, whereas the empirical regularity required $\lambda^{-1/2}$. Numerically, Nicholson's formula decayed much faster than Rybczyński's, yet Austin-Cohen decayed much slower than Rybczyński's. Which one was correct? Rybczyński justified his predictions by comparing the three formulae with a selected set of measured data from Austin's paper in 1911—seven daytime and four nighttime data points within a range between 400 and 1,000 miles for a wavelength of 3,750 meters (these data came from the voyage in 1909 instead of that in 1910). For these cases, Rybczyński's predictions were closest to the experimental data. Nicholson's predictions were too low, the Austin-Cohen formula's too high (figure 4.1).

Rybczyński's paper in 1913 was the diffraction theorists' first formal response to the quest for evidential support. It aimed to develop a surface-diffraction theory *both* free from mathematical inconsistencies *and* with support from experimental data. He almost succeeded. His theory met with approval from some researchers, especially the Germans. Even though it was not a complete physical picture of long-distance radio-wave propagation, it characterized wave propagation along the earth approximately correctly when the atmospheric effects were absent. Sommerfeld wrote to Wilhelm Wien in November 1913:[6]

> Poincaré gave a lecture in Göttingen on some mistake about the wireless telegraphy that he corrected in *Rendiconti di Palermo*. Here he is correct in having the exponential damping, and that is in agreement with Nicholson. For the comparison with experience, it depends on the numerical scale, which is difficult. Rybczyński found (following March's method, which actually had an incorrect result; in other words, the absence of exponential damping) a smaller damping factor [rate] than Nicholson did. (Poincaré did not give a numerical value.) This smaller damping factor

6. Sommerfeld to Wilhelm Wien (letter) (my translation), 29 Nov. 1913, Arnold Sommerfeld Papers (NL 056), 010, also available at http://sommerfeld.userweb.mwn.de/KurzFass/00079.html (last accessed on 3 December 2012).

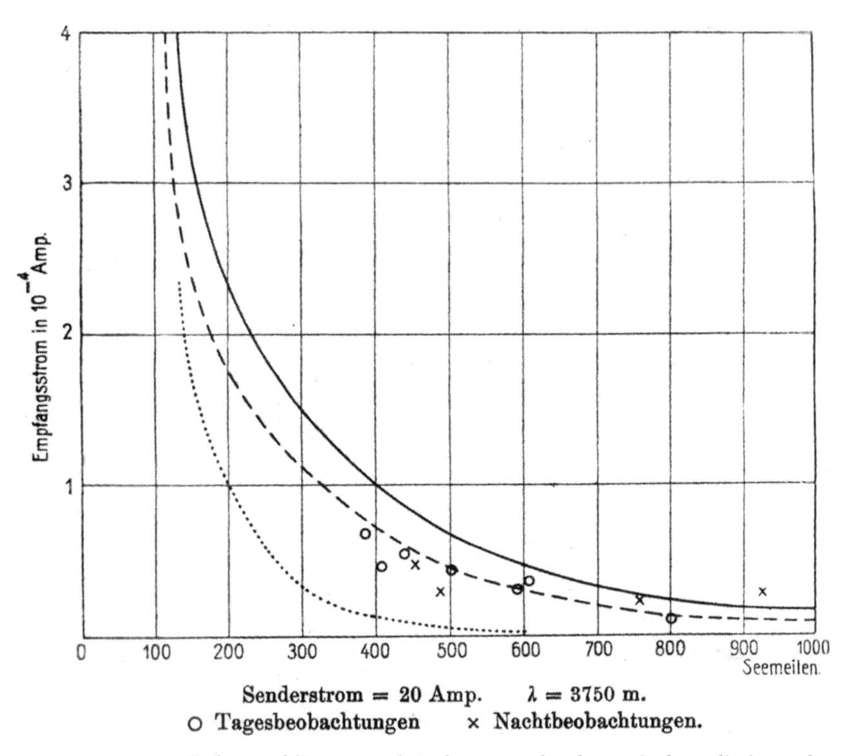


FIGURE 4.1. Rybczyński's comparison between the theoretical predictions, the empirical formula, and the experimental data. The vertical axis is incoming antenna current in microamperes, the horizontal axis is distance in miles. The wavelength is 3,750 meters. Legend: circle: daytime data; cross: nighttime data; solid curve: the Austin-Cohen formula; dashed curve: Rybczyński's formula; dotted curve: Nicholson's formula. Rybczyński ,"Über die Ausbreitung der Wellen" (1913).

> is in good agreement with Austin's observations. But I am not misled [to conclude] that Rybczyński's damping factor agrees with the reality for all long distances. One must make the case possible that the reflection from the well-conducting upper layers is responsible for the [wireless] telegraphy at quite long distances. To sum up, Poincaré (following an attempt that failed at first) discovered the truth in principle; his negative conclusion that the electrodynamic telegraphy at long distances is impossible might not be maintained, and it depends on the numerical details of the questions.

Thus Sommerfeld believed that Rybczyński's theory was analytically right and gave a rate of attenuation that somewhat matched experimental data. But, he continued, reflection from conducting layers in the upper atmosphere (we see more about this below) was probably crucial for very long-distance propagation.

Nevertheless, Rybczyński's work did not fare as well elsewhere. His ad hoc procedure of approximation lacked mathematical rigor. More seriously, he did not select the most reliable data to compare. He obtained his seven daytime data points from the voyage of 1909, when the instruments were still under adjustment and the weather severe. And four of his data points were from night, when wireless signals were notoriously unstable. In a paper of 1914, Austin cited these problems and suggested that Austin-Cohen was still a valid empirical law, citing evidence from a new experiment by the U.S. Navy.[7]

## *A New Experiment*

The U.S. Navy reorganized in the early 1910s. It dissolved the Bureau of Equipment in 1910, transferred the Wireless Telegraphic Laboratory to the Bureau of Steam Engineering, and in 1912 renamed the wireless facility the Naval Radio-Telegraphic Laboratory.[8] In conjunction with reorganization, the navy kept updating its communications technology, constructing its first transoceanic radiotelegraphic station in Arlington.

The results of the test voyages of the *Birmingham* and the *Salem* in 1909 and 1910 did not fulfill the requirements of the contract (the maximum distance under test was 1,000 miles rather than 3,000, and atmospheric disturbances often interrupted transmission). But wireless communications using the Fessenden rotary-spark discharger proceeded apace when the navy decided to continue use of the NESCO apparatus. The navy started construction on the buildings and the antenna tower at Fort Myer in Arlington, Virginia, in 1911 and completed work in December 1912. It then installed the transmission equipment, including the main antenna proper (350 feet high) and the 100 kw synchronous rotary-spark transmitter, and commissioned the station on 13 February 1913.[9]

Prior to completion, the Naval Radio-Telegraphic Laboratory was testing the transmission equipment, preparing for another long-range measuring voyage. Two days after the commissioning, the *Salem* set sail from the League Island Navy Yard in Philadelphia for Gibraltar, 3,970 miles from Washing-

7. Austin, "Quantitative experiments" (1914), 69–86.

8. Allison, *New Eye for the Navy* (1981), 13–14; Douglas, "Technological innovation" (1985), 120–30.

9. Howeth, *History of Communications-Electronics* (1963), 178–83.

ton, DC.[10] Austin's first assistant, George Clark, several technicians from the laboratory, and a few engineers from NESCO were onboard. The voyage planned the final test of the Arlington station before the navy incorporated the facility into its operations schedule. Perhaps as important, and certainly more critical for future naval procurement and operations, the voyage would compare the efficacy of the existing rotary-spark transmitter with that of an alternative transmission technology—the arc transmitter.

The Australian-born engineer Cyril F. Elwell had introduced the arc transmitter to the U.S. Navy in 1912. The invention of the Dane Valdemar Poulsen, the device produced a much less damped waveform than did the spark-gap.[11] In 1911, Elwell had formed the Federal Telegraph Company in California to develop arc machines; after his firm opened a long-distance circuit between San Francisco and Honolulu in 1912,[12] he tried to interest the navy. He had two supporters in the navy's Radio Division: Stanford Hooper and A. J. Hepburn. Through their efforts, the chief of the Bureau of Steam Engineering, H. I. Cone, agreed to let Federal Telegraph install a 30 kw arc transmitter at Arlington.[13] At this point the navy had not decided to adopt the device; it planned field tests before deciding between spark and arc. The operational test for Arlington permitted systematic measurements of the arc machine. Furthermore, because the cruiser on the voyage received signals from both the 100 kw spark transmitter and the 30 kw arc transmitter simultaneously, the navy could compare them directly at the same time and under the same conditions. The test voyage offered a "trial of strength" between the two competing technologies.

In the 1913 test, the naval experimenters faced a doubly novel situation. This time, unlike in 1910, two transmitters (spark and arc), not one, were in play. The detector technologies had evolved, too. In 1910, a rectifier-galvanometer and a shunted telephone with electrolytic barretter received electromagnetic waves. In 1913, the tests used an alternative detection technology, a heterodyne detector that Fessenden invented.[14] The new device was more sensitive at long distances than were the instruments of 1910.

10. Ibid., 183.

11. Hong, *Wireless* (2001), 119–55.

12. Aitken, *Continuous Wave* (1985), 122–40.

13. Ibid., 90–91.

14. A heterodyne detector mixed an incoming electromagnetic wave with a continuous sinusoidal wave from a local oscillator. The two signals combined nonlinearly in a device such as an earphone. The combination produced a "beat" whose frequency was the difference between the local oscillator's frequency and the frequency of the incoming signal. The opera-

Experimenters from the Naval Radio-Telegraphic Laboratory and NESCO began taking measurements onboard the *Salem* and at the Arlington station shortly after the cruiser left Philadelphia on 15 February 1913.[15] The round-trip voyage to Gibraltar took one and a half months.[16] Both Austin from the laboratory and John L. Hogan, a NESCO engineer, published the results.[17]

These reports had two major implications, one relevant to the navy, the other to wave-propagation theorists. The findings did tentatively close the trial of strength: the 30 kw Federal arc transmitter outperformed the 100 kw spark machine. According to Austin, both devices yielded comparable signal attenuations below 900 miles, but "at distance over 1000 miles the arc waves appear to begin to show advantage over the spark waves."[18] Austin added, "Messages were continuously received with both arc and spark in the daytime up to 2100 miles (3900 km). Occasional signals were heard much farther, the arc being heard on one day in the daytime even in the harbor of Gibraltar. Both arc and spark were heard at night at all times during the voyage."[19] The captain of the *Salem*, Commander E. T. Pollock, vividly illustrated the team's impression of the arc transmitter:[20]

> On the return trip, the N.E.S.Co. operator, an exceptionally keen worker for his company, received messages on the arc from Arlington in regard to the health of his child, which messages were not heard on the spark, and he

tional procedure of the heterodyne detector was similar to that of the shunted telephone—the operator tuned a variable resistor until the beat he heard from the earphone was barely audible and then recorded the resistance. The 1913 experiment used a heterodyne detector along with an electrolytic barretter.

15. Note that no representative from Federal Telegraph was on board, which reflected its minimal influence. In 1913, NESCO had established a good connection with the administrative officers in the navy, while Federal Telegraph was just starting to build that kind of relationship. Aitken remarked that when the Arlington station allowed Elwell to set up his arc transmitter, it forbade him to "put any nail holes in the floors, walls or ceiling." Aitken, *Continuous Wave* (1985), 90–93.

16. George H. Clark to the Chief of the Bureau of Steam Engineering (letter), 14 April 1915, Bureau of Ships Records (RG 19), E 988, 841(24), box 1926.

17. Austin, "Quantitative experiments (1914)," 69–86; and Hogan, "Quantitative results" (1913), 720–23.

18. Louis W. Austin to the Chief of the Bureau of Steam Engineering (letter), 3 Apr. 1913, Bureau of Ships Records (RG 19), E 988, 841(24), box 1926.

19. Austin, "Quantitative experiments" (1914), 82.

20. E. T. Pollock to the Secretary of the Navy (letter), 31 March 1913, Bureau of Ships Records (RG 19), E 988, 841(24), box 1926.

> stated rather profanely that he had to admit the superiority of the arc signals, especially when his own company's apparatus couldn't give him any news.

The spark machine consumed more power and occupied more space but performed more poorly than its arc competitor beyond 1,000 miles, and the navy decided to build its next few high-power stations with arc transmitters.

Among those doing research on radio-wave propagation, the Gibraltar test voyage offered a chance to verify the Austin-Cohen formula under new conditions. There were now two distinctly different transmitters. The wavelengths had changed from 1000 and 3750 m in 1910 to 2000 and 3800 m in 1913. And the Gibraltar voyage covered a much greater range; in 1910, the farthest reach was 1,000 miles, and in 1913, 3,970. Researchers obtained measured data for distances as great as 2,480 miles. If the data from these varying conditions were consistent with Austin-Cohen, then that formula's epistemic status would increase. Both Hogan's and Austin's reports suggested that the new daytime data did nicely match the formula. Hogan plotted the measured signals from the *Salem* (2000 m wavelength) up to 1,860 miles, the measured signals from Arlington (3,800 m wavelength) up to 2,640 miles, and the corresponding predictions from Austin-Cohen. There were several anomalies in the *Salem*-Arlington data, which Hogan waved away by attributing them to "a period of high absorption." But the signals from Arlington to the *Salem* were "in exceptionally close agreement" with the formula.[21]

Working on the same set of data, Austin went one step further. By the time he wrote his report, he was well aware of Rybczyński's challenge. With the new data in hand, Austin had ammunition to respond. Before delving into a detailed data analysis, he stated that in choosing the measured results of December 1909 instead of those from July 1910, Rybczyński had not picked the most accurate data.[22] He then plotted Rybczyński's formula, Austin-Cohen, and the 1913 data (figures 4.2 and 4.3).[23] Most data points were closer to the

21. Hogan, "Quantitative results" (1913), 720–21. Hogan also discussed nighttime data but found it difficult to make sense out of them. Strong static simply precluded accurate measurement. Hogan did attempt to fit the nighttime data with a series of exponential-absorption curves in order to calculate the absorption levels to which they corresponded, but without apparent success. Ibid., 721–22.

22. Austin stated: "Von Rybczyński in his article gives a curve in which the Brant Rock observed values are shown to agree more perfectly with the theoretical formula than with the empirical formula. Unfortunately, however, the observations chosen by him were among the earliest and least accurate in the series of experiments." Austin, "Quantitative experiments" (1914), 71–72.

23. Ibid., 77–79.

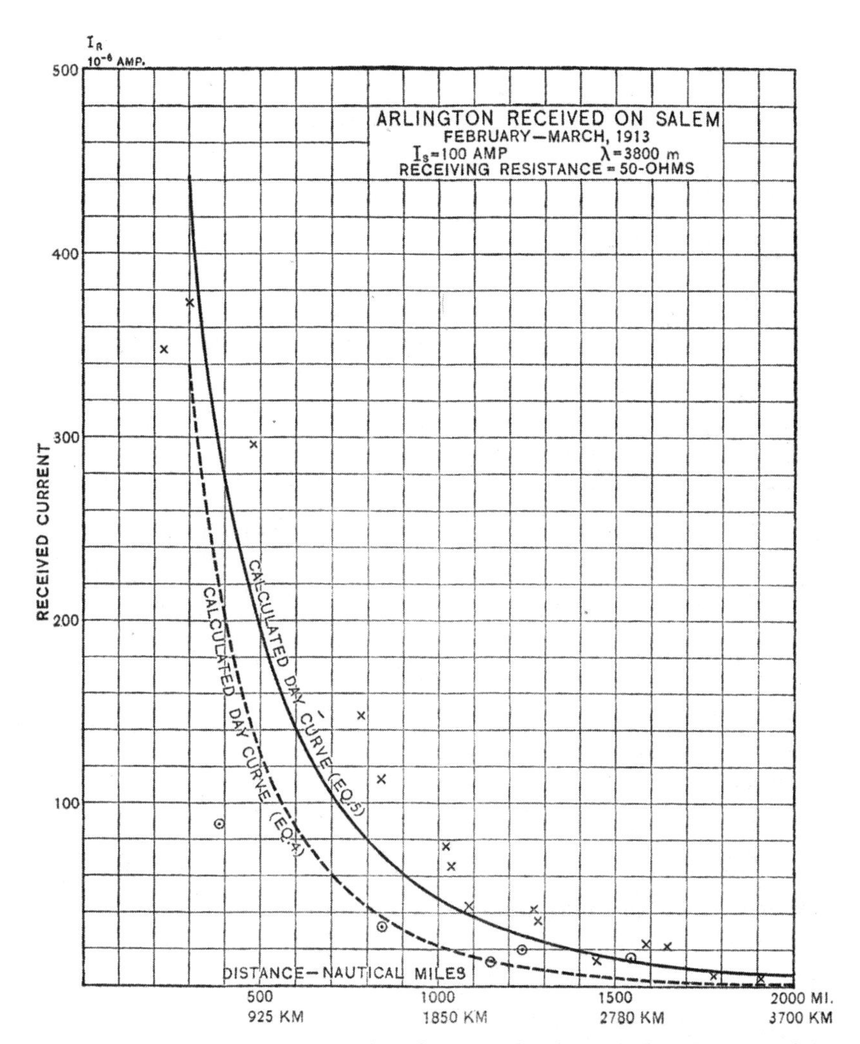


FIGURE 4.2. Austin's comparison between the theoretical and empirical formulae and the data from 1913: signals from Arlington to the *Salem*. Legend: solid curve "(EQ.5)": the Austin-Cohen formula; dashed curve "(EQ.4)": Rybczyński's formula; cross: data from outward trip to Gibraltar; circle: data from return trip. Austin, "Quantitative experiments" (1914), Figure 2.

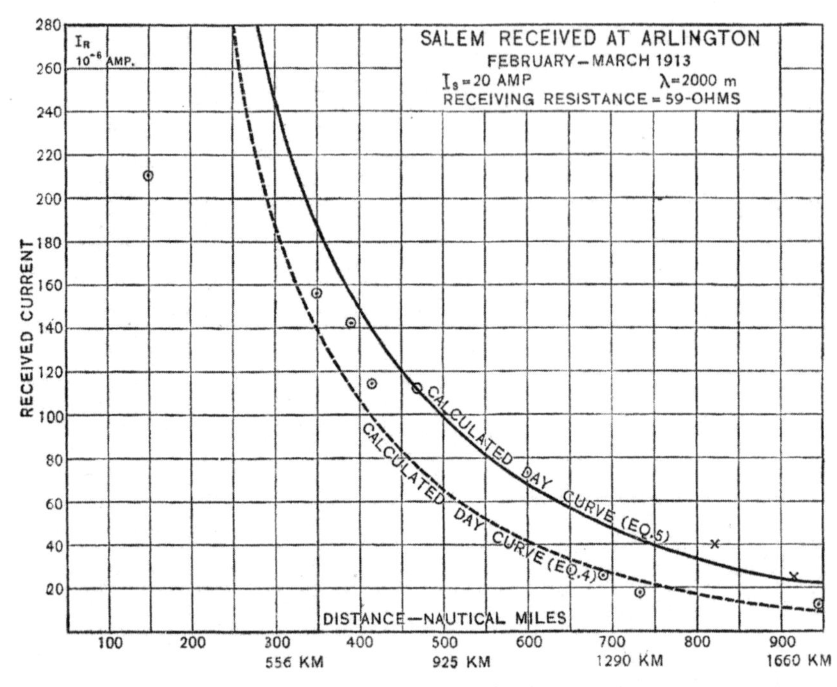


FIGURE 4.3. Austin's comparison between the theoretical and empirical formulae and the experimental data from 1913: signals from the *Salem* to Arlington (same legend). Austin, "Quantitative experiments" (1914), Figure 3.

predictions from Austin-Cohen than to Rybczyński's, particularly the signals from Arlington to the *Salem* (which was also consistent with Hogan's prior remarks). Most of the data points lay either above the Austin-Cohen curve or between it and Rybczyński's, but were in any case much closer to the former than to the latter. Austin concluded, therefore, "There can be no doubt from these results that [Rybczyński's formula] gives values too low to be reconciled with the observations, but that they are in very fair agreement with the [Austin-Cohen]."[24]

Both Hogan and Austin's experimental reports provided convincing further evidence for Austin-Cohen's generalizing power. Moreover, Austin's explicit comparison of the predictions from his and Rybczyński's formulae with the data provided more support for a formula deriving from an empirical basis than for one from the diffraction theory. Rybczyński's claim that his

24. Ibid., 78.

predictions were in good agreement with Austin's observations clearly did not hold.

Nevertheless, this did not destroy altogether the diffraction theory. Even the experimenters, supposedly adamant defenders of the empirical law, did not think that way. Austin wrote to Zenneck in 1916:[25]

> I am becoming quite convinced that the theoretical transmission formula, given in your book, represents approximately the weakest signals observed; while our Navy formula gives a fairly good average. Although I have taken a great many observations, I am still somewhat doubtful regarding the power to which the wave length should be raised in the exponential term. The observations are exceedingly discordant, apparently due to selective reflection.

The experimenter believed that the diffraction theory could predict the "minimum" signal strength when there was no chaotic reflection. Nor did he trust too much the $\lambda^{-1/2}$ dependence of the decay rate in the empirical formula, since "the observations are exceedingly discordant," because of "selective reflection." Meanwhile, the diffraction theorists did not give up, either. Regardless of the discrepancy between the theoretical predictions and the empirical law, the British mathematicians Macdonald and Augustus Love worked to resolve the theory's mathematical contradiction.

## *The British Response*

For Hector Munro Macdonald, life went on after he crossed swords in 1902–4 with Rayleigh on long-distance wave propagation. In 1905, he returned from Cambridge to become professor of mathematics at Aberdeen; two years later, he joined the university's administration. While switching to the "pure" mathematics of special functions and burying himself in administrative work, Macdonald did not forget the unnerving logical issues of the surface-diffraction theory. He continued to think of a solution to the approximation problem for the diffraction integral. In 1914, eleven years after his groundbreaking work, Macdonald proposed a new method to approximate his infinite series in equation (2.2).

What he did was to introduce a new series, much easier to sum up, that ap-

25. Austin to Zenneck (letter), 14 Sept. 1916, Jonathan Zenneck Papers (NL 053).

proximated the original series quite well when the spherical harmonic's order $n+½$ was close to $ka$. Macdonald argued that the new sum well approximated the old one because the dominant contribution to both series came from the same neighborhood around $ka$ (i.e., the terms in which $n$ was close to $ka$). Within this neighborhood, he replaced the Hankel function of order $n+½$ and its derivative by the Hankel functions of order 1/3 and 2/3 in the new series, using mathematical identities. One could interpret this sum, which contained the Hankel functions of orders 1/3 and 2/3 only, as a Riemann sum of an integral. Macdonald thus evaluated the integral using Cauchy's residue theorem and found that its value was the sum of the integrand's residues, which equalled the integrand's poles and, equivalently, the zeros of the Hankel function of order 2/3. When $ka$ was large, only the pole with the minimum imaginary part dominated. Macdonald found from calculating the value of the dominant pole that the resultant field intensity had an exponential decay in the form of $\exp[-\gamma(ka)^{1/3}\sin(\theta/2)]$.[26]

The functional form of Macdonald's exponential decay $\exp[-\gamma(ka)^{1/3}\sin(\theta/2)]$ differed from Nicholson and Rybczyński's factors, both having the form of $\exp[-\beta(ka)^{1/3}\theta]$. Whose diffraction theory was correct in terms of mathematical consistency or fitted with experimental data better? The Englishman A. E. H. Love tackled this question as he became familiar with Macdonald's work.

Augustus Edward Hugh Love was another Cambridge-trained mathematician. At St John's College, he became second wrangler in 1885. He obtained his master's degree in 1889 and left to take the Sedleian Chair of Natural Philosophy at Oxford. Love came to the problems of electromagnetic diffraction via his preoccupation—solid mechanics, particularly elasticity and geodynamics. A topic that had interested him was the propagation of elastic waves inside the earth and its applications to seismic phenomena. His major achievement in geodynamics came in 1911 as he identified a transverse, distortional wave (the Love wave) inside the earth. This surface disturbance propagated mainly along the boundaries between homogeneous layers and often caused more damage in an earthquake than the longitudinal, solid waves.[27]

The research on geodynamic waves required the same mathematical skills in real and complex analysis as acoustics and electromagnetic diffraction. Since the 1880s, Love had been corresponding with Rayleigh about the techniques

26. Macdonald, "The transmission of electric waves" (1914), 50–61.
27. Milne, "Love" (1939–41), 467–82.

for solving wave equations.[28] Moreover, the propagation of elastic waves along geological boundaries and the propagation of radio waves along the earth's surface resembled each other closely. In 1913, Cambridge physicist Joseph Larmor "lured" Love "into the quagmire of radiation theory." Within weeks, Love grasped the current diffraction theories and developed his own view.[29]

In a paper of 1915, Love gave a comprehensive overview of research on long-distance propagation of radio waves. He used a novel approach to compute the numerical values of the diffraction series in equation (2.2), compared his calculations with values from Macdonald's new formula, found them in agreement, and declared Macdonald's theory the best of the lot. And finally he reinterpreted the long-distance experimental data by correcting the asserted relation between the audibility factor (the results from the shunted-telephone measurements) and the antenna current.[30] This improved the fit of the data with his (and hence Macdonald's) predictions.

Love pointed out that there were three theories of wave diffraction along a large spherical conductor: Macdonald's, Nicholson's, and Rybczyński's. Using distinct schemes of analytical approximation, they predicted field intensity proportional to different factors:[31]

$$(4.2\text{a})\quad E(\theta)\propto\frac{1}{\lambda}\sqrt{\frac{\cos^2(\theta/2)}{\sin(\theta/2)}}\exp[-\mathbf{47.89}\lambda^{-1/3}\sin(\theta/\mathbf{2})]\quad\text{(Macdonald)}$$

$$(4.2\text{b})\quad E(\theta)\propto\frac{1}{\lambda}\sqrt{\sin\theta}\exp(-\mathbf{23.8}\lambda^{-1/3}\theta)\quad\text{(Nicholson)}$$

$$(4.2\text{c})\quad E(\theta)\propto\frac{1}{\lambda}\sqrt{\frac{1}{\theta\sin\theta}}\exp(-\mathbf{11.3}\lambda^{-1/3}\theta)\quad\text{(Rybczyński)}$$

$$(4.2\text{d})\quad E(\theta)\propto\frac{1}{\theta\lambda}\exp(-\mathbf{9.6}\lambda^{-1/2}\theta)\quad\text{(Austin-Cohen).}$$

28. Love to Rayleigh (letters), 1888–1915, Rayleigh Papers, Rare Book Collection, Air Force Research Laboratory, Hanscom Air Force Base, Lexington, MA.

29. Love to Larmor (letter), 1 May 1913, Joseph Larmor Papers, Special Collection, St John's College Library, Cambridge University.

30. Love, "Transmission of electric waves" (1915), 105–31.

31. Ibid., 125. In Love's paper, the coefficient in Nicholson's formula was $[\sin(\theta/2)]^{1/2}$. Yet Balthasar van der Pol showed that Love was incorrect and the coefficient should be $[\sin\theta]^{1/2}$. van der Pol, "On the energy transmission" (1918), 868.

In the above equations, $\lambda$ was in kilometers, $\theta$ was in radians, and the radius of the sphere was the earth's radius (6370 km). Love criticized March and Rybczyński's integral approach by noting that the spherical harmonics in their integrand—equation (4.1)—went to infinity, as their orders were not integers plus ½: when the order $\alpha$ was not an integer and the Legendre function $P_{\alpha-1/2}(\cos\theta)$ was finite at $\theta = 0$, it was infinite at $\theta = \pi$. Thus, expressing the diffracting field in terms of an integral over spherical harmonics was illegitimate, since it artificially introduced a singularity at the antipode $\theta = \pi$. (See figure 4.4 for a numerical comparison.)

Love developed his own method to calculate the diffraction series in equation (2.2). His approximation was numerical instead of analytical. Like Nicholson and Macdonald, he approximated the terms of the series in the neighborhood of $n+½ = ka$. But he did not approximate the series' dominant contribution with a closed form. Rather, he computed the numerical values of a "sufficient" number of terms, added them up, calculated this truncated finite series, and compared the numerical results with those he obtained from Macdonald's method at several separation angles. The predictions from these two different methods agreed well. So Love confidently declared that his results *confirmed* Macdonald's.[32]

Finally, Love discussed the discrepancy between theoretical predictions and experimental data. The Austin-Cohen formula made the field intensity proportional to $(1/\theta)\exp(-9.6\lambda^{-1/2}\theta)$, which was quite different from all the diffraction formulae, including Macdonald's. Love saved the phenomena by reinterpreting the empirical results. He compared data from John Hogan's report in 1913. He noticed that Hogan used a "device law" to convert the data from measured audibility factor to antenna current: audibility factor was proportional to the square of antenna current. From Austin's 1911 report, however, this device law did not always hold. Austin's measurements for calibration suggested that audibility factor was proportional to antenna current proper, not to the square of antenna current for weak signals. Because most data came from periods when signals were weak, Love argued that the proper device law should be direct proportion. Thus he modified Hogan's data accordingly, from which he made the "calibrated" data fit Macdonald's predictions better than the Austin-Cohen formula. The reinterpreted empirical evidence supported Macdonald's diffraction theory.

Was Love's conclusion justified? No. His numerical method used assumptions about the infinite series identical to Macdonald's; it was not a surprise

32. Love, "Transmission of electric waves" (1915), 116–23.

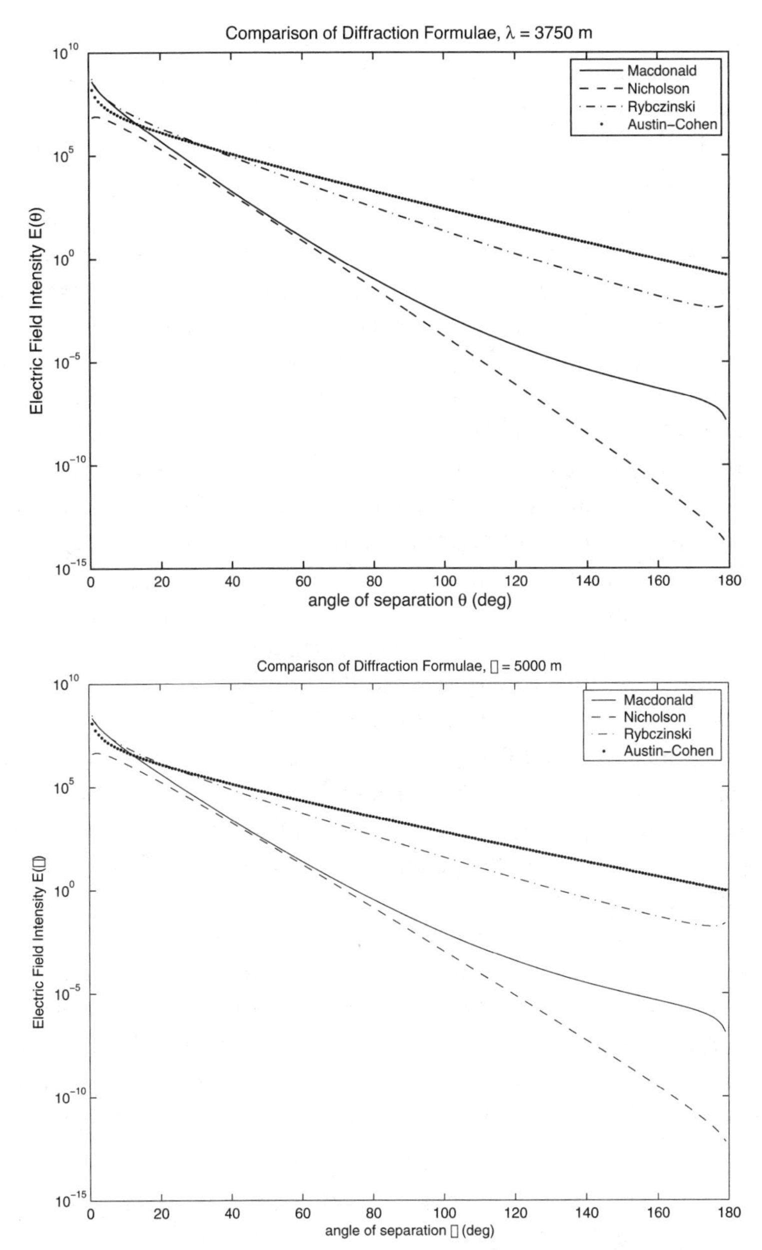


FIGURE 4.4. Comparison of Macdonald's, Nicholson's, Rybczyński's, and Austin-Cohen's formulae. Upper panel: wavelength 3750 m; lower panel: wavelength 5000 m (the ordinate's scale is relative).

that the two theories gave similar predictions. The real issue—whether the underlying assumptions were legitimate—remained. Moreover, Love's reinterpretation of the data was problematic. He relied on the device law that Austin developed in 1911. Austin reported the same experimental results as Hogan did in 1913, but he did not discover any significant deviation of the measured data from the Austin-Cohen formula.

Whether Love's reinterpretation of data was correct is still a question. Some contemporaries were aware of his problem. English electrical engineer G. S. O. Howe, for instance, commented that he based his device law on an "unfortunate misunderstanding" of a statement that Austin made concerning the resistance of the shunted telephone.[33] Historically, diffraction theorists did not acknowledge in publications that Love's approach resolved the long-distance wave-propagation problem. They remained silent on his paper.

In fact, not just Love's work went unheeded. The surface diffraction theory dwindled in the mid-1910s. Although the unresolved problem of approximation was still a worthy topic for mathematical inquiries, this approach seemed less and less empirically relevant, as the Austin-Cohen formula gained more and more technical followers. The political atmosphere did not help, either. The outbreak of war in Europe forced the mathematical physicists to pull away from more theoretical work towards duty to their nations. From 1915 to 1918, Sommerfeld and his disciples in Munich concentrated on research with more direct military applications, such as the gyroscope, whereas the Cambridge wranglers took up various kinds of war service. With Poincaré dead, Rayleigh aging, and the other active participants busy in war, surface diffraction did not thrive.

As the surface diffraction model faded, however, an alternative theory for long-range propagation gained increasing support in the 1910s. It presupposed a novel geophysical entity in the upper atmosphere; it promised to explain not only long-range propagation but also the diurnal change of signals and the existence of irregular noise; and it became popular especially among people practicing wireless technology such as engineers, inventors, and technicians.

## ATMOSPHERIC REFLECTION THEORY

Although Marconi's transatlantic wireless was inevitably the most conspicuous empirical observation on wave propagation at the turn of the century,

33. G. S. O. Howe, "Note on Radio Research Board Sub-Committee A Paper 13," 1921, DSIR 10/134, Public Record Office, Kew, London, England.

it was by no means the only one. As soon as the radio pioneers moved the Hertzian-wave set outdoors and exploited its telegraphic applications, they started to note how environmental conditions affected the quality of signals. The outdoors, after all, was not a simple extrapolation of laboratory space. Landscape, climate, and even air humidity influenced how far their signal could reach and how much noise contaminated it. And they had to tilt the antenna from vertical to enable optimum transmission.

Among the major wave-propagation phenomena they discovered in the 1890s and 1900s, two effects stood out, for they did not depend directly on terrain or weather (the most obvious factors to affect wave propagation): First, wave propagation experienced diurnal change. The maximum effective distance at night exceeded that in daytime. Marconi discovered this in his second successful transatlantic wireless experiment and reported it in 1902.[34] Second, noise, probably from the atmosphere, interfered with wireless communications links. This "static," "atmospheric," or "stray" wave reduced the rate of correct recognition of incoming Morse code. Static was more serious at night than in the day, during summer than during winter, and in low-latitude regions than in high-latitude regions. A number of wireless telegraphers identified the existence and characteristics of atmospheric noise. Feriyi, Popoff, and Turpain made early observations of atmospheric noise resulting from thunderstorms or other meteorological disturbances.[35] British naval officer and radio explorer Henry Jackson noted the day-night and summer-winter static differences as he experimented with wireless telegraphy while serving on Royal Navy warships between 1899 and 1902.[36]

To some scientists, the possibility of transoceanic propagation, the diurnal change of signals, and the existence as well as variations of static noise seemed to have the same cause, which had to do not with the ground but with the region above it. About the same time as European mathematical physicists were articulating the theory of surface diffraction, a few electrical researchers attempted to explain all these phenomena in terms of an electrically active atmospheric layer.

Unlike the Cambridge and Munich surface diffraction theorists or the U.S. naval experimenters who established the Austin-Cohen formula, advocates of atmospheric reflection did not form a coherent, specific social unit. Only two

34. Dunlap, *Marconi* (1937), 122–27; Marconi, "On the effect of daylight" (1902), 344–47.

35. Fleming, *Principles* (1916), 851–52.

36. Jackson, "On the phenomena" (1902), 254–72. For Jackson's life and work, see Weaver, "Jackson" (1930), 448–50.

individuals published articles on the model in the 1900s; three people at most were seriously working on it in the 1910s. Yet even though the formal papers were scarce, support was growing—mostly from electrical engineers, engineering professors, radio technicians, and wireless inventors, many of them from Britain and the United States. They liked a model that could intuitively explain a variety of phenomena they encountered in their technical practice in wireless. Rather than being diehard fundamentalists, they held a pragmatic view about theory.

### *Kenelly-Heaviside Layer*

The idea of long-range propagation via an atmospheric layer is straightforward: instead of diffracting along the earth's curvature, radio waves reflected back and forth between the earth and the upper layer. In other words, the earth and the atmospheric layer—both electrically conductive—served as the two boundaries of a gigantic waveguide for radio signals. Waves did not creep along the surface like ground waves; they shot high in the sky and returned to the earth, making them "sky waves" (figure 4.5). While André Eugène Blondel and C. E. Guillaume in France, Oliver Heaviside in England, and Arthur Edwin Kennelly in the United States all independently considered atmospheric reflection a plausible explanation for long-distance wave propagation, only Heaviside and Kennelly elaborated the model and put it in print.[37]

Oliver Heaviside was a peculiar figure in the history of science. Growing up in a working-class family in London, he learned and did physics by himself, published valuable works on electromagnetism in an obscure trade journal, refused academic affiliations, and remained a bitter, disillusioned maverick despite his central contribution. Historians have called him "Maxwellian," one of the post-Maxwell physicists who elaborated on the master's paradigm of electromagnetism. He achieved reputation for his new formulation of Maxwell's electromagnetic theory (which we use today) and his technique of operational calculus for solving linear ordinary differential equations. To many observers, Heaviside represented the first generation of engineering scientists.[38]

Unlike some other Maxwellians, Heaviside never seriously investigated wireless technology. But his work on wired signal transmission inspired his thought on propagation of radio waves. Once a telegraphic operator, he had worked on the theory of how electromagnetic energy flowed through a tele-

37. Russell, "The Kennelly-Heaviside layer" (1925), 609.

38. For Heaviside's biography, see Nahin, *Heaviside* (1987), and Yavetz, *Heaviside* (1995).

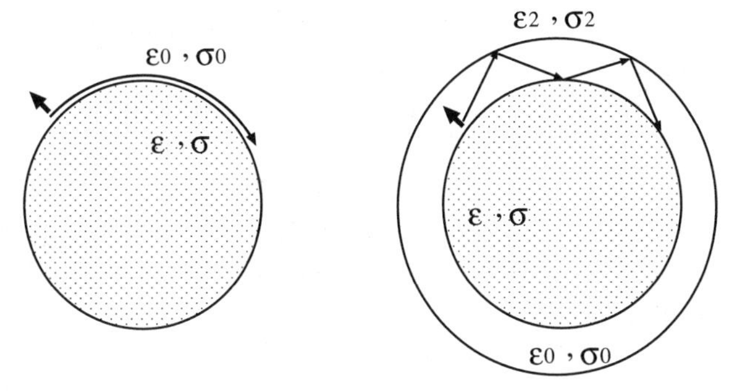


FIGURE 4.5. The surface-diffraction model (left) versus the atmospheric-reflection model (right).

graph line and developed a "telegraph equation" to describe the variations of voltage and current. In 1901, he was studying the electromagnetic field patterns that emerged when an electrical signal passed through a coaxial cable consisting of two concentric cylindrical conductors. He obtained the field by solving Maxwell's equations for the boundary condition that confined free space between the two coaxial conductors. Then he made various geometric metamorphoses of the cable to other mathematically tractable structures—for example, transforming the cylindrical boundary condition into a spherical condition, in which another hemisphere surrounded a point source on top of a large hemisphere. He found that this structure could support propagation of waves along the curving surfaces and conjectured that the same mechanism *might* account for Marconi's transatlantic wireless signals.[39]

In 1902 Heaviside contributed an introductory article on telegraphy to the *Encyclopaedia Britannica*, presenting in it his theory of atmospheric reflection.[40] The short paragraph presented a two-step argument. First, because both sea and land have nonvanishing conductivities, radio waves may travel along the earth's surface in the same manner as radio waves pass along a conducting telegraph wire. Second, it is also *possible* that the earth's surface and also a conducting layer (like a huge reflecting board) in the upper atmosphere bound traveling radio waves. That was all. He did not describe the physical mechanism generating the conducting layer, evidence for the layer's existence, or the qualitative or quantitative properties of the propagating waves therein. Without these discussions, Heaviside's hypothesis remained an interesting yet highly incomplete concept.

39. Nahin, *Heaviside* (1987), 279–81.

40. Heaviside, "Electric telegraphy" (1902), 215.

While Heaviside announced his theory, another person independently entertained a similar idea. Arthur Edwin Kennelly's career path epitomized the beginning of American electrical engineering. Born in India as the son of a minor British colonial official, Kennelly matriculated in engineering at the University of London but soon became an itinerant telegraph operator. After years of travel, he moved to the United States, joined the research staff at Thomas Edison's West Orange Laboratory in New Jersey, worked at General Electric, and then opened a consulting firm with electrical industrialist Edwin Houston. Eventually, he became a professor of electrical engineering at Harvard, where his main research interests were electric circuits and power systems. He popularized the theories and technologies of wireless by providing simple physical explanations of his observations. He emphasized physical intuition rather than complicated mathematical theory in dealing with wave-propagation problems.[41]

In 1902, Kennelly proposed and published a model similar to Heaviside's. He also argued that reflections from the conducting layer in the upper atmosphere made long-distance radio-wave propagation possible. Unlike Heaviside, however, he provided a physical cause for the layer. His explanation rested on J. J. Thomson's discovery in 1892 that the air at low density had a nonzero electric conductivity. The more dilute the air, the higher the conductivity. Kennelly assumed that the air density was proportional to air pressure and evoked the standard dependence of atmospheric pressure on height and Thomson's extrapolated experimental formula to deduce that, 80 kilometers up, the air conductivity was 20 times that of seawater! After securing this causal explanation, he contended further that if the atmospheric layer and the earth's surface confined the space, then the propagating wave's energy density would diverge as fast as a cylindrical wave. In the absence of the layer, the energy density would diverge as fast as a spherical wave, which decayed faster than a cylindrical wave. To Kennelly, the slower energy divergence resulting from the upper layer's existence explained why the radio wave could still maintain a detectable intensity at long distances.[42]

Kennelly's paper was more specific than Heaviside's paragraph: it explained the conducting layer and estimated its height. But both models differed in kind from the diffraction theory. While the diffraction theory represented radio waves' field intensity *quantitatively*, Heaviside and Kennelly explained a puzzling wireless phenomenon *qualitatively* and in a much less

41. Wagner, "Kennelly" (1931), 1.

42. Kennelly, "Electrically-conducting strata" (1902).

mathematical manner. Moreover, while diffraction theorists built mathematical representations on fundamental principles (Maxwell's equations) and a straightforward boundary condition (the earth), supporters of the conducting layer invoked a hypothetical entity. As the reflection model developed further in the 1910s to connect a variety of other radio phenomena relating to atmospheric conditions, the distinction between quantitative representation and qualitative explanation sharpened, and the demand to know more about the hypothetical layer became more intense.

## *Eccles's Ionic Refraction*

As Heaviside and Kennelly applied their model to explain long-distance propagation, they also raised difficult questions about the layer's nature. What was it? What physical process caused its formation? How could it reflect radio waves? Was it really like a metallic surface that blocked and bounced back electromagnetic energy?

In the early 1910s, English engineering scientist William Henry Eccles started to address these questions. His answer, in brief, portrayed the upper layer as a body of electrons and positive ions that emerged as sunlight decomposed air. Moreover, the layer deflected electromagnetic waves not through blunt reflection but by gradual refraction resulting from variation of electron density over height. Eccles's model of ionic refraction marked the first effort to turn atmospheric reflection into a more complete physical theory. His richer assertions about the layer's characteristics led to the explanation of more empirical facts. And his formal expression for the layer's refractive index represented an attempt to mathematize the theory.

But Eccles's model did not yet compete directly with surface diffraction. While the Cambridge and Munich mathematical physicists focused on the relationship between field intensity and distance and hence felt uneasy about the discrepancy between the Austin-Cohen formula and their theoretical predictions, Eccles and his followers did not bother to fit their model with that empirical law. They sought rather to connect various wireless phenomena with distinct features of the upper layer. They drew on the intellectual tradition of microphysics since the late nineteenth century, which studied propagation of electromagnetic waves in various media in order to reveal the internal molecular structures of materials.[43]

Although Heaviside and Kennelly wrote on the upper conducting layer in

43. Buchwald, *From Maxwell to Microphysics* (1985), 177–268.

1902, it was through Eccles's work in the 1910s that the wireless community began to acknowledge the empirical ground of the atmospheric-reflection hypothesis. The child of a Lancashire mechanical engineer, Eccles was curious about steam engines and thermodynamics. He obtained his bachelor's degree in physics from the Royal College of Science, London, in 1898. In 1899, he joined Marconi's research team on wireless telegraphy, which shaped his later career in radio. After leaving Marconi's company, Eccles taught first at a local college in Chelsea and then at University College, London, beginning in 1910. He was a pioneer of radio technology in Britain. He would be an advisor to the War Office during World War I, president of the Radio Society of Great Britain, and a consultant for Britain's Radio Research Board in the 1920s. He was famous for his inventions in measuring techniques (such as audibility) and vacuum-tube electronic circuits (such as the flip-flop oscillator) that went into broad use in communications engineering.[44]

Eccles began to work on radio-wave propagation in the late 1900s because of his interest in static. An experienced wireless engineer, he was familiar with the omnipresent natural noise that produced sharp clicks, grinding sounds, or hissing tones in telegraph operators' earphones. Since the late 1890s, a number of researchers had been correlating static with local weather disturbances. Eccles wondered whether static was a merely local phenomenon and set up an experiment in 1910 with his colleague Morris Airey at University College in London to measure simultaneously the static at a wireless station at the college and at another station in Newcastle, 270 miles (432 km) away. They found that if static was noticeable at one station, then it was very likely to be so at the other. Eccles deduced that the static at two places resulted from the same cause, involving a long-distance mechanism, perhaps discharge of atmospheric electricity hundreds or thousands of miles from the receiving stations.[45]

This was only the beginning of Eccles's research into static. In 1912, he developed a model to explain its most notable feature—the diurnal variation of its intensity.[46] Since both the transatlantic telegraph signals and static could travel long distances, they might propagate in a similar manner. Hence a correct physical model for long-distance wave propagation might explain not only the data on transatlantic wireless telegraphy but also those on static.

In Eccles's model, a permanent conducting layer (he dubbed it the "Heavi-

44. Ratcliffe, "Eccles" (1971), 195–214.

45. Eccles and Airey, "Note on the electrical waves" (1911), 145–50.

46. Eccles, "On the diurnal variations" (1912), 79–99.

side layer") in the upper atmosphere surrounded the earth, as did another concentric layer between it and the earth. This new layer corresponded to a region of air with gradually changing physical properties instead of a sharp boundary. When sunlight penetrated the Kennelly-Heaviside layer to enter the middle region, its ultraviolet component ionized the air there and produced free microscopic charged particles.[47] The lower the height, the more dissipation sunlight experienced as it penetrated deeper into the atmosphere, and the fewer ions it generated. The number of charged particles per unit volume thus increased with height.

The idea of an ionized atmospheric layer was not entirely new. Nineteenth-century students of terrestrial magnetism such as Carl Friedrich Gauss, Arthur Schuster, and Balfour Stewart had speculated on a similar hypothesis to account for variations in the geomagnetic field.[48] They contended that diurnal and seasonal changes in the ionized layer would affect the earth's magnetic field. Under Eccles's framework, the century-long tradition of geomagnetic research was beginning to connect with emerging studies of radio-wave propagation.

The effect of the ionic layer on long-distance transmission of wireless signals was a function of how radio waves propagated in such a medium. Eccles evoked a simple microphysical model to describe wave propagation in the ionized air: the ionized air consisted of free ions (including electrons and light positive ions like hydrogen ions) and massive neutral particles (gas molecules). When an electric field was applied, the ions moved. Because the much heavier positive ions responded much more sluggishly than the electrons, the electrons incurred the dominant effect on the electric field. Eccles derived the average induced ionized current from Newton's second law of motion, ignoring the statistical energy distribution of ions. Incorporating the induced current into Maxwell's equations, he expressed the ionized air's refractive index (indicating the reduction of electromagnetic wave speed) in terms of the ions' charge, density, mass, and number. From this theory, Eccles demonstrated that the phase velocity of an electromagnetic wave increased with the ion-number density (more details in chapter 6).

Eccles used this theory to explain three wireless phenomena: the possibility of wave propagation along the earth's curvature, the radiation directivity of Marconi's tilted aerials, and the diurnal variation of static. First, the theory

47. Ibid., 88–89.

48. Stewart, "Terrestrial magnetism" (1875–89), 181–84; Schuster, "Terrestrial magnetism" (1889), 467–518.

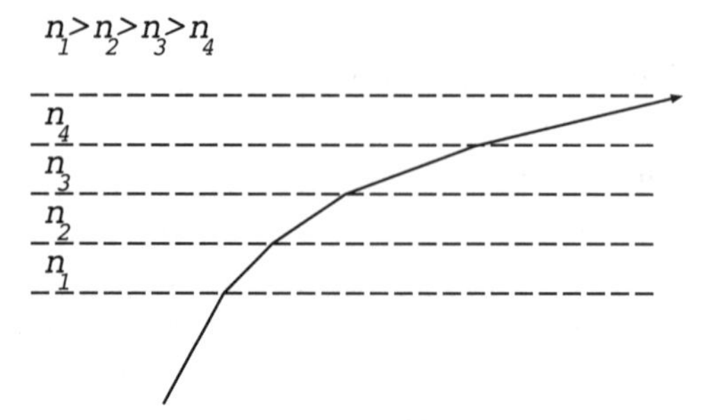


FIGURE 4.6. Wave refraction in a stratified ionized atmosphere.

suggested that the greater the ion density, the higher the phase velocity. From Eccles's physical model, ion density increased with height, as therefore did the wave's phase velocity, which implied a smaller refractive index at a higher altitude. According to Snell's law, one of the oldest principles in optics, an upgoing radio wave propagating in this ionized medium moved gradually downward because of the refraction of the medium that was gradually becoming, in effect, optically looser with height. That is, the wave path curved. If the curvature of the refracted wave path equaled that of the earth, the wave could run naturally along the earth's surface *without any action by the ground material* (figure 4.6). Unfortunately, these quantitative predictions disagreed with Austin-Cohen.[49]

Second, Eccles explained the directivity of a tilting antenna. His theory suggested that the amount of refraction decreased with frequency. When the frequency fell low enough, the curvature of the refracting wave path became larger than the earth's curvature. In this case, the wave path was obliquely incident to the ground and the direction of wave polarization no longer remained vertical. To match the polarization as much as possible, technicians at the receiving antenna must tilt it towards the direction of the transmitter. Thus Eccles reached the same conclusion as Zenneck's without using the "surface wave."

Third, Eccles dealt with the effect of sunlight. The ionized layer resulted from sunlight and thus did not exist at night. In England, the major source of long-distance static came from tropical Africa, where thunderstorms and other electrically disturbing weather processes were more severe. By day, refraction of the ionized layer in the middle and lower atmosphere directed long-distance static. At night, the ionized layer was absent, and reflection from the

49. Eccles, "On the diurnal variations" (1912), 91.

conducting Heaviside layer in the upper atmosphere directed long-distance static. The waves refracting through the absorptive ionized atmosphere lost more energy than the waves that a conducting surface reflected. Hence static was weaker during the day than at night.

Eccles's theory of atmospheric refraction differed in essence from Kennelly's and Heaviside's atmospheric reflection theory. In Kennelly's and Heaviside's model, the concentric spheres of the earth and the permanent conducting layer in the upper atmosphere guided radio waves in accordance with the conductivities of air and ground. In Eccles's, the refractive condition of the ionized atmosphere alone directed radio waves, which had no relation to ground conditions. For Eccles, long-distance wave propagation was not a result of the boundary condition that the conducting ground and the atmospheric layer together formed; ionic refraction *alone* explained the bending of the waves. This difference was clearly revealed in a letter from Eccles to Heaviside in 1912. Eccles learned from his former student working at the Marconi Company that "day signals from Clifden become much fainter on rounding Newfoundland and entering St. Lawrence." So the land did produce a great effect. Eccles observed,[50]

> This fact, of course, supports your view of guidance by the surface of the globe and is at first sight against my view that the solar ionization causes helpful bending of the rays round the globe—for on your guidance theory intervening land may reasonably be supposed to weaken sig[nal]s by its low electric conductivity relative to seawater and by its irregular contour.

Eccles did not give up, however—he suggested that the air above land might absorb more waves.

Eccles's 1912 work changed atmospheric reflection from a tentative hypothesis to a sophisticated model. The Heaviside-Kennelly hypothesis explained only one fact. In contrast, Eccles covered several apparently unrelated phenomena—long-distance wave propagation, tilted polarization, and the diurnal variation of static. It was essentially a theory of qualitative wireless phenomena. Eccles did attempt to develop a mathematical theory of ionic refraction but achieved only partial and preliminary results, inconsistent with the empirical law—the Austin-Cohen formula. In addition, his model could

50. Eccles to Heaviside (letter), 27 Nov. 1912, Oliver Heaviside Papers (UK0108 SC MSS), 005/I/6/10, Institute of Electrical Engineers Archives, London, England.

account for the maximum or minimum of static in daily records but could not predict static levels.

Eccles's theory caught the attention of English physicist John Ambrose Fleming, who had collaborated with Marconi on high-power wireless transmission and invented the Fleming valve (an early vacuum tube). Fleming thought that the ionized layer could perhaps account for the diurnal variations of wireless signals. But the static patterns seemed too complex for him to work out a consistent explanation. At the annual meeting of the British Association for the Advancement of Science in 1912, Fleming organized a discussion on "the scientific theory and outstanding problems of wireless telegraphy" and suggested "formation of a British Association Committee to guide and formulate research on some of these unsolved problems." The association formed a committee that included "physicists, mathematicians, radio-telegraphists, and meteorologists," and it arranged for systematic observations on static. Unfortunately, that body produced no significant results, and the outbreak of war terminated its activity.[51]

Attendees at Fleming's radio-telegraphic session included Arthur Kennelly. Along with Fleming, he had been among the first people to accept Eccles's theory of atmospheric ionization; he believed that the boundary between the sunshine and the shadow regions formed a reflecting surface crucial to diurnal variation in wireless telegraphy. At the association meeting, he proposed a model for the variation of wireless signals at about twilight. The model concerned two wireless stations sitting along an east-west direction. The eastern facility experienced both darkness and sunshine prior to its western counterpart. About the time the sun rose at the eastern station, the western facility was still in darkness, and the transitional band moved gradually towards the eastern station. Before the band crossed it, the two stations were on the same side of the blocking curtain at the transitional band, so signal transmission between them remained unaffected. When the band moved right behind the eastern station (shortly before its sunrise), the blocking curtain became a reflecting surface to bounce the overshooting waves from the west back to the east, and vice versa. Hence the incoming signals' strength intensified. After sunrise at the east station, the band moved between the two facilities, blocking wave propagation, and incoming signals weakened significantly. When the sunrise had just passed the western station, the band functioned again

51. Fleming, *Principles* (1916), 860. Fleming's later interest in the refraction theory shifted from ionized media to nonionized ordinary air. See Fleming, "On atmospheric reflection" (1914), 318–33.

as a reflecting surface behind it to strengthen incoming signals. As it moved farther away, the reflective enhancement also waned. The same pattern happened about sunset.

Kennelly's mechanism predicted that a wireless signal moving from east to west had a maximum just before twilight at the eastern station and right after twilight at the western station, and a minimum in between. The pattern of the signal variation from Kennelly's model agreed with observations at the wireless stations in Nova Scotia and Amesbury, Massachusetts.[52]

These investigations typified the work of the atmospheric-reflection theorists. They paid more attention to a broader realm of propagation phenomena than to the mathematical relations between signal intensity and distance, and they emphasized qualitative characteristics of experimental data. Their theoretical work involved sophisticated model building and simple mathematical theories of wave propagation deriving from microphysics.

If we judged the popularity of a theory in terms of the number of publications about it, then surface diffraction would clearly outperform atmospheric reflection in the 1900s and 1910s. While the former yielded dozens of papers and lured the leading European mathematical physicists into high-profile debates, the latter generated fewer than five articles, only one of them in a leading scientific journal. Yet, despite the scarcity of research literature, the Kennelly-Heaviside layer and atmospheric reflection actually gained support among engineers in the 1910s, while surface diffraction lost adherents. Researchers studying the diurnal variations of signals and noise generally accepted Eccles's and Kennelly's models. Another case in point: some American wireless inventors employed the Kennelly-Heaviside layer to explain their fading experiments about 1912. In addition, a British researcher working for the Royal Engineers during World War I explained the errors in the army's radio direction finders in terms of atmospheric reflection (more details in chapter 7). Even Sommerfeld had to acknowledge in his 1913 letter to Wien that "perhaps we must make the reflection from the well conducting upper layers responsible for the possibility of wireless telegraphy at quite long distances."[53] Despite the lack of specific knowledge about and concrete evidence for the upper layer, the theory of atmospheric reflection gradually became common sense for radio technologists.

Long-distance propagation, however, remained problematic. Regardless

52. Kennelly, "The daylight" (1913), 12.

53. Sommerfeld to Wien (29 Nov. 1913).

of its explanatory power, atmospheric reflection did not reconcile with the empirical law of propagation. Eccles's quantitative prediction was still preliminary, and it contradicted Austin-Cohen. It seemed that his pure ionic refraction without the land effect was wrong and that Heaviside's propagation between ground and layer was right. But if that was the case, what was the mathematical solution of the wave equation corresponding to Heaviside's boundary condition?

In the late 1910s, a young experimental physicist tried to bring the mathematical tools from the surface diffraction theory into the atmospheric reflection model, with the hope of yielding the Austin-Cohen formula. Born in Utrecht, The Netherlands, Balthasar van der Pol worked at the Cavendish Laboratory of Cambridge University as an assistant between 1917 and 1919. He was keen about radio. The scientific and practical natures of the long-distance wave propagation problem intrigued him. During his Cavendish years, he worked on refraction theories, diffraction theories, and experiments on ionized media,[54] and after returning to Utrecht he wrote a doctoral dissertation on radio-wave propagation.[55] In 1918, he tried to tackle the discrepancies among different diffraction formulae (Macdonald, Nicholson, and Rybczyński). He wished to find out which formula was mathematically rigorous. If any one of them turned out to be logically correct, then diffraction could not account for the Austin-Cohen formula. In other words, van der Pol intended to figure out whether it was possible *at all* to account for the Austin-Cohen formula using the wave theory. The mathematics was too difficult for him, so he turned to a Cambridge-trained mathematician, George Neville Watson, for help.[56]

## WATSON'S SYNTHESIS AND ITS ACCEPTANCE

George Neville Watson was another veteran of the Cambridge mathematical regime. Growing up in Oxfordshire, he entered Trinity College, Cambridge, in 1904. He became a senior wrangler, Smith's Prizeman, and fellow of Trinity College, where he remained until 1918, when he took a professorship at Birmingham.[57] In the 1910s he studied complex-variable theory, seeking mathe-

54. "Publications of Balthasar van der Pol," Philips Research Public Relations Department, Eindhoven, The Netherlands.

55. Van der Pol, "De Invloed van een Geioniseerd Gas" (1920). Van der Pol's synopsis of wave-propagation studies up to 1918 appeared in van der Pol, "On the energy transmission" (1918), 858–76.

56. For van der Pol's life, see Casimir, "Introduction" (1960), vi–viii.

57. Whittaker, "Watson" (1966), 521; Rankin, "Watson" (1966), 551–65.

matical properties of Bessel functions via the techniques of complex analysis.[58] Meanwhile, he also worked on theories of approximation; a closed-form solution did not suit him, and he was sensitive to numbers and computability.[59] This technical qualification allowed him to complete the jigsaw puzzle (with a mathematical piece) on long-distance propagation.

### *The 1918 and 1919 Papers*

In early 1918, van der Pol asked Watson to investigate long-distance wave propagation. The results came out as two papers, in 1918 and 1919.[60] The first presented a rigorous mathematical proof that the field intensity diffracting along the curvature of a large conducting sphere had an exponential decay rate proportional to $\lambda^{-1/3}$, rather than $\lambda^{-1/2}$ (as Austin-Cohen suggested). The second showed that the field intensity diffracting in a space bounded by a large conducting sphere and a conducting surface exterior and concentric to the sphere had the $\lambda^{-1/2}$ dependence. Thus atmospheric reflection, which modeled the earth and the atmospheric layer with two concentric spheres, could explain the empirical law. This conclusion integrated the findings of the surface diffraction theorists, atmospheric reflection model makers, and signal-transmission experimenters and resolved the two-decades-long controversy over the reason for long-distance radio-wave propagation. The importance of Watson's work warrants an explication of its technical details.

First, Watson sought a rigorous solution to the diffraction problem avoiding earlier mathematical pitfalls. He started by focusing on the (German tradition's) Hertzian potential instead of the (British tradition's) magnetic field intensity. But following the Cambridge approach, he expanded the Hertzian potential in terms of a *discrete sum* of spherical harmonics instead of an *integral*. The result was an infinite diffraction series similar to equation (4.1) but with a discrete sum rather than a continuous integral over the index; this sum

58. Watson, *Treatise* (1952), In Cambridge, Watson received training on modern analysis from mathematicians E. W. Barnes, G. H. Hardy, and E. T. Whittaker. Lecture draft to Cambridge University (undated), George Neville Watson Papers, Special Collections, University of Birmingham Libraries, Birmingham, England.

59. Whittaker, "Watson," 522. In the 1930s, Watson participated in the machine computation of mathematical tables for the British Scientific Computing Service. L. J. Comrie (Scientific Computing Service) to George Neville Watson (letter), 17 Oct. 1937, George Neville Watson Papers, Special Collections, University of Birmingham Libraries.

60. Watson, "The diffraction of electric waves" (1918–19), 83–99; Watson, "The transmission of electric waves" (1918–19), 546–63.

differed from Macdonald's, but with a similar angular dependence and pole structure:

$$\text{(4.3)}\quad \Pi = -\frac{1}{ka^2}\sum_{n=0}^{\infty}(2n+1)P_n(\cos\theta)\frac{\varsigma_n(ka)}{\varsigma'_n(ka)}.$$

As Love pointed out, the problem with the integral expansion was that the Legendre polynomial in the integrand blew up at the antipode when its order was a noninteger. Watson's central innovation was to convert the series expansion of the Hertzian potential in equation (4.3) into an integral expansion without such a problem. Key to accomplishing that was interpreting the term in the diffraction series $(2n+1)P_n(\cos\theta)\zeta_n(ka)/\zeta'_n(ka)$ as the residue of the function

$$2s\pi\frac{P_{s-1/2}(-\cos\theta)\varsigma_{s-1/2}(ka)}{\cos(s\pi)\varsigma'_{s-1/2}(ka)}$$

at the pole $s = n+½$ (in contrast to the Germans' formula, the Legendre function $P_{s-1/2}(-\cos\theta)$ in this new expression was finite at $\theta = \pi$). So the series in equation (4.3) was the sum of the above complex function's residues on the positive real axis, which, as Watson showed, equaled a contour integral

$$\oint s\pi\frac{P_{s-1/2}(-\cos\theta)\varsigma_{s-1/2}(ka)}{\cos(s\pi)\varsigma'_{s-1/2}(ka)}ds$$

around the righthand side of the complex $s$ plane (figure 4.7). This technique—later known as Watson's transformation—converted the Hertzian potential in equation (4.3) from a series into a complex integral that one could evaluate by Cauchy's residue theorem, and the integrand's pole—the zero of $\zeta'_{s-1/2}(ka)$—with the smallest imaginary part dominated its value. After computation, Watson showed that the contribution from the dominant pole had an approximate form with an exponential decay $\exp(-23.94\lambda^{-1/3}\theta)$, which was quite close to Nicholson's $\exp(-23.8\lambda^{-1/3}\theta)$.

Watson's 1918 work confirmed that the intensity of the field diffracting along the earth's curvature was significantly weaker than Austin-Cohen predicted. Watson's exponential decay rate was proportional to $\lambda^{-1/3}$, and the Austin-Cohen was proportional to $\lambda^{-1/2}$. The diffraction theory was now mathematically consistent, but empirically inadequate.

Watson located the problem in the physical model. All the diffraction theorists assumed that the earth's surface *alone* diffracted the field radiating from the dipole oscillator. Since surface diffraction could not account for the em-

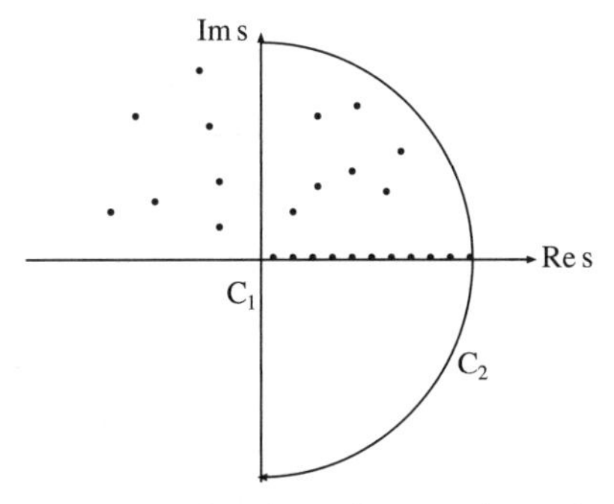


FIGURE 4.7. Watson's integral path on the complex *s* plane. The points are the poles of the integrand. Those on the positive real axis are s = 1/2, 3/2, 5/2, and so on.

pirical observations, the upper reflective regions of the atmosphere might be decisive in delivering radiation at long distances. The question was: could the diffraction theory incorporate the physical assumption of an atmospheric reflective layer and thereby yield predictions consistent with the empirical law? Or, to express the question in a different way: could the atmospheric reflection model appropriate the mathematical apparatus of the surface diffraction theory to generate the Austin-Cohen formula?

Watson took up this question in his paper of 1919. He took the earth to be a conducting sphere and the atmospheric reflective layer to be a spherical surface exterior and concentric to the earth. He assumed no structure for the layer and no gradient of refractive index. Both the Hertzian dipole and the receiver sat on top of the inner sphere. Waves radiating from the dipole propagated in the free space between the two spheres. The atmospheric reflection model became in effect a new and only slightly more complex boundary condition.[61]

To evaluate the Hertzian potential in this new boundary condition, Watson expanded it in terms of a series of spherical harmonics again and matched it with the boundary condition to determine the coefficients in this series. Then he applied "Watson's transformation" to convert the series into a complex integral, analyzed the pole structure of this integrand, and evaluated the integral in terms of these poles. He found that when both inner and outer conductors were perfect, there were about 20 to 80 poles on the real axis. Each real pole corresponded to a pure oscillatory mode. Thus the field intensity was a superposition of numerous oscillatory modes periodic with respect to distance. That is, the field did not decay at all! This result was consistent with physical

61. Watson, "The transmission of electric waves" (1918–19), 547.

intuition, since the two perfectly conducting spheres formed a perfect cavity resonator that could retain electromagnetic energy indefinitely. Yet a perfect cavity resonator never existed in nature. Both the earth and the upper atmospheric layer (if any) should have finite conductivities to dissipate propagating electromagnetic waves.

Watson therefore considered the case when both inner and outer conductors were imperfect. The integrand's pole structure changed accordingly, and all the real poles now moved away from the real axis and gained nonzero imaginary parts. Nonetheless, the locations of the new poles shifted only slightly from their original locations at perfect conductivity, when the conductivities were still quite high so that the wave-number ratio $k / k_c \cong \sqrt{\omega\varepsilon_c / \sigma_c}(1+i)/\sqrt{2}$ was small ($\omega$ was angular frequency, $k$ was free-space wave number, $k_c$, $\varepsilon_c$, and $\sigma_c$ were the conductor's wave number, dielectric constant, and conductivity, respectively). Watson employed a perturbation analysis to the leading orders of $\delta_e \equiv \sqrt{\omega\varepsilon_e / \sigma_e}$ and $\delta_a \equiv \sqrt{\omega\varepsilon_a / \sigma_a}$ ($\delta_e$ represented the effect of the earth and $\delta_a$ the effect of the atmospheric layer) and found that the dominant pole contained an imaginary part proportional to $\delta_e+\delta_a$. This pole contributed an exponential decay $\exp[-a(\delta_e+\delta_a)\theta/2h]$ ($a$ was the earth's radius, and $h$ the height of the atmospheric conducting layer). So the field intensity had an exponential rate of decay proportional to $\lambda^{-1/2}$, since $\delta_e+\delta_a$ was proportional to $\omega^{1/2} = (2\pi/\lambda)^{1/2}$. This result, in contrast to all the previous diffraction theories, had the same $\lambda$-dependence as the Austin-Cohen formula. By tuning the ionized layer's conductivity at $1.67 \times 10^7$ CGS units and height at 100 km, Watson could match the numerical value of the decay rate from his theory and that from Austin-Cohen.[62]

Watson's accomplishment was twofold. His 1918 paper developed a rigorous mathematical technique—Watson's transformation—to evaluate the diffracting field. This method represented the spherical-harmonic series in terms of a complex integral over a properly chosen path that was approximated by the dominant-pole contribution. Watson thereby confirmed the mathematical consistency of Nicholson's predictions and hence demonstrated the empirical inadequacy of the surface diffraction theory. His 1919 paper employed the same technique to calculate the diffracting field between two concentric spherical conductors. He showed that when the conductivities were high but finite, the field intensity had an exponential decay rate proportional to $\lambda^{-1/2}$. By adjusting the upper conductor's height and conductivity, he reproduced Austin-Cohen numerically. The theory of atmospheric reflection now gained

62. Ibid., 551–56.

major empirical support from long-distance propagation experiments. It no long remained an ad-hoc hypothesis to explain (away) some qualitative traits of observed phenomena.

## *Consolidating the Watson-Austin-Cohen Paradigm*

Appreciation of Watson's achievement did not come immediately. The first publications to mention his work all emphasized its mathematical rather than its empirical implications. Van der Pol discussed Watson's mathematical contribution to clarifying the controversy over different approximation methods for wave diffraction above a spherical conductor; but he ignored his theoretical prediction of the Austin-Cohen formula.[63] Macdonald commented on the mathematics of Watson's work without mentioning its empirical adequacy. He concentrated on extending the approach to the general case where single-frequency time-dependence did not hold.[64] In Germany, Sommerfeld's student Otto Laporte used Watson's results to reconcile the British representation of the diffracting field in terms of a series and the German representation in terms of an integral, without addressing Watson's empirical implications in their own right.[65] To these early responders, Watson's principal breakthrough was to develop a mathematical technique—Watson's transformation—that could generate rigorous and accurate closed-form approximations to contrived complex series or integrals and hence solve the longstanding mathematical debate regarding different versions of surface diffraction. That Watson's theory was the first one to match the major empirical law of propagation they did not stress.

Most scientists working on atmospheric reflection in the 1920s studied wave refraction through ionic media. They downplayed Watson's sharply defined Kennelly-Heaviside layer and complex-variable theory of diffraction. In a classic paper in 1924, Joseph Larmor mentioned Watson only at the end of a list of diffraction theorists beginning with Macdonald, and then as the man who demonstrated that surface diffraction could not account for long-distance propagation of waves around the earth.[66] Among these responses to Watson's papers, one thing stood out: after Watson had proved the mathematical rigor

63. Van der Pol, "On the propagation" (1919), 365–80.

64. Macdonald, "The transmission of electric waves" (1920), 52–76; "On the determination" (1925), 587–601.

65. Laporte, "Ausbreitung elektromagnetischer Wellen" (1923), 595–616.

66. Larmor, "Wireless electric rays" (1924), 1025–36.

of Nicholson's predictions, no one any longer attempted to reconcile the experimental data with the surface diffraction theory. The diffraction theory as a physical model for long-distance wave transmission was dead.

A direct empirical confirmation of Watson's theory appeared in 1925. Like the experiments leading to the Austin-Cohen formula, the measurements offering observational support of Watson's work were a by-product of a pragmatic project, too. If the former derived from a navy's renovation of its communications systems, then the latter was the result of a corporation's efforts to broaden its market coverage. In the early 1920s, the British Marconi Wireless Telegraph Company, the leading provider of long-distance wireless services on the Atlantic, aimed to extend its network. It made plans to build new high-power transmitting stations in South Africa and Australia.[67] In order to design stations or just to ascertain whether they were commercially possible, the company decided to organize a global expedition of signal measurements.

The endeavor took place in 1922–23. Captain Henry Round—the firm's leading engineer and an expert on direction finders who helped Britain's intelligence work during World War I—was the chief planner. The actual measurements were the task of Marconi's long-time collaborator K. W. Tremellen and his assistant Mr. Allnutt. The key apparatus was a receiving set consisting of a tuning circuit, a vacuum-tube amplifier, and a heterodyne circuit. On 28 January 1922, Tremellen and Allnutt boarded an English commercial liner, SS *Dorset*, in Liverpool. The ship crossed the Atlantic to Newport News, Virginia, sailed south to Panama, passed through the canal, traversed the South Pacific, and landed in Auckland, New Zealand, on 26 March 1922. The men unpacked the receiving set and took it to Sydney, Australia, where they stayed for a year. On 18 May 1923, they began the return trip. Boarding an Australian liner, SS *Boonah*, in Sydney, they sailed around the Australian coast to Melbourne, Perth, and Wyndham. Then they traveled across the Indian Ocean to Colombo, passed Port Sudan on the Red Sea, crossed the Mediterranean, exited past Gibraltar, and arrived in London on 8 December.[68]

On this two-year trip, Tremellen and Allnutt regularly measured signals from almost all the major high-power stations in Europe and the Americas (Chelmsford, Bordeaux, Nauen, Rocky Point, Panama, and San Francisco; wavelengths ranged from 10 km to 30 km). Unlike the U.S. Navy's 1909–13 propagation experiments from the Arlington station, therefore, the Marconi's expedition involved many transmitters. In addition, the latter's signal-

67. Jolly, *Marconi* (1972), 247.

68. Round et al., "Report on measurements" (1925), 933–34.

reach distances (14,000 km maximum) were much longer than the former's (4,000 km). Thomas Eckersley, a Marconi research engineer, clearly understood the journey's significance as he analyzed the vast trove of data that Tremellen and Allnutt gathered. He compared measured signal strength and distance with a formula from Watson's main result in 1919:

$$E \propto \frac{1}{\lambda\sqrt{\sin\theta}} \exp(-\alpha\lambda^{-1/2}d),$$

where $d$ was transmitter-receiver distance and the subtended angle $\theta$ derived from $d$ and the earth's radius $R$ via $\theta = d/R$; note the slight difference from the Austin-Cohen formula (3.1).[69]

The data that Eckersley and Round selected to compare with Watson's theory were from daytime (atmospheric noise made the nighttime data fluctuate greatly), and transmitter-receiver distances did not exceed 12,000 km (at longer distances, the receiver picked up the same signal from opposite sides of the earth). These data were highly consistent with Watson's main result. Eckersley and Round found especially good fits between theory and experiment for measurements from the routes Liverpool–Newport News and Panama–Auckland. (In both cases, most signals carried over seas, not land.)[70] The numerical value of the coefficient $\alpha$ differed with geographical regions and seasons and hence had to come from measurements. But Watson's theory of atmospheric reflection now had empirical corroboration from a set of experimental data that was more systematic and abundant than that which led to the Austin-Cohen formula. More confident about the reality of the Kennelly-Heaviside layer, Eckersley and Round gave an estimate of its height at 30–40 km.[71] Round reported their findings at the Institution of Electrical Engineers in May 1925.

The Marconi Company's long-range wireless experiment marked a turning point for Watson's work. After 1925, the wave-diffraction theory of the Kennelly-Heaviside layer gradually gained in stature not only for its mathematical consistency but also for its empirical adequacy. In 1928, George W. Kenrick at the physics department of the University of Pennsylvania reinvestigated the mathematical and empirical significance of Watson's theory. He pointed out that there had been much recent work to explain short-wave propagation by reflection and refraction, but "less attention has been given

69. Ibid., 977.

70. Ibid., 973–83.

71. Ibid., 996. This estimate turned out to be too low.

to modifications produced in the classical Hertzian solution for the field at a distant point due to an oscillating doublet."[72] He called for reexamination of Watson's work from the angle of the short-wave analysis. He calculated the electromagnetic field radiating from a Hertzian dipole applying the same boundary condition that Watson had used. Instead of working on the residue waves, he computed the multiple reflecting rays bouncing through the space between the earth and the atmospheric layer and summed all the reflective terms. The outcome of this approach reproduced Watson's mathematical formula and hence the Austin-Cohen formula. (That is, the diffraction integral converged to its geometric optics approximation.) Watson's work became significant not only for disproving the diffraction model, but also for its quantitative predictions of observed phenomena.

In the 1930s, some scientists and engineers working on radio-wave propagation began to extend their scope from ionic refraction to diffraction between the earth with composite material and geometry and the heterogeneous atmospheric layers. Watson's transformation proved a useful technique for analyzing these problems. In 1937, Harvard professor of physics Harry R. Mimno reviewed the literature on the physics of the ionosphere.[73] He highlighted the significance of the discrepancy between the $\lambda^{-1/3}$ dependence that all the diffraction theorists before Watson predicted and the $\lambda^{-1/2}$ dependence from Austin-Cohen, and he stressed Watson's invaluable theoretical account of the empirical $\lambda^{-1/2}$ dependence. Van der Pol's student H. Bremmer at the Philips Company in Holland also wrote:[74]

> The pioneering work clearing the way for further investigations was done by Watson in 1918. By a transformation with the aid of an integral in the complex plane, this author succeeded in transforming the rigorous series of zonal harmonics into a new series converging rapidly enough to be of use in the radio problem. As a matter of fact, almost all of the later literature is based upon this transformation of Watson.

Since then, radio scientists and engineers have held that wave reflection between the earth and the upper layer is responsible for long-distance radio transmission, and Watson's transformation has been a useful mathematical tool for analyzing propagation of electromagnetic waves.

72. Kenrick, "Radio transmission formulae" (1928), 1040–50.
73. Mimno, "Physics" (1937), 1–43.
74. Bremmer, *Terrestrial Radio Waves* (1949), 7.

Often in the history of science two mutually exclusive theories have competed to answer the same set of questions. In radio-wave propagation studies in 1901–19, however, two mutually exclusive theories addressed different types of questions. Surface diffraction and atmospheric reflection differed essentially in epistemic status—the nature and scope of the questions at which they aimed—rather than in their physical models.[75] Surface-diffraction theorists began by asking why long-distance propagation was possible. To answer this question, they constructed a straightforward physical model and attempted to solve it mathematically. But they soon discovered a technical difficulty in obtaining an accurate approximate solution. So they switched to a mathematical question—what was an accurate approximation of the diffracting field's intensity above a large conducting sphere? Almost all their effort involved proper approximations of the diffracting field's analytical form.

The atmospheric-reflection theorists asked for causal explanations for a broader realm of propagation phenomena. To do so, they constructed more elaborate physical models. They offered reasonable explanations for observed phenomena. But they failed to develop a mathematical theory for making systematic quantitative predictions before 1910, succeeded to a very limited extent after Eccles, and began to make steady progress only after Watson. The reflection theorists had difficulty formulating answers to questions regarding numerical prediction. Thus the diffraction theorists and the reflection theorists had different agendas: the former tried to resolve a mathematical problem of approximation, the latter to explain new radio-wave propagation effects. This explains why some mathematical physicists kept working on the diffraction theory regardless of its dubious empirical adequacy, and why an experimenter strongly supporting reflection models still gave credit to diffraction models.

The epistemic distinction corresponded to the social identities of both theories' promoters. The proponents of surface diffraction were mathematical physicists in Cambridge, Munich, and Paris. They had inherited a research and pedagogical tradition of theoretical physics that tended to represent complicated physical problems in terms of differential equations with pertinent boundary conditions and used various mathematical techniques to solve those equations. The advocates of atmospheric reflection were Anglo-American electrical engineers, radio inventors, and experimental physicists. They sought above all to make sense of the wireless phenomena they ob-

75. For more about the question-answering nature of scientific theories, see Bromberger, *On What We Know We Don't Know* (1992).

served in practice. The mathematical rigor or logical completeness of theory was not a top priority for them. These two groups, along with the American naval experimenters who produced the Austin-Cohen formula, constituted the major communities of historical actors in early propagation studies.

Each of the three communities left an unexpected legacy out of its research on long-distance propagation. The diffraction theorists found that approximation was crucial in physical problems in which analytic solutions could not give meaningful quantitative information and direct numerical computation was intractable. The new condition that the wavelength was much shorter than the scatterer's dimension forced them to develop a repertoire of advanced mathematical tools other than those for acoustic scattering for dealing with approximations of diffraction series or integrals. Sommerfeld's integral and Watson's transformation initiated the studies of problems that would become classical in mathematical physics: propagation of waves from a vertical or horizontal dipole oscillator above or below a homogeneous or layered horizontal plane or sphere, and so on. These problems did not necessarily correspond to real physical ones. But the mathematicians still investigated the complex-variable techniques to solve the problems for their own sake. The mathematical theory of complex series and integrals that prospered from the 1930s to the 1950s owed much to the heritage of the German and the British diffraction theorists in the 1900s and 1910s.[76]

Perhaps the most notable legacy came from advocates of atmospheric reflection. They started from the puzzles of radio operations and ended up introducing a new entity in the upper atmosphere. The Kennelly-Heaviside layer was originally a hypothesis to explain radio-wave propagation. The layer itself was never the focus of study in the 1900s and 1910s. Despite Eccles's and Kennelly's intelligent speculations on solar ionization, scientists did not have concrete ideas about the layer's geochemical and geophysical characteristics, and they did not know why solar ionization yielded a layer, how it formed, or what constituted it. They barely began to understand the behavior of electromagnetic waves in ionized plasma from the new electron-inspired microphysical perspective. And they did not possess more "direct" evidence for the layer's existence. But these problems did not remain difficulties for

76. One can find examples in a classical textbook on dipole radiation from the 1960s: Baños, *Dipole Radiation* (1966). The mathematical techniques that developed from studies of radio-wave propagation found application even in other areas of physics. For instance, Watson's transformation became part of the standard tool kit for the phenomenological S-matrix approach to theoretical particle physics in the 1960s. I owe this point to Dr. David Kaiser.

long; they became research opportunities later. From the 1920s on, scientists showed increasing interest in the upper atmospheric layer. Radio waves thus shifted from being the object of research to providing the key to investigating the geophysical entity. This brought a paradigmatic change to atmospheric science.

The original goal of the American wireless experimenters was to test equipment for the U.S. Navy's first long-distance radio station. They had a practical engineering problem to solve, but in pursuing it they made a priceless contribution to science. The navy's questionable decision to settle for NESCO's transmitter, the missionary agenda of the National Bureau of Standards, and Austin's technical training in German experimental science brought meticulous instrument design, systematic operational procedure, and mathematical representation of data into the investigation. These factors broadened the types of questions that the naval experimenters asked (from designs of apparatuses to nature's underlying regularity) and transformed the epistemic status of the experimental results from engineering data to scientific evidence. Their outcome was an empirical law that served as the only quantitative check on theories of long-distance radio-wave propagation in the 1910s.

What to physicists became a bedrock law for testing mathematical theory was for engineers a reliable design rule. Since the Austin-Cohen formula had proved to hold across various wavelengths, transmitters, and receivers, it could serve as a general criterion for designing radio apparatuses. To wireless engineers in the 1910s, Austin-Cohen guided the design of long-distance transmitting stations, for it stipulated the quantitative dependence of incoming signals' strength on antenna height, distance, transmitter power, and wavelength. Thus, it could determine the amount of power, the height of the antenna, and the size of the wavelength for a station to satisfy a minimum signal-level requirement at a given distance.[77]

77. For instance, in a letter to the secretary of the navy, the acting chief of the Bureau of Steam Engineering addressed a proposal of cooperation on wireless telegraphy by R. C. Caletti, who built a high-power radio station in France and intended to send signals to North America for experimental purposes. The acting chief provided him with the Austin-Cohen formula to specify appropriate antenna height and transmitting power. Acting Chief of the Bureau of Steam Engineering to Secretary of the Navy (letter), 13 Jan. 1911, Bureau of Steam Engineering, General Correspondences 1911–22, E 988, 840–841, box 1924, Bureau of Ships Records (RG 19). In his textbook on wireless telegraphy, John Ambrose Fleming stated that wireless engineers used Austin-Cohen to design technical specifications for transmitting stations. He also gave several examples of how to produce numerical specifications from the formula. Fleming, *Principles* (1916), 787–91.

The Austin-Cohen formula's central engineering consequence was its domination of long-wave (low-frequency) radio. The formula predicted longer propagating distances at longer wavelengths. In the 1910s, builders of long-range wireless stations followed this mandate by lowering their operating frequencies as much as possible and erecting giant antenna towers to have their signals reach wider areas (antenna size should match wavelength to transfer maximum energy). Long-range wireless communications hence became a business that only rich firms and the state could afford.

Nevertheless, the Watson-Austin-Cohen paradigm became obsolete as soon as it consolidated at the end of the 1910s. Contrary to that doctrine, radio amateurs and wireless engineers in the late 1910s and early 1920s found that short waves could also propagate over very long distances, and with just moderate transmitting power. This not only opened the era of short-wave communications but also led to discovery of the ionosphere and to radio studies of it.

* 2 *

# *Discovering The Ionosphere: 1920–26*

CHAPTER 5

# Radio Amateurs Launch the Short-Wave Era

No one visiting the District of Columbia in 1912 could miss the spectacle of the U.S. Navy's new, record-breaking wireless transmitter. "Topping the summit of an irregular, red-clay, bush-dotted hill immediately south of the Government military reservation at Fort Meyer, Virginia, and within eye range of the western front of the Capitol in Washington," a reporter from the *New York Times* wrote, "are the triplet peaks of the most powerful radio station in the world." Following the reporter, we would cast our immediate gaze on the 600-foot-high steel-lattice tower (even taller than the Washington Monument, the area's highest building) and the two supplementary 450-foot-high towers. The construction hosted $400,000 worth of equipment for a soundproof operations room and a 100-kilowatt spark-gap transmitter that covered up to 4,000 miles. The Arlington naval station was not the only one, however. Six others, as the reporter declared, were on the drawing boards for Guam, Honolulu, Manila, Panama, Samoa, and San Francisco. On completion, they would "girdle the earth" and keep the navy "in direct touch with its vessels wherever they may be."[1]

The United States was not the only nation to build gigantic wireless stations. Before Arlington, the German firm Telefunken had erected a 200-meter-tall transmitter in Nauen near Potsdam that could reach as far as North America. Soon after the U.S. undertaking, the French army built a high-power station on top of the Eiffel Tower to cover all of Europe. The British Marconi Com-

1. "First of a chain of seven which will girdle the earth and keep the Department at Washington in direct touch with vessels of our fleet wherever they may be." *New York Times* (20 Oct. 1912).

pany's huge wireless senders had dominated maritime communications on both sides of the Atlantic, and the firm was extending its wireless chain to all the British territories in Africa, Asia, and Oceania. Responding to increasing demand, manufacturers improved spark-gap senders tremendously and developed new transmitting machines such as arcs and alternators. If every decade after the Industrial Revolution had a technological emblem, then the high-power radio transmitter was definitely the one of the 1910s.

These transmitters, which achieved an outstanding performance in range, signal quality, and reliability, operated with waves longer than 300 meters. The scientific theory of radio-wave propagation that the Austin-Cohen formula and Watson's mathematical work represented showed that propagating range increased with wavelength. For great distances, long-wave radio was the way to go. As a result, wireless companies and governments were building transoceanic and transcontinental sending stations with higher and higher antenna masts. (To maximize power-transfer efficiency, the antenna length had to be comparable to wavelength.) Giant aerials constituted landmarks around the world, and long-range wireless became an exclusive business of the powerful and the rich who could afford such aerials.

Short-wave radio was much humbler. The difficulty of generating continuous waves—key to reducing interference—at high frequencies (i.e., short wavelengths) was a major technical barrier. Continuous-wave transmitters with frequencies above 1 MHz (1 million cycles per second) did not prevail until thermionic tubes such as John Ambrose Fleming's valve and Lee de Forest's audion became stable industrial products in the late 1910s. Short-wave radio's propagating range was too short, and its output power from tube oscillators was far less than that from alternators, arcs, or spark gaps, which unfortunately had limited frequency.

The common belief that long waves were much more valuable than short waves shaped the U.S. government's spectrum policy. Assuming that waves shorter than 200 meters were unable to travel along the earth's curvature beyond several hundred miles in normal circumstances, American policy makers left the "useless" high-frequency end of the spectrum to radio amateurs. The Radio Act of 1912 dedicated waves longer than 200 meters (frequencies lower than 1.5 MHz) to commercial and military uses. Amateurs could only use waves shorter than 200 meters for transmitting power below one kilowatt.[2]

2. An Act to Regulate Radio Communication (13 Aug. 1912), section "regulations," subsection "general restrictions on private stations," in the U.S. Department of Commerce (Bureau of Navigation), *Radio Laws and Regulations of the United States* (1914).

World War I changed the situation. The military authorities of both the Allies and the Central Powers discovered the advantages of short waves in compactness and bandwidth for applications in short-range communications in battle. They deployed short-wave sets for ground troops, warships, and airplanes and built their own laboratories or collaborated with industry to improve high-frequency wireless electronics. Trench soldiers' telegraph machines, airmen's communications kits, and early military radiotelephony operated at higher frequencies than long-range commercial wireless.[3] Thus short-wave radio made great strides during the war. For the armed forces and industry, however, it was only a *complement* to long-wave wireless telegraphy. As the natural laws of propagation restricted it, it was applicable only to short-range communications. Although the Austin-Cohen formula applied only in daytime and there were occasional reports of successful high-frequency long-range communications at night, the communication quality at night, because of noise and fluctuations, was too low to sustain a regular link.

Nevertheless, the technology eventually thwarted the scientific theory, not vice versa. In the early 1920s, a series of signal-transmission experiments demonstrated that, contrary to Austin-Cohen, radio waves shorter than 200 meters could propagate across the Atlantic and even beyond. Moreover, such high-frequency propagation seemed stable enough to warrant regular transoceanic or transcontinental links. This result overturned the two-decades-long technological paradigm of long-wave radio. From then on, long-range wireless was no longer under the monopoly of gigantic antenna towers and power plant–like transmitters that only the state or big corporations could afford. A small private sending station with moderate antenna height and output power would do. The era of short-wave radio began.

This discovery transformed propagation research. The physics of electromagnetic waves shorter than 200 meters turned out not to be a mere extrapolation to the physics that European mathematicians, Anglo-American engineers, and U.S. naval experimenters had struggled to attain since 1900. The homogeneous Kennelly-Heaviside layer, Watson's theory of wave propagation between the earth and the uniform layer, and the Austin-Cohen formula were valid only at low frequencies. Above 1.5 MHz, scientists and engineers found, the ground ceased to be a central factor—as Eccles had predicted a decade earlier. Rather, the structure and material constitution of the atmospheric layer shaped propagation of radio waves. The old theory's inability

3. See Nebeker, *Dawn of the Electronic Age* (2009), chapter 1; Hartcup, *The War of Invention* (1988), 118–45; Fagen, *Bell System* (1975).

to account for some other novel short-wave phenomena that emerged in the 1920s—fading, direction-finding errors, fluctuations of wave polarization, and skip zones—forced this turn of direction. A new theory of propagation was necessary; and it should rely more on Eccles's ionic refraction than on Heaviside and Watson's model of ground-sky "waveguide."

This chapter deals with the groundbreaking experiments in the early 1920s that established the feasibility of long-distance signal transmission at wavelengths shorter than 200 meters. These experiments were extraordinary not only because of their outcome, but also because of the people who carried them out. The folks who tested short-wave communications across the Atlantic were not academic scientists, corporate engineers, or governmental researchers. They were radio amateurs in Britain, France, and the United States. First surfacing about 1900 but really becoming more numerous before the First World War, these hobbyists shared a culture that valued establishment of communication links under difficult conditions. This goal encouraged them to take the part of the spectrum that policy makers deemed useless and to attempt what many professionals ridiculed as next to impossible—sending regular wireless signals above 1.5 MHz across thousands of miles. Their success marked one of the most unusual episodes in the history of radio.

Entering the story of propagation research about 1920, this community introduced a different form of experimentation. Wave propagation experiments, by definition, occurred outdoors, involved large scales, and defied laboratory-level control. The U.S. Navy's engineering tests leading to the Austin-Cohen formula and the British Marconi Company's corroboration of the formula were good exemplars of such field experiments. Yet what wireless hobbyists did provided an alternative possibility. Unlike the highly centralized military or corporate expeditions with a dozen or so technicians boarding a couple of battleships or commercial liners, short-wave trials in the early 1920s involved hundreds of loosely federated "hams" (radio amateurs' nickname) all over North America and Western Europe transmitting or receiving signals at the same time. Experiment became a highly *distributed* practice.

Furthermore, while Austin and Round and their colleagues relied on the state or capital to coordinate long-distance actions, arrange logistics, and supervise schedules, the wireless hobbyists resorted to civil groups to perform these tasks. Associations of radio amateurs in the United States (and to a lesser extent in Britain and France) were the main executors of the transatlantic tests. They worked out timetables, determined signal formats, organized overseas delegates, and mobilized hobbyists' participation through magazines. The aim of these endeavors was neither to answer a research question (as most scientific experiments did) nor to test the performance of machines (as Austin's and Round's teams did), but rather to break the record (and its public-relations implications for the amateur societies' lobbying for their part of the spectrum). Variable control, precise measurement, and analysis of mathematical data gave way to meticulous check for credibility, heroic accounts of expeditions, and sensational coverage in mass media. The radio amateurs' "experiments" on short-wave long-range propagation were far from the term's traditional sense.

## AMERICAN RADIO AMATEURS IN THE EARLY TWENTIETH CENTURY

For a young technology such as wireless at the turn of the century, the demarcation between professional and amateur was ambiguous. Most wireless pioneers—Armstrong, de Forest, Fessenden, Marconi, and so on—were amateurs before they joined companies or established enterprises of their own. Strictly, no "professional" radio engineers existed before Marconi set up the first wireless operating firm in the late 1890s. The category of radio scientists was also absent, since many wireless experimenters did not consider themselves scientists (nor did other people view them as such). Even after wireless telegraphy had become a regular business by the early 1900s, laypeople still continued to "play with," not just use, this technology. A person with some elementary electrical and mechanical knowledge could easily assemble a receiving set and, with slightly more effort, a transmitter.[4]

It is thus difficult to make a claim about when radio amateurs started to appear. Suffice it to say that they already existed in 1900. Their numbers grew rapidly as wireless technology progressed and the market for electrical components expanded. The United States had the largest population of radio amateurs; Britain and France followed closely. These people were technology enthusiasts who loved new gadgets, tinkering with machines, and the idea of talking remotely to people with no mediation but ether. They could be telegraph operators, engineering students, science teachers, factory technicians, affluent teenagers, even lawyers and shopkeepers.

Early radio amateurs' principal activities included assembling transmitters or receivers, "working" (communicating with) other amateurs on air using their homemade devices, and eavesdropping on commercial wireless telegraph. Their sense of achievement came from reaching as many amateurs as

4. Yeang, "Between users and developers" (2007).

possible from all over the world or making novel improvements on designs for transmitting or receiving devices. In the early phase, radio amateurs operated individually. As their numbers grew, they organized clubs. The first American radio amateur organization, the Junior Wireless Club, emerged in 1909 in New York City. Within a few years, such groups boomed around the country; by 1912, the *New York Times* estimated that there were 122 of them.[5]

At first, wireless clubs were peer communities for hobbyists to get to know one another, to exchange information, and to coordinate interesting activities. Some radio amateur organizations that professional engineers and academic scientists formed, especially those in Europe, more resembled amateur scientific societies. As the groups became larger, their functions extended to political and commercial aspects. Politically, the large bodies aimed to defend amateurs' legal right to use the spectrum, to lobby against armed forces and wireless companies' request for further restrictions on the wavelengths and power of amateur radio, and to advocate in public media the legitimacy of amateurs' right. Commercially, they planned to open the business of relaying long-range messages via the existing amateur networks.

The American Radio Relay League (ARRL) was one such hobbyist society. Its precursor, the Radio Club of Hartford, started in January 1914 in Hartford, Connecticut.[6] Hiram Percy Maxim, son of the machine-gun inventor Hiram Maxim, helped set it up. A graduate of MIT, he was an inventor and engineer too. He invented an equally controversial rifle silencer and designed an automobile featuring a front-mounted engine and a steering wheel on the left, which set the standard for American cars.[7] He began to take an interest in radio through his own son and built a one-kilowatt radio station in his home on Prospect Avenue, Hartford. Other members of the Hartford group, such as the secretary, Clarence D. Tuska, were local young hobbyists.[8] Maxim's Radio Club resembled hundreds of other societies around the country. But he had a different vision: he considered short-wave communications over long dis-

5. DeSoto, *Two Hundred Meters* (1936), 24; Douglas, *Inventing American Broadcasting* (1987), 205.

6. DeSoto, *Two Hundred Meters* (1936), 38.

7. Garraty and Carnes, *American National Biography* (1999), 751–52. Hiram Maxim also competed with Thomas Edison on inventing incandescent light; see Bazerman, *The Languages of Edison's Light* (1999), 202–10.

8. "LaSalle road links the whole 'ham' world," *West Hartford News*, 3 Nov. 1954, from Hiram Percy Maxim Papers, RG 69:12, box 2, Connecticut State Library, Hartford, CT. Tuska was the club's first secretary and launched the magazine *QST*.

tances a challenging task with great technical interest and commercial potential. His idea of long-range communications was to transmit messages across the continent by relaying with amateur stations. Such efforts required nationwide coordination among amateurs. To pursue this goal, Maxim changed the name of the Radio Club of Hartford to the American Radio Relay League (ARRL) in March 1914.[9]

Maxim organized ARRL in a much more formal manner than most radio clubs, and he attempted to engage amateurs from all over the United States. He divided the leagues into districts and, by emulating telephony, built wireless "trunk lines" between major relay points via high-performance transmitters and receivers. ARRL published an annual call book containing the names, addresses, call numbers, power, ranges, receiving speeds, and operating hours of stations around the country.[10] Starting in December 1915, the league published *QST* (acronym for "general attention" in amateurs' code) as a forum for wireless amateurs to exchange information and discuss issues of common interest. *QST* would later become an important vehicle for hobbyists to share their nitty-gritty technical knowledge. It was also ARRL's organ to invite participation in large-scale experiments and a channel to publish experimental data.

A goal of magazines such as *QST* was to help amateurs to design and build their own radio sets without infringing the government's spectrum regulations.[11] Radio amateurs *had to* work on short waves; the Radio Act of 1912 prohibited nonprofessional stations from operating at wavelengths longer than 200 meters and transmitting at power greater than one kilowatt. European governments had even stricter spectrum regulations: for instance, Britain's maximum power for hobbyists was 10 watts. Amateurs had to focus on higher frequencies than did wireless operating companies, and they had to master the art of short-wave radio.

World War I offered amateurs a major opportunity to hone such talents. Although martial law in the U.S. suspended their hobby activities, they continued radio operations in the military, as many enlisted in the armed forces. In their wartime service, radio amateurs experienced the new wireless technologies that military communication was using: audion tubes, crystal detectors, feedback amplifiers, Fleming valves, heterodyne circuits, and loop an-

9. DeSoto, *Two Hundred Meters* (1936), 41.
10. Douglas, *Inventing American Broadcasting* (1987), 296.
11. Yeang, "Between users and developers" (2007).

tennas. Many of these devices applied to high-frequency sets. After the war, the amateurs built more advanced radio stations applying their new technical knowledge and thereby started pursuing ambitious short-wave experiments.

### FADING EXPERIMENTS

Fading—irregular variations in intensity of incoming radio signals lasting from a tenth of a second to a few minutes—was a crucial wave-propagation property that affected the quality of short-wave communications. Rapid and intensive fading prevented wireless telegraph operators from recording correct Morse-code sequences. Even today, fading is still a nuisance to wireless users. (Just think how often the signal quality of our cellular phones fluctuates over time!) But fading rarely occurred at long waves or short distances; it was serious only for long-distance short-wave communications. This explains why early wireless technologists did not identify this apparently pervasive phenomenon until the 1910s.

Lee de Forest and Leonard Fuller first documented fading at California's Federal Telegraph Company in 1912. While Cyril Elwell (the firm's head) was arranging the arc-transmitter test in the U.S. Navy's transatlantic experiment of 1912–13 (see chapter 4), he was expanding his business in long-distance telegraphy from San Francisco to Los Angeles and Honolulu. De Forest (the inventor of the thermionic tube "audion") and Fuller (an expert on arc oscillators) joined Federal Telegraph to help Elwell carry out his plan. As they were building and testing these wireless stations, they found fading. In 1912, de Forest reported to the periodical *Electrician* that a 3,260-meter wave from Los Angeles faded out in San Francisco while a 3,100-meter wave remained at full strength. At the same time, the 3,260-meter wave was observed in Phoenix, Arizona, with normal amplitude. De Forest cited Harvard professor George W. Pierce's 1910 textbook on wireless telegraphy to explain the effect, with the interference between the direct wave going along the earth's surface and the wave reflecting from the Kennelly-Heaviside layer. Later, Fuller also observed and measured the intensity variation of the San Francisco–Honolulu signals at wavelengths between 3,000 meters and 10,000 meters. He reported the results at the meeting of the American Institute of Electrical Engineers in 1915.[12]

12. Villard, "The ionospheric sounder" (1976), 850–51; Tuve, "Early days" (1974), 2079–80; Lee de Forest, "Absorption (?) of undamped waves" (1912), 369–70; Pierce, *Principles of Wireless Telegraphy* (1910), 139; Fuller, "Continuous waves" (1915), 809–27.

Fading actually became a concern only in the late 1910s, when the use of short waves became popular among radio amateurs and the military. Signal fluctuation certainly damaged communication quality, but radio technologists knew very little about fading. They knew that it was unusual during the day or within reliable ranges of transmitting stations; other than that, it was irregular. It could come and go suddenly in an unpredictable manner. Two stations in the same region could experience different levels of fading at the same time. The signal fluctuations did not follow any regular pattern.

Also, the explanation of fading was far from conclusive. De Forest and Pierce shared the same account of fading as the consequence of interference between ground waves and sky waves. But they disagreed on the height of the reflective layer that caused the Los Angeles–San Francisco fading: Pierce estimated about 196 miles, and de Forest proposed *three* layers at heights of 17, 27, and 37 miles. Worse, the simple reflective model explained different yet *fixed* signal levels at different places or wavelengths, but not time variations of signal levels.

Engineers had sought to explain radio signals' time variations in terms of variations in transmitting devices' characteristics, sudden changes in propagating distances, or fluctuations of the Kennelly-Heaviside layer. Yet no one obtained direct evidence for any of these theories. Some people believed that fading was the product of mechanisms occurring on the path of radio-wave propagation in the atmosphere. If so, its occurrences and patterns must relate to specific conditions of the atmosphere's physical parameters. Unfortunately, scientists did not know what physical parameters related to fading.

Because fading seemed to have connections with many unknown atmospheric variables, its experimental study was tricky. A single measuring instrument operating during a short period might not adequately observe variations of atmospheric conditions: New England's climate differed from the Midwest's; winter differed from summer; temperature, humidity, and pressure could all matter. Without a specific theory, scientists had to consider all aspects of atmospheric conditions. Thus fading experiments could not use the traditional laboratory setup—a small group of investigators using a few measuring devices on a handful of controlled variables. They needed many observations from various geographical regions, seasons, and weather conditions.

Laurens E. Whittemore and S. Kruse at the Radio Division of the U.S. National Bureau of Standards first proposed such an experiment in 1919. They believed that a fading experiment should involve many observers covering a large geographical area and for a long period.[13] Neither NBS nor any single university, military, or corporate laboratory could conduct this large-scale endeavor alone. The bureau had to find a wide network of short-wave radio stations to carry out the task collectively. It turned to radio amateurs.

Kruse and Whittemore proposed collaboration with the American Radio Relay League. On 2 April 1920, a conference took place in the NBS building in Washington to discuss the fading experiment. Attendees included Kruse, Whittemore, and John H. Dellinger from NBS, ARRL's Maxim and Kenneth B. Warner (the organization's secretary and *QST*'s editor since 1919), Albert Hoyt Taylor of the U.S. Naval Air Station, and S. J. Mauchly and A. Sterling of the Carnegie Institution of Washington.[14]

They plotted an experimental scheme: within several days each season, a few transmitting stations would send test signals every night according to a predetermined schedule. Hundreds of receiving stations would record the fading characteristics of the test signals. A central organization would collect and analyze their data. ARRL was responsible for arranging transmitting and receiving stations, coordinating signal-sending schedules, publishing schedules, and collecting data. NBS designed the experimental procedures and the data format and analyzed the resulting data.

The ARRL-NBS fading experiment took place in 1920 and 1921. In 1920, there were summer tests from June to August and fall tests in October; in 1921, winter tests in January and spring tests in April. The sending network consisted at first of six transmitters[15] but soon grew to seventeen—in New England, along the east coast between New York and Virginia, and in Illi-

13. Kruse, "The Bureau of Standards—ARRL tests" (1920), 6; Dellinger et al., "A study of radio signal fading" (1923), 196.

14. For Warner, see "It seems to us," *QST* (Nov. 1948), 10, box C4-1, folder 21 "Kenneth B. Warner," American Radio Relay League (ARRL) Archives, Newington, CT. For the Washington conference, see Radio Laboratory, National Bureau of Standards, "Conference, April 2, 1920, to plan test schedule for the investigation of the swinging of radio signals on short wave lengths" (conference record), 7 April 1920, 1–12, in box C23-5, folder "Fading Tests," ARRL Archives. Participants discussed possible causes of fading. Mauchly suspected sunspots. Maxim and Mauchly conjectured that it happened when propagating conditions changed abruptly from abnormal to normal. Taylor blamed interference between ground waves and sky waves. Whittemore held that short waves propagated along isobaric lines. Maxim proposed a fading experiment to correlate the variations of radio signals to atmospheric, geomagnetic, and weather conditions.

15. They were the Naval Air Station NSF in Washington; Maxim's 1AW in Hartford; D. J. Coette's 2JU in Woodhaven, NY; Westinghouse's 8XK run by Frank Conrad in Pittsburgh; R. H. Mathews's 9ZN in Chicago; and J. A. Crowdus's 9ZV in St Louis. See NBS to par-

nois, Indiana, and Ohio.[16] The transmitting stations operated at wavelengths of 200, 250, and 375 meters. ARRL initially planned to secure 150 receiving stations but later expanded to 200. Like Gauss et al.'s "Magnetic Union" and Sabine et al.'s "Magnetic Crusade," this fading experiment required a lot of participants from vast regions.

How did ARRL and NBS mobilize so many radio stations? This was where amateur magazines came in. These publications spread recruiting information. Before the fading experiment, ARRL traffic manager J. O. Smith announced in *QST* the motivation, plan, and schedule of the tests, and invited participation.[17] Recruiting volunteers from mass media was an unusual way to reach experimenters. (Imagine what would happen if General Electric's Schenectady Laboratory called for experimenters in *Scientific American*!) This large-scale experiment consisted of sending, receiving, and recording radio signals according to an open schedule; in principle, anyone who knew the schedule could join in by tuning his or her receiver at the appropriate times. Yet there was selection going on, and not every volunteer was eligible. When a volunteer asked ARRL secretary Kenneth Warner in a letter why the fading experiment rejected him, he received the reply: "These tests are not being participated by all the members of the ARRL. Out of our total membership, we are selecting about 150 men of our best qualified operators as the work requires men of experience and ability. Our Division Managers were required to select the men in their different territory."[18] To ARRL, selection of participants was a necessary condition to ensure data quality.

Radio amateurs liked the idea of large-scale experiments. Since fading had been a threatening problem for their hobby, they had a practical motivation to obtain more specific information about it. But "experiment" in this collective, long-range format was also a kind of *activity* that posed a challenge they were glad to face. Amateurs loved making wireless connections under all kinds of situations. The complex coordination of the project would make it noteworthy among their peers; in other words, its success would "break a record." From ARRL's viewpoint, arranging such activities was crucial to holding its members' attachment, no matter how much actual scientific value

---

ticipants starting June 1920 (letter), box C23-5, folder "Fading Tests," American Radio Relay League Archives.

16. Dellinger et al., "A study of radio signal fading" (1923), 198.

17. Smith, "Variation of strength" (1920), 17; "The ARRL QSS test" (1920), 5–6.

18. Warner to E. N. Fridgen (letter), 29 Sept. 1920, box C23-5, folder "Fading Tests," ARRL Archives.

the activities would generate. We can see ARRL's intention clearly in a letter Warner wrote to Kruse:[19]

> This fading business is getting my cast-iron goat. You know the ARRL, on its own and apart from Bureau activities, is going to extend the fading test idea all over the country. These tests have caused intense interest and there were many requests to get in on them, but we were actuated more by the desire to hold things together thru the summer than anything else . . . This we hoped to do away with if we could "invent" something which would *keep the stations going* during the summer season.

When the activities indeed yielded interesting results, a different rhetoric emerged—a novel image of radio amateurs as contributors to knowledge "of high scientific value."[20] The hobbyists soon attached "scientific value" to their large-scale tests. Amateurs had already justified their legal right on the radio spectrum in terms of their experiments' actual and potential contribution to the art of radio. With the propagation experiments, they gained more symbolic capital: they were no longer mere tinkerers, but rather investigators of natural phenomena with scientific significance.

But how could such an activity be an *experiment*? The project incorporated hundreds of volunteers with different training and education. A lot of them did not have any laboratory working experience. These participants used all kinds of different instruments for receiving and recording data. How did the organizers ensure the quality of the measurements and control the operational variables? More fundamentally, what were the quantities to measure, and what was the evidence the experiment aimed to obtain?

### *Standard Experimental Procedure and Data Format*

ARRL and NBS came up with a solution to these problems: *QST* announced a standard experimental procedure and a data-recording format before the experiment.[21] It designated in advance transmission dates and time schedules for the tests in every season.[22] The message from a transmitting station was a

19. Warner to Kruse (letter), 26 June 1920, ibid. (my emphasis).

20. *QST* editor, "The ARRL QSS test" (1920), 5–8.

21. Smith, "Variation of strength" (1920), 17; *QST* editor, "The ARRL QSS test" (1920), 5–8.

22. For instance, the July tests took place each week on Tuesday, Thursday, and Saturday; Dellinger et al., "A study of radio signal fading" (1923), 203. In each test, the first transmitting

simple sequence of alphabets starting and ending with information about the station and the test.[23] Audibility measured signal intensity in a discrete nine-level "Eccles scale" ranging from "very strong" to "nothing."[24] Participants could use a standard data sheet to make records on a table: the alphabets' audibility represented signal intensity, and the alphabetic sequence represented a discrete time series. So a table recorded the variation of signal intensity over time—the characterization of fading by definition. In addition, the sheet also asked for background information, which a receiving operator filled in.[25] Participants were to mail the data sheets to the ARRL's Divisional Fading Committee.

The fading experiment began in June 1920 and continued to mid-1921. The 243 receiving stations that participated all over the U.S. Northeast and Midwest as well as in Canada produced thousands of records.[26] Thanks to the standard format, most data were processible and analyzable. But their accuracy was controversial because of the measuring techniques and data format. Some critics thought amateur operators' aural perception an inaccurate measure of signal intensity. Before the experiment, NBS considered measurement

---

station began to broadcast signals at 10:10 p.m. Eastern Standard Time. It stopped sending at 10:20 p.m., and the second station immediately following continued for ten minutes, and so on.

23. A typical sequence was: "1AW 1AW 1AW - . . . - AAAAA BBBBB CCCCC (*etc.*) YYYYY ZZZZZ YYYYY XXXXX (etc.) BBBBB AAAAA . . . - . 1AW." Here 1AW was the call number of the transmitting station. NBS, "Proposed letter to sending stations," 27 Sept. 1920, box C23–5, folder "Fading Tests," ARRL Archives. The alphabetical sequences went out at a constant speed of 90 letters per minute.

24. The nine levels of the Eccles scale were very strong, strong, good, fair, rather faint, faint, just readable, very faint, just audible, and nothing. *QST* editor, "The ARRL QSS test," 5. The choice of Eccles's scale took place at the conference on 2 April 1920; see NBS, "Conference, April 2, 1920" (7 April 1920), 8–9.

25. The background information included the call number of the receiving station, its location, the date, the beginning time of observation, the general reception condition, the general character of static noise, the transmitting station's call number, the wavelength, the weather condition (clear, cloudy, rain, snow, sleet, fog, or lightning), the wind direction, and the wind strength (calm, light, medium, strong, storm). Kruse, "The Bureau of Standards—ARRL tests" (1920), 6. When Sterling questioned at the conference on 2 April 1920 whether it was too much for an operator to obtain and to document so much information during a short measuring period, Maxim replied, "No, he is accustomed to being quick else he is not a radio operator." NBS, "Conference, April 2, 1920," 10.

26. Dellinger et al., "A study of radio signal fading" (1923), 198. The number of data reports in October exceeded 2,200. *QST* editor, "The fading test" (1921), 12.

by audibility meter, galvanometer, and phonographic recorder as a method of comparison.[27] Unfortunately, a typical radio amateur could not afford any of these automatic measuring devices.

Another problem was the design of data presentation. In the data-recording table, the resolutions of both signal intensity and time were too coarse to capture the subtle and rapid variations of fading: the time resolution was the duration of a five-letter alphabetic group (3.33 seconds), and the intensity scale had only nine discrete levels. Soon after the summer tests began, NBS and ARRL found that marking entries on a data table with checks was inadequate. They needed continuous curves that followed closely the variations of signal intensity, not a set of discrete check marks. So they changed the recording scheme. Instead of revising the table, they asked receiving operators to change their way of recording data. The operators were to draw continuous waveforms on the table; it was their responsibility to capture the variations within the resolution and to produce continuous curves (figure 5.1).

Mapping data in this manner required some skill. An operator had to know how to *disobey* the table's format, how to follow continuously and precisely the variation of signal intensity with his or her perception, and how to avoid smoothing the curves too much. Although Kruse found the new method feasible after he did a "laboratory test,"[28] not all radio amateurs had these measuring skills, which differed from the normal wireless operator's know-how. A good operator was able to recognize a signal's content no matter how weak it was, so he or she might pay too much attention to the content instead of to the variation of audibility within the duration of the alphabet. Dellinger and his colleagues observed that the best recorders were the amateurs who had some training in measurement work and had undertaken lab experiments. In general, the best radio operators were not good recorders: their curves were flat and lacked details. Because they concentrated on content, they apparently "followed the variation in signal strength without being aware that any variation is taking place."[29]

The large amount of data went to NBS for analysis. At first, the analysts

27. Dellinger et al., "A study of radio signal fading" (1923), 205.

28. Radio Laboratory, National Bureau of Standards, "Conference held at Bureau of Standards, Saturday, July 24, 1920, on A.R.R.L.—Bureau of Standards fading tests" (conference record), 5 Aug. 1920, 3, box C23–5, folder "Fading Tests," ARRL Archives.

29. Dellinger et al., "A study of radio signal fading" (1923), 208; *QST* editor, "Fading test" (1921), 13–14.

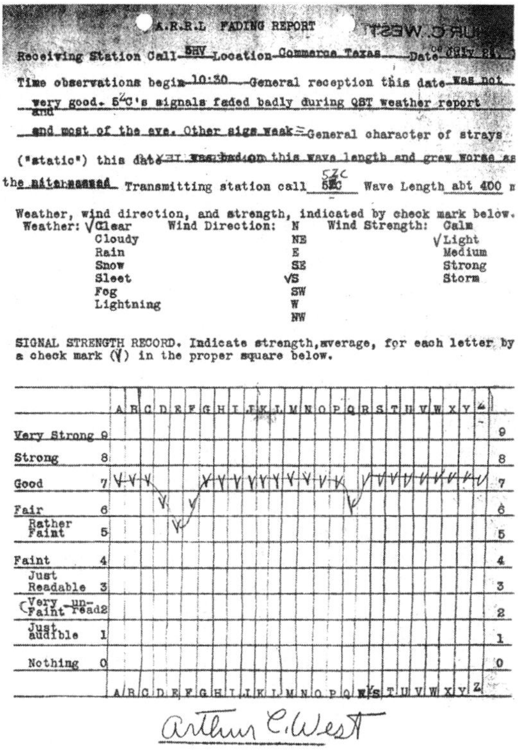

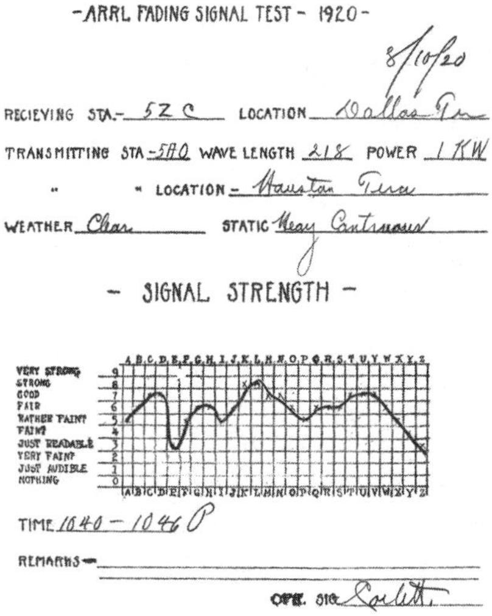

A.R.R.L FADING REPORT

Receiving Station Call 5HV Location Commerce Texas Date [illegible]

Time observations begin 10:30 General reception this date was not very good. 5ZC's signals faded badly during QST weather report and most of the eve. Other sigs weak. General character of strays ("static") this date was bad on this wave length and grew worse as the nite passed. Transmitting station call 5ZC Wave Length abt 400 m

Weather, wind direction, and strength, indicated by check mark below.

Weather: √ Clear, Cloudy, Rain, Snow, Sleet, Fog, Lightning

Wind Direction: N, NE, E, SE, √ S, SW, W, NW

Wind Strength: Calm, √ Light, Medium, Strong, Storm

SIGNAL STRENGTH RECORD. Indicate strength, average, for each letter by a check mark (√) in the proper square below.

Very Strong 9, Strong 8, Good 7, Fair 6, Rather Faint 5, Faint 4, Just Readable 3, Very Faint unread 2, Just audible 1, Nothing 0

A B C D E F G H I J K L M N O P Q R S T U V W X Y Z

Arthur C. West

-ARRL FADING SIGNAL TEST - 1920-

8/10/20

RECIEVING STA.- 5ZC LOCATION Dallas Tex

TRANSMITTING STA - 5AQ WAVE LENGTH 218 POWER 1 KW

" LOCATION - Houston Tex

WEATHER Clear STATIC Heavy Continuous

- SIGNAL STRENGTH -

VERY STRONG 9, STRONG 8, GOOD 7, FAIR 6, RATHER FAINT 5, FAINT 4, JUST READABLE 3, VERY FAINT 2, JUST AUDIBLE 1, NOTHING 0

A B C D E F G H I J K L M N O P Q R S T U V W X Y Z

TIME 1040 – 1046 P

REMARKS –

OPR. SIG. Corlett

FIGURE 5.1. Two examples of fading data from the ARRL-NBS experiment. Box C23-5, folder "Fading Tests," courtesy of American Radio Relay League.

tried to compare directly all the curves to identify those with similar shapes and then to examine their common conditions. This method was cumbersome for numerous data. So instead they assigned every data curve a degree of fading and registered it along with receiving stations' locations on weather maps to identify the atmospheric conditions relating to fading. Finally, they input the record on punch cards and used the tabulating machines at the Bureau of Census to conduct statistical analysis.[30]

The analysis confirmed the longstanding belief that fading at short wavelengths was more serious than at long wavelengths. It also showed that waves propagating along an isotherm or isobar had weaker fading and that clouds at the receiving station caused stronger fading, stronger atmospheric noise, or stronger signal strength.[31] But none of these emerged as clear scientific evidence. The claims relating to atmospheric conditions did not lead to a theory of fading. Nor was the conclusion very credible to the experimenters at NBS.[32]

The importance of the ARRL-NBS fading experiment lay more in its process than in its results. By participating in the tests, radio amateurs learned how to take part in a large-scale scientific experiment. ARRL gained invaluable experience in how to incorporate participants, arrange schedules, standardize communications format, coordinate operations, and collect data. Individual hobbyists had fun and gained a sense of achievement—ARRL held a contest for stations with outstanding performance during the test (receiving many signals or having many receiving stations pick up their signals). They also published the names of the award winners in *QST*.[33] Amateurs learned new measuring techniques, too. More important, American radio amateurs created a new type of activity relevant to the technical community at large—a collaborative "mass" experiment. As they continued this new type of activity, they would eventually produce results of scientific significance.

30. Dellinger et al., "A study of radio signal fading" (1923), 211–21, Kruse, "The Bureau of Standards—ARRL tests (part 2)" (1920), 13–22.

31. Dellinger et al., "A study of radio signal fading" (1923), 222–24.

32. Ibid., 221–22. Dellinger, Whittemore, and Kruse proposed that fluctuations of the Kennelly-Heaviside layer caused fading, but they offered no experimental results to support this theory.

33. Kruse, "Station performance" (1920), 11–14; *QST* editor, "Performance of January QSS recorders" (1921), 14–15.

## SETTING THE STAGE FOR TRANSATLANTIC EXPERIMENTS

Many contemporaries believed that wireless amateurs were unable to communicate across the Atlantic Ocean. From the Austin-Cohen formula, the propagating distance of a wave shorter than 200 meters with transmitting power less than one kilowatt could not exceed 1,000 miles. Even though an increasing number of scientists and engineers began to doubt the formula's validity beyond the long-wave regime and normal daytime conditions, they still believed that short-wave transoceanic propagation occurred, if at all, only under anomalous and unstable atmospheric conditions.

But from early on, wireless hobbyists entertained the possibility of regular transoceanic short-wave communication. Before World War I, Maxim and Louis Pacent of the Radio Club of America conceived of transatlantic tests.[34] After the war, radio amateurs in Britain, France, and Holland occasionally claimed reception of signals from American amateurs. Were these reports credible? Claimants did not have "reliable" and "impartial" witnesses, which radio amateurs considered crucial. (Later amateurs developed a "notarized" procedure of witness in their propagation experiments.) Worse, their receptions were not repeatable.[35] Even if the reports were true, the highly unstable circumstances implied that a short wave propagated across a long distance only under randomly abnormal atmospheric conditions. The only way to refute these critiques was to obtain positive results from carefully designed experiments.

In 1921, American radio amateurs were at the right moment to conduct such experiments. On the one hand, they had accumulated adequate technical knowledge to handle the effort. Wartime service had familiarized many of them with the new electronic and antenna technologies at short wavelengths. Some amateurs who were also engineers even invented new devices: the Armstrong feedback amplifier, the Beverage antenna (long horizontal wires hanging between two vertical frames), and the Reinartz tuning circuit (a single-tube tuning circuit whose feedback structure included the tube's in-

34. DeSoto, *Two Hundred Meters* (1936), 72.

35. For instance, amateurs Hugh and Harold Robinson in Keyport, New Jersey, learned from amateur George Benzie in Aberdeen, Scotland, in November 1921, that he heard their transmitting speech and phonographic music at his own station. But the Radio Club of America's investigation proved this claim false. *QST* editor, "2QR's transatlantic claim disproved" (1922), 8.

nate capacitance). These novel designs circulated among amateurs through personal contacts and hobbyist magazines and had been common in American private stations since the late 1910s. Moreover, amateurs also learned from the fading tests how to conduct large-scale experiments with volunteers. Like their Humboldtian predecessors, they were now versatile with organizing and logistical skills such as distributing information, coordinating participants, imposing measuring standards, and collecting data.

On the other hand, the amateurs were competing fiercely with the emerging commercial broadcasting for the legal right to use the short-wave spectrum. As soon as radio broadcasting began about 1920, large electric manufacturers stepped into the business. They formed interest groups to lobby for regulations to reserve a wider short-wave spectrum for commercial broadcasting, which would force radio amateurs to yield a portion of their original spectrum and further limit their transmitting power to prevent interference with broadcasting. This raised a conflict between radio amateurs and commercial broadcasters and launched a publicity battle. ARRL endeavored to promote a positive image of amateurs in mass media. It secured the support of Secretary of Commerce Herbert Hoover, who in 1922 organized the First National Radio Conference in Washington, DC, inviting representatives from corporations and private station owners to discuss spectrum allocation and related regulatory issues. At this moment, a sensational achievement with significant scientific value would give the amateurs much power in any negotiations.[36]

Americans were not the only actors in transatlantic experiments. Wireless hobbyists in Britain and France faced different situations from the Americans; European governments imposed much stricter limits on radio amateurs. U.S. authorities, for a variety of reasons, tolerated wireless hobbyists, but the Europeans considered amateur radio's only legitimate goal to be "scientific experimentation." In Britain, for instance, the six biggest manufacturers of radio sets (Marconi Wireless was the largest) formed a British Broadcasting Company (the predecessor of today's British Broadcasting Corporation) in 1921 and gained a monopoly from the government. Consequently, a person could legally possess a radio receiver that the company did *not* make and *not* for listening to its broadcasting programs *only if* he or she had a license to "conduct experiments."[37]

36. DeSoto, *Two Hundred Meters* (1936), 74–76, and Herbert Hoover to Hiram Percy Maxim (letter), 10 April 1930, Hiram Percy Maxim Papers, RG 69:12, box 2.

37. Johns, "The great oscillation war" (2005). For the formation of the British Broadcasting Company, also see Baker, *A History of the Marconi Company* (1970), 191–97. The six big-

Under the same rationales, radio transmitters posed an even more serious threat to both broadcasting order (by interfering with commercial programs) and national security (by serving as a tool for espionage). Thus transmission in Europe was under much tighter control than reception. The maximum allowable power of a British private transmitting station was 10 watts, one percent of that in the United States. And again, only people planning "scientific experiments" received transmitting licenses, which were harder to obtain than receiving licenses.[38] These regulations profoundly affected radio amateurs: they worked much more on less-restricted receivers than on transmitters, and they used private stations to do electric experiments (or at least to claim to do them).[39] Because governments granted amateurs space on the spectrum for their potential to generate new technological or scientific knowledge, a principal way to claim amateurs' rights was to conduct experiments.

Two amateur federations—in Britain and in France—were organizing wireless experiments. In 1913 René Klein launched the London Wireless Club in Hampstead in north London. It was similar to many local wireless amateur groups in Britain. In September 1913, F. Hope-Jones advocated for "a Society founded on a more ambitious scale than that of the Wireless Clubs then springing up all over the country."[40] So the group became the Wireless Society of London and later the Radio Society of Great Britain, a mirror image of the American Radio Relay League. Both national groups started as local hobbyist clubs and later went national. Both issued magazines to connect nationwide amateur communities: ARRL's *QST* and the British Society's *Wireless World and Radio Review*.

But Hope-Jones and his colleagues, unlike ARRL, did seek "men of eminence in the science of wireless telegraphy" as administrators. The first president was the Scottish electrical engineer A. A. Campbell Swinton, who was a pioneer of medical radiography and electronic television, a fellow of the Royal Society beginning in 1915, and later an adamant opponent of the British Broadcasting Company's monopoly. The society also attracted other notables for its board, such as William Duddell, William Henry Eccles, John

---

gest manufacturers were British Thomson-Houston, General Electric, Marconi, Metropolitan Vickers, Radio Communication, and Western Electric.

38. Child, "5WS" (1923), 877; Deloy, "A letter from France" (1920), 52.

39. A French amateur, Léon Deloy, commented on his visit to the United States: "Your principal aim seems to be traffic handling while our only aim is experimenting." Deloy, "My impressions" (1923), 17.

40. Anonymous, "History of the Wireless Society of London" (1922), 257.

Ambrose Fleming, Henry Jackson, Oliver Lodge, and Sylvanus Thompson.[41] The body was closer to the tradition of gentlemen's learning societies that had existed in Britain since the seventeenth century. Thus the demarcation between wireless amateurs and professionals in Britain was more ambiguous, probably because experimenting was the only legal wireless hobby in the country.

In France, Gustav Ferrié and Alfred Pérot established la Société des Amis de la T.S.F. (Telegraphie sans fils, or wireless telegraphy) in 1922.[42] Like its counterpart in London, it was more like a society where scientists, engineers, and nonprofessionals exchanged knowledge and publicized discoveries. Its magazine, *l'Onde électrique*, was a major scholarly journal for radio engineers. And, unlike the American and British amateur groups, la Société collaborated closely with the government. The background of its cofounder explains its distinct orientation—General Gustav Ferrié was perhaps the most important pioneer of radio technology in France. A graduate of the École Polytechnique, he had served in the French army for over twenty years. He was famous for introducing wireless to the army, administering the development of military communications during World War I, and coordinating (with Poincaré and others) the campaigns for longitude measurements using radio. At the time when la Société began, he was the inspector of the army's Military Telegraph Services.[43] From his viewpoint, la Société was not a mere venue for hobbyist gatherings; it should be an organ to facilitate research on radio that complemented the state and industry's investigations. Therefore, it should and would maintain close relations with the government.

## TRANSATLANTIC EXPERIMENTS

### *The First Trial*

In 1919, M. B. Sleeper in Britain proposed the first plan for transatlantic experiments. As the editor of the hobbyists' *Everyday Engineering Magazine*, Sleeper intended to carry out his program with the magazine's resources. In 1920, a financial crisis at the periodical forced him to take his plan to ARRL.[44]

41. Ibid., 258–63.

42. Gutton, "Dix années de T.S.F." (1932), 397.

43. Ibid., 397–99, Jouaust, "Ferrié" (1932), 45–52. For a biography of Ferrié, see Amoudry, *Ferrié* (1993).

44. *QST* editor, "Transatlantic sending test" (1921), 20.

ARRL's operating department decided to take over the project and began preparation in 1921.

The experiment was to test one-way transatlantic communications from North America to Europe, for the United States had more high-power transmitters than Europe. ARRL managed to recruit several transmitting stations at home. As for receivers in Europe, it asked the Wireless Society of London to help and arranged for Philip Coursey, the assistant editor of *Wireless World and Radio Review*, to locate receivers. Coursey and ARRL concluded a transmitting schedule: on the nights of 1, 3, and 5 February 1921, 24 American amateur stations transmitted predesignated signals to the British Isles, where 250 receiving stations were listening.[45] The experiment failed. Coursey found no evidence for receiving signals in participants' reports.

The Britons and the Americans thought differently of the experiment's failure. Coursey believed that it failed because the test lasted only three nights. Successful signal transmission did not happen every night because of randomly fluctuating atmospheric conditions. The chance for signals to get through was slimmer as the test duration was shorter.[46] The Americans held that the British amateurs' inferior receiving devices and techniques were part of the cause. They found two lethal effects—interference from commercial stations and interference from other participating receiving stations; the major source of the latter was radiation from self-regenerative amplifier-detectors. If the British experimenters, as many American amateurs did, had used the heterodyne receivers whose oscillating frequencies lay below or above the operating frequencies, the chance for success would have been significantly higher. Besides, although British amateurs had done much work on receiving techniques, the know-how for this experiment was novel to them. A *QST* editor commented, "Such reception is a new field for British experimenters and they hardly can be expected to show the same performance as an American dye-in-the-wool ham who has learned how to get amateur DX only after years of patient struggle."[47]

## *The Second Trial*

In response to comments from both sides of the ocean, the second transatlantic experiment proceeded differently. The number of test dates trebled. The

45. DeSoto, *Two Hundred Meters* (1936), 72. At night short waves usually propagated much farther.

46. Coursey, "Report on receptions" (1922), 23.

47. *QST* editor, "Failure of the transatlantic test" (1922), 15–16.

experiment was to take place every night between 8 and 17 December 1921. ARRL attempted to incorporate more qualified transmitting stations. Its Operating Department first posted a request in the September 1921 issue of *QST* to invite all American amateur transmitting stations to participate. "Fellows," the note stated, "our good old A.R.R.L. is calling to you amateurs with your excellent transmitters, inviting you to enter this contest in the name of good sportsmanship and in the interest of the advancement of Amateur Radio . . . We want the Atlantic Ocean spanned on schedule by an amateur station and we want definite proof that it has been done" (figure 5.2).[48] Seventy-eight stations were willing to enter. A preliminary test took place between 1 and 5 November for these stations. The transmitted signals had to reach 1,000 miles over the continent. Following the preliminary test results, ARRL chose twenty-seven transmitting stations for the final tests.[49]

To further increase the number of transmitters, ARRL decided to make the experiment "democratic," i.e., open to all transmitting stations. It divided each night in the final tests between 8 and 17 December into two parts: it opened the time from 7 to 9:30 p.m. to all American amateurs who were willing to transmit and reserved the time from 9:30 p.m. to 1 a.m. for the transmitting stations that had passed the preliminary test.[50] The wavelengths of the stations and all the free-participating amateurs should be about 200 meters.[51]

Additionally, ARRL's traffic manager, Fred Schnell, a former radio electrician in the Naval Reserve from Chicago,[52] proposed that the organization send to Britain an experimenter of its own, Paul Forman Godley, with the receiving devices and the operating techniques that the Americans trusted. A native of Garden City, Kansas, Godley had been a technician in the railroad business and telegraph firms and a member of ARRL's advisory technical

48. Schnell, "Transatlantic sending tests" (1921), 12.

49. *QST* editor, "The story of the transatlantics" (1922), 7–9.

50. Ibid., 10. The free-to-all part had further divisions into ten segments (fifteen minutes each), corresponding to nine geographical districts of the United States and Canada. An amateur could transmit only within the fifteen minutes for his or her district. The order of transmission for all districts was different each night. For example, in the first night the order was 1, 2, 3, . . . , 9, C (Canada); in the second night, 2, 3, 4, . . . , 9, C, 1, and so on. See *QST* editor, "Godley to England" (1921), 29. The division reduced the flow of instant information for the convenience of traffic handling. The schedule rotated to allow each station to transmit in different parts of the night; many experts thought that the sun's position affected the efficacy of long-distance wave propagation.

51. Schnell, "Transatlantic sending tests" (1922), 20.

52. See anonymous, "Schnell" (1919), 29.

it used to be, from which the QRA can be determined in the call book:

Calls 1AA to 1WZ are not changed. Calls 1AAA to 1WAZ have the first two letters transposed. For example, old 1RAY is now 1ARY. Now notice that in the erroneous way in which the calls were first issued, the combinations having the common letter A had to conclude at 1WAZ, calls beginning with X, Y and Z being specials; but when corrected by transposing the A to become the first letter this difficulty no longer holds. Consequently the calls 1ABA to 1CBZ have been appropriated to fill in this blank at the end of the A scale, and this throws the middle letter of the rest of the B system three letters behind. Therefore, calls 1DBA to 1WBZ have had (1) the first two letters transposed; (2) the resulting middle letter slid back three letters in the alphabet. For example, old 1GBC is now 1BDC. Similarly, the calls 1ACA to 1FCZ were necessarily appropriated to fill in the blank, now twice as large, at the end of the B scale, and this throws the middle letter of the rest of the C bunch six letters behind. For example, old 1GCN is now 1CAN.

# Transatlantic Sending Tests

## By The Traffic Manager

THE Operating Department wishes to announce that the second attempt to span the Atlantic Ocean will take place December 8th to 17th, inclusive. Until the complete list of transmitters is known the exact time cannot be given, but the tests will start about 8 p.m. Eastern Standard Time and continue until about midnight, giving each transmitter a fair chance to accomplish this almost unbelievable feat.

In order to have only the very best and most far-reaching transmitters in this test, preliminary tests will be held November 7th to the 12th, inclusive. The preliminaries will be over land and probably will specify that 1000 miles air-line must be covered in order to qualify for the finals. Details will be announced soon.

Mr. Phillip R. Coursey, B.Sc., F.Inst.P, A.M.I.E.E., assistant editor of "The Radio Review" (London), will have complete charge of the receiving stations in England and other countries where amateurs will listen for our signals. Should we be successful in our attempts, Mr. Coursey will decide the winners after he has received all the data from his receiving stations. It was thru the untiring efforts of Mr. Coursey that we received the splendid co-operation of the English amateurs in the tests of last February, and Mr. Coursey assures us that they are keen to try it again.

Whether or not we shall have prizes rests entirely with our manufacturers and dealers. If the fascination of this idea of getting 'cross-seas is such that they wish to donate apparatus to be given the winners, we will be glad of the opportunity to give them full credit and announce their prizes in this magazine.

Fellows, our good old A.R.R.L. is calling to you amateurs with your excellent transmitters, inviting you to enter this contest in the name of good sportsmanship and in the interest of the advancement of Amateur Radio. We know you will answer as you have in the past. We want the Atlantic Ocean spanned on schedule by an amateur station and we want definite proof that it has been done. Full credit will be given the amateur or engineer who has anything to do with the transmitter that succeeds. The only requirements are those of the U.S. Radio Communication Laws. The power input must not exceed 1000 watts and the wave length must be 200 meters. The laws permits transmission on waves below 200 meters but since the English stations will be tuned for reception on 200, we ask you to use that wave.

This announcement is for the purpose of getting entrants. If you have a good DX transmitter or contemplate having one by November 8th, then send in your name. Applications will be accepted up to and including October 12th. Use the form below, or make up one similar thereto if you wish to avoid mutilating your copy of QST.

Traffic Manager, A.R.R.L.,
1045 Main St., Hartford, Conn.

Please enter my station as a transmitter in the Transatlantic Sending Tests, December 8th to 17th. I will be ready to transmit in the preliminary tests on November 7th to 12th, and if I fail to cover the specified distance in the preliminary tests I shall relinquish my rights to transmit in the final tests.

Name.................... Call Letters..........

Street..............City and State..............

Power of transmitter..........................

Type (CW or spark)..........................

Greatest distance heard (give three records)..........................

FIGURE 5.2. In its periodical *QST*, ARRL called for radio amateurs to participate in the second transatlantic experiment. Schnell, "Transatlantic sending tests," *QST* (September 1921), 12, courtesy of *QST*.

committee and the Radio Club of America. In these amateur societies, he was famous for having taken "the Armstrong circuits, then considered impractical for short-wave work, and adapted them to amateur work."[53] His expertise on heterodyne receiving circuits explained why ARRL chose him for the mission. Godley departed New York for England on 15 November, carrying a super-heterodyne regenerative receiver with five-stage amplification.[54] His original plan was to set up a receiving station in London, but strong static interference there forced him to move to Ardrossan, Scotland.[55] By December, Godley had secured operating permission from the British Post Office, set up a Beverage antenna, hooked up the receiving set, found the best tuning, and was ready to proceed.

Credibility was a key issue for experimentation. How did the American and British radio amateurs make their experimental results credible to others? Like the Humboldtian experiments of the nineteenth century, the transatlantic wireless tests involved a lot of participants with different degrees of experimental skill and technical capability. Thus some sort of standard was necessary. Unlike many earth-science experiments, however, these tests did not concern precise *measurements* of quantitative features of certain physical phenomena that *had existed* in nature. Instead, the phenomena resulted from experimenters' sending radio waves as messages. And obtaining accurate numerical data was not the issue. The actual goal was ensuring that the messages arrived with little or no error. In other words, they were *communications* experiments. Therefore, the approaches that usually established scientific credibility—accurate instruments, visual or literal presentation of data, and error analysis, among others—were less relevant here. Rather, credibility demanded a procedure that authorized the witnessing of successful message exchanges. Such a procedure, which the communications protocol laid out, resembled a notary.

This was exactly what the sponsors arranged: every U.S. transmitting station passing the preliminary tests received a sealed secret cypher combination just before the experiment. The operators would open the seal and transmit the encrypted signals on 8 December 1921. In Britain, no one except the main organizer, Philip Coursey, knew the secret combinations, which Godfrey took to him. All the incoming messages that British amateurs claimed should go

53. *QST* editor, "Godley to England" (1921), 29.

54. *QST* editor, "The story of the transatlantics" (1922), 10; Godley, "Official report" (1922), 39.

55. Godley, "Official report" (1922), 17–22.

to Coursey. By comparing these messages with his secret combinations, he could identify *authentic* successful communications. Moreover, since Godley's work was a hallmark of this experiment, he had to be under stricter supervision. M. D. Pearson from the Marconi Company in Glasgow constantly watched him during the test and verified the reception of every signal at his station.[56]

The second transoceanic trial was much more fruitful than the first one: the radio amateurs' short waves did cross the Atlantic. On page 1 of the January 1922 issue of *QST*, the editor enthused, "Oh, Mr. Printer, how many exclamation points have you got? Trot 'em all out, as we're going to need them badly, because WE GOT ACROSS!!!!!!" (figure 5.3).[57] Godley's first successful reception of American signals occurred on 7 December. Sitting in a tiny tent engulfed by Scotland's chilly winds, the American ham wrote in his log: [58]

> At 1:33 A.M. picked up a 60-cycle synchronous spark at about 270 meters, chewing rag. Adjusted for him, and was able to hear him say "C U L" and sign off what we took to be 1AEP; but atmospherics made sign doubtful! That this was an American ham there was no doubt! I was greatly elated, and felt very confident that we would soon be hearing many others! Chill winds and cold rains, wet clothes, and the discouraging vision of long vigils under most trying circumstances were forgotten amidst the overwhelming joy of the moment—a joy which I was struggling to hold within! I suggested hot coffee at once, and Pearson volunteered to warm it on our stove. He had pot and bottle in his hands when I called sharply to him to resume watch! Our welcome American friend was at it again with a short call for an eighth district station! His signal had doubled in strength, and he was booming through the heavy static and signed off clearly 1AAW, at 1:42 A.M.!

But this was just the beginning. In the ten-day test period, Godley received signals from nineteen American and Canadian continuous-wave (CW) stations and nine American and Canadian sparkgap stations. Also, eight British

56. *QST* editor, "The story of the transatlantics" (1922), 11.

57. *QST* editor, "Transatlantic tests successful" (1922), 7.

58. Godley, "Official report" (1922), 23. Note that a radio wave of 270 m is now classified as a medium-length wave, not a short wave. But to the radio amateurs and engineers in the early 1920s, this wavelength was clearly shorter than the common wavelengths for long-range communications. The historical actors thus would not hesitate to designate it as a short wave. To them, the numerical definition of short waves was not available yet.

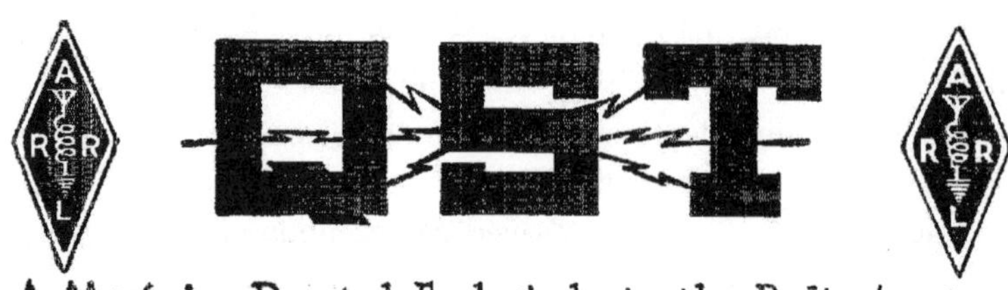


A Magazine Devoted Exclusively to the Radio Amateur

## *Transatlantic Tests Successful*

OH, Mr. Printer, how many exclamation points have you got? Trot 'em all out, as we're going to need them badly, because WE GOT ACROSS!!!!!!

As we prepare the copy for this issue of QST our Transatlantic Tests are in progress and we have the highly gratifying news from Paul F. Godley, our special listener in Scotland, that the A.R.R.L. has spanned the Atlantic! For the first time in history the signals of United States and Canadian amateur stations have been heard across the ocean on schedule.

Mr. Philip R. Coursey, in charge of arrangements in Great Britain, radioed us on Dec. 13th as follows:

"*Many your stations heard by British amateurs. Details later.*"

We are most impatiently awaiting receipt of Mr. Coursey's detailed report, the compilation of which necessarily will have to await the collection and examination of the individual logs from the British listeners. It is this phase of the tests in which we are particularly interested—we want the British amateurs, with their normal receiving apparatus, to hear our signals if they can, so that we may hope to move amateur traffic to them on schedule. We trust that Mr. Coursey's report will be received in time for our next issue.

Paul F. Godley, special representative of the A.R.R.L., with special American equipment, located his station at Ardrossan, a small fishing village some twenty miles to the west of Glasgow, Scotland, after experimenting with various locations, and there listened for our signals thruout the ten day period, reporting nightly via radiogram from MUU which was repeated on this side by WII. To date twenty-six stations have been reported by him, as listed on the cover of this issue—six sparks and twenty-two C.W. stations. These are mostly in the eastern part of the country, rather contrary to expectations, the westernmost one being in Cleveland, Ohio. There is but one Canadian reported, 3BP, Rogers of Newmarket, and on his spark at that, but Mr. Coursey's report may show more of our cousins in the Dominion.

Station 1BCG at Greenwich, Conn., was reported on two consecutive nights and indications are that it had the greatest signal strength of any heard. This station was especially erected for the tests and was jointly owned and operated by Messrs. Minton Cronkhite, E. H. Armstrong, George Burghard, John Grinan, Ernest Amy, and Walter Inman. In its testing it has been reported from the Pacific Coast and must have kicked up considerable of a rumpus. Encouraged by the report of their signals, these men attempted to transmit an actual message, and to their credit be it said that they succeeded in putting across the ocean the first private radiogram ever transmitted across this span by an amateur station. The message was transmitted on the night of Dec. 11th, and acknowledged by a cablegram to A.R.R.L. Headquarters by Godley, reporting its reception at 3 a.m. G.M.T. on the 12th. The message read as follows:

"*Nr 1 NY ck 12 to Paul Godley, Ardrossan, Scotland. Hearty congratulations. Burghard Inman Grinan Armstrong Amy Cronkhite.*"

Thus not only have amateur signals been heard overseas in astounding number, but a coherent message has been put over by the same means.

---

This is all the news we can give you at this writing, fellows. We got over, as we said we would, and our A.R.R.L. did it. It opens the door to big things and the scientists of the world are of course gasping and marvelling that such small powers on such short wave lengths could cover such distances. It will take some weeks to get the official story of the Transatlantics in final form, as we must now await Godley's return and Coursey's detailed report, but we will present it just as quickly as possible. And there will be some more call letters in the British report, you bet!

FIGURE 5.3. ARRL announced the success of the transatlantic experiment in *QST*. *QST* editor, "Transatlantic tests successful," *QST* (January 1922), 7, courtesy of *QST*.

receiving stations succeeded in copying messages from eleven North American CW transmitting stations.[59] Almost all the signals that came in were from stations transmitting during the second parts of the nights. Amateurs in The Hague and Nice[60] received the American signals too.

The success of the transatlantic test was great news to radio amateurs. Godley received a grandiose welcome on returning to his homeland: friends, press correspondents, and photographers. To American amateurs, Godley was the kind of hero who would anticipate Charles Lindbergh a decade later. Two electrical engineers remembered that when they were "radio boys" in New York City in the 1920s, every friend of theirs dreamed of seeing Lee de Forest and Paul Godley in person.[61] Yet—even more important—not only a single hero but also quite a number of radio amateurs succeeded in sending and receiving transatlantic signals in this collective large-scale experiment. The outcome did not just demonstrate that short waves could reach long distances, despite the Austin-Cohen formula. It showed the repeatability of this effect and the possibility of *regular* long-range short-wave communications.

The next step, the American and British amateurs thought, would be to make short-wave communications across the Atlantic regular. So they decided to organize another experiment with even more participants.

## *The Third Trial*

The third amateur transatlantic experiment followed in 1922. The test used the basic procedure from the previous two experiments, but on a broader scale. More transmitting and receiving stations joined. ARRL and the Radio Society of Great Britain invited amateurs in Western Europe to participate. Experience indicated that radio signals from North America had a good chance to reach not only the British Isles, but also Continental Europe. Three French amateur societies—the Radio-Club de France, la Société des Amis de la T.S.F., and la Société Française d'Étude de T.S.F.—formed a Comité Français des Essais Transatlantiques to take care of the experimental issues.

59. DeSoto, *Two Hundred Meters* (1936), 73; *QST* editor, "The story of the transatlantics" (1922), 11–12. The British receiving station with the best performance among the eight successful ones belonged to W. R. Burne from Manchester. He later won a prize that local electric manufacturers donated. See anonymous, "Presentation of prizes" (1922), 10–12.

60. *QST* editor, "The European transatlantic results" (1922), 20; DeSoto, *Two Hundred Meters* (1936), 74.

61. Interview with Albert Wallace Hull, part IV, no. 109, and with Lloyd Espenshied, part IV, no. 72, Radio Pioneers, Columbia University Oral History Collection.

Pierre Corret, a member of all three bodies, chaired the committee. Gustav Ferrié also supported the project. Individual amateurs in Belgium, Holland, and Switzerland joined the tests.[62] The experimenters planned to try both eastward and westward communications across the ocean.

Selection of transmitting stations was under way in October on both sides of the Atlantic. In North America, Schnell organized a preliminary test between 25 October and 3 November, in which transmitting stations sent signals over distances beyond 1,200 miles.[63] In Britain, in the November 1922 issue of *Wireless World and Radio Review*, Coursey called for transmitting stations capable of sending signals 400–500 miles or more.[64] Because laws severely limited maximum transmitting power in Europe, westbound communication was more difficult than eastbound. To compensate, the Radio Society of Great Britain built for the experiment a transmitting station—call number 5WS, in Wadsworth, London—and the Post Office gave it a special license to use up to one kilowatt of power.[65]

To prepare for the test, both American and British organizers educated amateurs by placing in hobbyist magazines information about the designs of short-wave radio. Coursey introduced heterodyne circuits to British amateurs and advocated their use in receivers.[66] Godley shared his experience in the 1921 tests in arranging receiver circuits, erecting antennae, selecting vacuum tubes, and choosing locations.[67]

The tests began in mid-December 1922. American and Canadian stations transmitted between 12 and 21 December, and British and French stations between the 22nd and the 31st. Organizers grouped U.S. and Canadian stations into ten districts, assigned each one a time segment to transmit, and changed the schedule of all time segments every night. British and French transmitting stations also divided nights: in the first night British amateurs took the first half, and French the second; in the second night they exchanged time slots, and so on.[68] Operating wavelengths were between 180 meters and 240 meters.[69]

62. Schnell, "Arrangements for 1922 transatlantics" (1922), 23.
63. Schnell, "The ARRL transatlantics" (1922), 11–12.
64. Coursey, "The transatlantic communication test" (1922), 185–88.
65. Coursey, "5WS" (1923), 785–88.
66. Coursey, "On heterodynes" (1922), 161–63.
67. Godley, "Listening for Europe" (1922), 33–35.
68. Schnell, "The transatlantic finals" (1922), 8–10; Coursey, "The transatlantic tests" (1922), 379.
69. *QST* editor, "The transatlantic triumph," *QST*, 6:7 (Feb. 1923), 15–16.

The results were both inspiring and disappointing: eastward communication was much more successful than westward. In the first half of the test, British amateurs heard 161 North American transmitting stations, and the French and Swiss heard 239; 315 American and Canadian stations reached Europe.[70] In contrast, only two European stations reached North America: 5WS and 8AB, the 500-watt CW station of the French amateur Léon Deloy in Nice.[71]

Since clearly short-wave communications were viable both ways across the ocean, it was natural to test two-way communications—two stations talking directly on air. But westbound traffic was far less satisfactory than its opposite. Radio amateurs had to improve this result before making transatlantic short-wave communications a regular link. They took action in 1923.

### *The Two-Way Test and the Fourth Trial*

The major European nexus testing two-way communications was Deloy's 8AB. A native of Paris, Léon Deloy developed an interest in electric apparatus through his astronomer godfather Camille Flammarion. In World War I, he served at the Eiffel Tower radio station, and the government dispatched him to Washington in 1917 to exchange radio-technology information with the U.S. Navy. His wartime service acquainted him with some influential future supporters of radio amateurs, including Gustav Ferrié and Edwin Armstrong. When peace came, Deloy established his private station, 8AB—one of the most advanced amateur stations in Continental Europe.[72] With the support of Ferrié's Military Telegraph Services, he obtained a special license for transmitting at 500 watts and adopted state-of-the-art transmitting and receiving technologies. In the transatlantic experiment in 1921, as we saw above, 8AB was the only station in France that received signals from North America. It was also one of the two stations that transmitted signals across the Atlantic in the 1922 experiment.

Deloy conducted a two-way communications test with station 1OKP in South Manchester, Connecticut, immediately after the 1922 experiment, but it failed. In March 1923, he made an announcement, through Kenneth B. War-

70. Ibid., 7–12.

71. Coursey, "5WS" (1923), 829. The *QST* editor reported reception of signals from another British station, 2FZ. But Coursey pointed out that the station with this call number was not operating at the reported time of reception. So the report was a mistake.

72. See anonymous, "Who's who in amateur wireless" (1922), 61, 63, 66.

ner, in *QST* calling for more participants.[73] That summer he visited the United States to study American amateurs' operating and design techniques. Returning to France in the early fall, he applied the knowledge to modify his station. He resumed the two-way transatlantic experiment in November, succeeding this time. The signals that he sent (with wavelength 100 meters)[74] on 25 and 26 November reached Warner and Fred Schnell at 1MO and John Reinartz at 1XAM, both in Hartford, Connecticut. On 27 November, Deloy was able to *converse* with Reinartz and Schnell on air. Between 9:30 and 10:30 p.m., both American stations received Morse-code messages from 8AB. Then they replied to Deloy. After a while, they received clear messages from Deloy acknowledging reception of their previous replies. It showed the possibility of real-time two-way communications across the Atlantic.[75]

Meanwhile, another large-scale experiment was under way. ARRL and the Radio Society of Great Britain decided to repeat the 1922 experiment in late 1923. This time they tested only westward transmission. Schnell announced the upcoming event in the December issue of *QST*. He reminded readers that although the unsatisfactory outcome of the last experiment might have resulted from the weak power of British and French transmitting stations, part of the problem lay with American amateurs' reception. A lot of U.S. participants had complained about serious jamming from their peers' transmitters when they were trying to receive. Too much radio traffic affected the quality of the receiving experiment. So Schnell strongly requested all American amateurs to remain absolutely quiet on air during the transatlantic test. If a station transmitted while the experiment was going on, then its owner would lose the right to win an award of up to $3,500 from domestic manufacturers and dealers, no matter how well he or she received.[76]

73. Warner, "Two-way tests with Europe" (1923), 13–15.

74. This wavelength was shorter than those in the previous transatlantic experiments. Some amateurs tried to push to shorter wavelengths in 1923. S. Kruse, the technical editor of *QST*, held in March a field test at 100 meters. Kruse, "Exploring 100 meters" (1923), 12–13. Deloy followed this trend.

75. For Deloy's two-way experiment with Reinartz and Schnell, see DeSoto, *Two Hundred Meters* (1936), 86–87; Deloy, "Première communication transatlantique bilatérale" (1923), 678–83, and "Communications transatlantiques" (1924), 38–42. Deloy received "la croix" in 1930 from the French magazine *Animateur des temps nouveaux* for his achievements in shortwave long-range radio; see "Chez nos confères—la 'croix' pour M. Léon Deloy," *TSF—Revue*, 8:307 (23 Nov. 1930), Special Collection, Libraries of Radio France, Paris.

76. Schnell, "The fourth transatlantic tests" (1923), 9–11.

The fourth transatlantic experiment lasted from 22 December 1923 to 10 January 1924. Dozens of British and French stations transmitted predesignated messages, and hundreds of American and Canadian stations attempted to receive them. The procedure followed the protocol from the previous tests. Afraid of expulsion, American amateurs remained silent. This did reduce interference significantly, and remarkable results emerged. North Americans picked up thirty-seven European amateur stations with assigned codes and five other unofficial stations, all operating at between 108 meters and 118 meters.[77] While there were only two European stations that succeeded in the previous experiment, this time there were forty-two.[78]

In the first half of the 1920s, American, British, and French radio amateurs performed a series of large-scale wave-propagation experiments. These tests embodied characteristics of amateur activities: competitive games, the struggle for records, many participants with heterogeneous backgrounds, and minimum control of variables. To many amateurs, these experiments may have seemed like expeditions to the Antarctic, transcontinental signal relays, and other fun-seeking activities of the period that reflected their masculine subculture. The amateurs did not organize these activities to seek answers to specific questions, nor did they care much about accurate measurements of well-defined physical quantities. Their main objectives were to communicate radio signals in various harsh environments, to connect with the maximum number of other amateurs on air, and to copy messages as correctly as possible. They developed their technological knowledge—circuit design, operational techniques, and skills of coordinating large-scale tests—principally for these objectives rather than for laboratory experimentation.

These features became a shortcoming in the ARRL-NBS experiments on fading. Organizers attempted to investigate the effects of atmospheric conditions on fading, but, aside from this generic direction, they did not formulate specific questions. The experimental design did not address issues from any theory of fading. Although a lot of data emerged, their quantitative analysis (except for the statistical correlation of certain qualitative factors) was inadequate. Nor did mathematical formulae represent the data. Moreover, a significant portion of the data was unusable, since many amateurs who followed radio operators' working style failed to reproduce the fine structures of signal

77. Warner, "The progress of transatlantic amateur communications" (1924), 15–16.
78. Warner, "Transatlantic tests report" (1924), 32–34.

intensity's variation over time. The ARRL-NBS fading tests left much less of a legacy to radio science than, for example, British physicist Appleton's fading experiments in the early 1920s (see chapter 7 below).

Nevertheless, the same features characterizing amateur activities contributed to the success of the transatlantic experiments, which sought straightforward evidence: they were to demonstrate the *possibility* and *stability* of a phenomenon, not to construct an empirical law, establish a scientific theory, or measure a physical quantity. Radio amateurs' energy and determination and their keenness to try to communicate in every kind of difficult situation sustained their collective effort. They were also good at maintaining the credibility of reports of communications, which was critical to the reliability of evidence. However, they did not have to pay much attention to the accuracy of measured quantities, the adequacy of data analysis, and the theoretical interpretation of results, for these were not their goals. The point for them was rather to demonstrate that an "impossible" effect was actually possible and stable.

Radio amateurs' sociopolitical status also helped. The legal barrier at 200 meters forced them to work exclusively on short-wave technology, making them experts on short waves and freeing them from concerns about long-distance long-wave communications, which preoccupied most government, corporate, and academic researchers. Their myriad numbers and highly interactive networks facilitated the well-coordinated large-scale tests and increased the chance for radio signals to travel far in unstable atmospheric conditions.

Consequently, radio amateurs' transatlantic experiments showed that the Austin-Cohen formula and its underlying theory of atmospheric reflection did not apply to short waves. Under normal, calm, nighttime atmospheric conditions, wireless waves shorter than 200 meters with overall transmitting power less than one kilowatt could travel more than 2,000 miles, much farther than contemporary physics predicted. Short-wave long-range communications no longer remained merely an idea, hearsay, or a peculiar case under anomalous or random geophysical circumstances. It became feasible. Of course, whether a reliable link of that kind using low-power short-wave radio would soon become possible was still a question. Some engineers and scientists were still skeptical about the practicality, if not the reality, of radio amateurs' transatlantic experimental results. P. P. Eckersley, an engineer of the British Broadcasting Company, commented in 1923:[79]

79. Eckersley, "5WS" (1923), 876.

> The result of these tests emphasises the extraordinarily small power used to cover enormous distances with success and many of you may wonder why the professionals build an aerial three miles long, putting in kilowatt after kilowatt[,] and then consider they are allowing only a safe margin to bridge the distance . . . The difference is in getting intelligible speech for a moment and getting good and lasting speech for twenty-four hours of the day . . . The attitude I emphasise is that the professional and the amateur are working from an entirely different point of view.

Nevertheless, conveying intelligible speech for a (repeatable) moment was already attractive enough for many professional radio users. Learning from radio amateurs' tests, the U.S. Navy speeded up research and development on high-frequency radio and began to seek collaboration with ARRL to test the propagation characteristics of short waves. The cooperative tests between the U.S. Navy and American amateurs led to another discovery in 1924 that shifted the understanding of short-wave propagation—the "skip zone" effect—as we see in the next chapter. Moreover, the U.S. Navy systematically collected the empirical data on the relationship between wave-propagating range and wavelengths shorter than 200 meters. The short-wave data—along with the skip-zone phenomenon—demanded a new theory of radio-wave propagation that depended much more heavily than the Kennelly-Heaviside model on the ionosphere.

CHAPTER 6

# From the Skip Zone to Magneto-Ionic Refraction

As soon as wireless hobbyists demonstrated the feasibility of long-range communications at frequencies higher than 1.5 MHz in 1920–22, governments, corporations, and academia jumped on the bandwagon of short waves to exploit their potential for applications. The state was on the verge of reversing spectrum legislation to take away amateurs' rights over high frequencies. Electronics manufacturers and communications firms hastened their development and deployment of high-frequency wireless telephones; and by taking over another innovation of amateurs'—broadcasting—at about the same time, they turned short-wave radio into a major instrument of modern mass media. To those who were following the science of propagation, the amateurs' finding was about to trigger an intellectual earthquake.

The more the professional engineers and scientists researched the high-frequency part of the spectrum, the more uncertain they became. Short-wave propagation not only disobeyed the Austin-Cohen formula, as the radio amateurs' experiments had shown; it did not behave monotonically as a "smooth" natural regularity should. A representative effect among the novel short-wave phenomena was the skip zone, which the U.S. Naval Research Laboratory (NRL) in Washington discovered in mid-1924, as we see below: radio signals diminished at certain distances from the transmitter but *reemerged* farther away. Neither a simple law of absorption nor Watson's theory of atmospheric reflection could account for such a peculiar fact.

The increasing attention to short-wave propagation led to revival of the ionic refraction theory in the early 1920s. Throughout the 1910s, the idea of radio waves refracting in a gradually ionizing atmosphere served mainly as a physical account of the Kennelly-Heaviside hypothesis in explaining long-distance propagation and atmospheric noise. Despite its visibility among engineers and experimenters, only Eccles did research on it. And the assumption underlying his theory—that only the atmosphere, not the ground, influenced wave propagation—appeared doubtful to supporters of Heaviside's waveguide-like boundary condition. Yet the radio experiments in the early 1920s implied that short-wave propagation, unlike long waves, seemed to involve *only* the atmosphere. For wavelengths shorter than 200 meters, the ground ceased to be a relevant boundary condition. This motivated scientists and engineers to reconsider Eccles's theory. In 1924–25, Cambridge Maxwellian Joseph Larmor gave a reformulation of ionic refraction, while English physicist Edward Appleton and AT&T researchers Harold Nichols and John Schelleng supplemented the theory with the effect of geomagnetism. The revitalized model—the magneto-ionic theory—quickly became a framework for understanding the new short-wave phenomena. In 1925–26, Albert Hoyt Taylor and Edward Hulburt at NRL explained the skip-zone effect and reproduced its quantitative variation by hypothesizing a refraction-inflicting ionized atmospheric layer. The skip zone, to Taylor and Hulburt, was a direct consequence of radio waves "escaping" from the finitely thick atmosphere.

Short-wave research in the early 1920s epitomized a transition of radio-propagation studies in both experimental practice and theory building. Experimentally, the discovery of the skip zone and the further accumulation of short-wave transmission data opened a new page of military involvement in empirical research on propagation. War had transformed the attitudes of the U.S. Navy, the major contributor in that area, towards research and development. While the navy had relied heavily on external contractors for inventions and innovations, when peace came it enacted its own research capacity.[1] The inception of the Naval Research Laboratory (NRL) was a representative move. Unlike the Naval Radiotelegraphic Laboratory, which functioned like a testing shop and obtained the Austin-Cohen formula accidentally from equipment testing, NRL now had a whole radio section for more scientific research on propagation.

Moreover, the staff at NRL was under the strong influence of radio amateurs' technical culture. People such as Taylor were engineers and hobbyists. They not only learned amateurs' way of large-scale, diffuse field experimenta-

1. Mindell, *Between Human and Machine* (2002), 19–30.

tion, but also sought their collaboration in implementing this style of practice. Their discovery of the skip zone was a result of military-hobbyist cooperation. Rather than setting up a radio transmitter and a few mobile measuring platforms (such as battleships), as Austin and his colleagues had arranged in 1909 and 1913, the skip-zone experiment involved a network of *fixed* short-wave stations, including NRL's NKF in Washington and a number of amateur stations in the eastern United States. This pattern foresaw the centrality of (hobbyist, commercial, and research) radio stations in the upcoming atmospheric propagation experiments.

Theoretically, Taylor and Hulburt's account of the skip-zone effect constituted a breakthrough for ionic refraction. It did not just extend the theory to cover one more empirical fact; it altered the theory's epistemic status. On its discovery, the skip-zone effect became a "p-predicament," in philosopher Sylvain Bromberger's term: it had not been conceived when the ionic refraction theory (or any other propagation theory) first emerged; it came as a surprise; and it was difficult to predict the form of any answer to the question "why are there skip zones?"[2] The ability to explain the skip zone thus indicated that ionic refraction was becoming a machine for generating novel empirical statements. In dealing with short waves, it appeared to be productive for many engineers and scientists by the mid-1920s.

The triumph of ionic refraction facilitated an even subtler epistemic transformation in radio propagation studies. While the old atmospheric reflection had built predictions and explanations on the geometry and very few material parameters of a featureless, homogeneous Kennelly-Heaviside layer, ionic refraction resorted to a much more structured, heterogeneous, and hence interesting atmosphere. The skip zone and other short-wave irregularities resulted from the upper layer's height, thickness, and electron-density profile or from the geomagnetic rotations of radio waves within it. This assertion had a flip side: the physical characteristics of the ionized atmosphere could account for various wave propagation phenomena, and those phenomena revealed the structure of the upper layer, too. The work of short-wave researchers in the early 1920s prepared the ground for a change of focus from the behavior of radio waves using the upper atmosphere as an explanatory tool to the properties of the ionized atmospheric layer using radio waves as a probing instrument. Propagation studies were beginning to evolve into atmospheric physics.

2. Bromberger, *On What We Know* (1992), 26–31.

## DISCOVERING THE SKIP ZONE AND SHORT-WAVE DATA

Short waves were strange. This was a primary lesson that wireless technologists learned in the early 1920s. The neat, simple relationships that governed long waves ceased to hold when the radio frequency exceeded 1.5 MHz. At high frequencies, quantities no longer increased or decreased monotonically with one another. Rather, the patterns of variations became more complex. Radio amateurs' finding against the Austin-Cohen formula alluded to one such complexity: the results from amateurs' transatlantic experiments suggested that, at short wavelengths, the intensity of propagating waves did not decrease exponentially with distance, as the empirical law had predicted. But this was just the beginning of numerous discoveries in the 1920s that questioned preconceptions about radio-wave propagation. What if the intensity of propagating waves did not decrease with distance *at all*? What if wave intensity did not decrease monotonically with wavelength? What if propagating waves could reach from one location to another only within a finite range of frequencies? What if this range of usable frequencies varied diurnally and seasonally?

The most notable discovery among the novel high-frequency phenomena was the "skip zone." A skip zone was a region where propagation of radio waves "skipped." Normally, the intensity of incoming wireless signals attenuated with separation between transmitter and receiver. When wavelength was shorter than 75 meters, however, signal intensity no longer decreased monotonically with distance. It diminished to a negligible level after a distance but *rose again* at a greater distance. No recognizable signal was perceptible in the intermediate region (skip zone), but radio signals exterior to the region were recognizable. In other words, radio waves jumped around; this ran against the intuition that an outgoing wave simply attenuated with distance as the wave front spread more and more widely over space.

The skip-zone phenomenon emerged when radio technologists, learning from amateurs' short-wave transatlantic tests, pushed towards higher and higher frequencies to see their effects on propagation. They performed experiments themselves or compiled data from wireless operating records, in an attempt to establish the relationship between wave intensity, distance, and frequency. In so doing, they found another interesting fact: a radio wave's range of propagation—the maximum distance at which it was detectible via state-of-the-art receiving technology—did not vary monotonically with its frequency. Instead, as the frequency moved from low to high (the wavelength from long to short), the range of propagation first decreased (following Austin-Cohen), *but then increased.* That is, the range of propagation was minimum at a certain frequency. Moreover, the empirical data indicated that the frequency corresponding to the minimum range of propagation was different in daytime, winter night, and summer night.

The skip-zone effect, the nonmonotonic variation of range of propagation with frequency, and the diurnal and seasonal changes dramatically affected short-wave wireless communications. On the one hand, these new facts reinforced what radio amateurs had shown from 1920 to 1924—an exciting possibility of long-range communications using much less power at higher frequencies. On the other hand, however, the short-wave phenomena pointed to the complexities in implementing such systems: careful choice of operating frequencies had to avoid skip zones as well as maintain maximum ranges of propagation. And these frequencies required adjustment vis-à-vis location, time of day, and month of year. Choosing proper operating frequencies in situ became a central operational practice of wireless communications. Because of the characteristics of short-wave radio channels, therefore, implementation of wireless systems involved not only design but also operation.

This practical imperative set the tone for short-wave propagation experiments in the mid-1920s. No wonder that the discoverers of the skip zone and the compilers of short-wave range data were not academic scientists. But they were not solely radio amateurs or corporate engineers, either. Once again, the military played an important part. The major experimenters behind these short-wave investigations were members of the new U.S. Naval Research Laboratory. Under Albert Hoyt Taylor, these naval researchers measured the range of radio-wave propagation at higher and higher frequencies. Their experiments resembled radio amateurs' fading and transatlantic tests in the early 1920s: being radio hams themselves, staffers at NRL also rallied volunteers (from ARRL and other amateur societies) on air to join the propagation measurements. Yet, Taylor and colleagues maintained a much clearer goal: to develop short-wave communications systems for the military. Eventually, their work led to the discovery of various short-wave phenomena that transformed the understanding of radio-wave propagation in the upper atmosphere.

### *Albert Hoyt Taylor and the U.S. Naval Research Laboratory*

Investigating nature, inventing machines, performing military tasks, playing at a hobby. Although these realms of activity were apparently disparate, they came together in specific contexts of American history in the late nineteenth and early twentieth centuries. Above all, American science had a pragmatic tradition that stressed tangible experience and applicability over theoretical sophistication. Rapid industrialization after the Civil War drove American Telephone and Telegraph, DuPont, General Electric, Standard Oil, Westinghouse, and other large companies to seek a "marriage between science and useful arts" by establishing laboratories and hiring PhDs in physics and chemistry to do applied research.[3] Moreover, the growing strength of industry and manufacturing encouraged Americans to view technology as a distinct cultural mark of their nation and thus motivated them to develop enthusiasm for technology. After the invention of airplane, automobile, and radio, more and more Americans cultivated the hobby of tinkering with these new gadgets.[4] Finally, World War I brought these trends together under the rubric of military need. The armed forces adopted the model of industrial laboratories, recruited scientists, and drafted amateurs to perform research, development, and operations regarding military technologies. As a result, the mobilization created a World War I generation of scientists and engineers.

Albert Hoyt Taylor was a representative of that generation, which moved into diverse realms of activity. Taylor was a physicist, engineer, naval officer, and radio amateur. Born in Chicago, he was enthusiastic about playing with electric devices at home from childhood on—a precursor of the "radio boy." He entered Northwestern University in 1896 to study physics and became an instructor at the University of Wisconsin in Madison. Like many ambitious fin-de-siècle Americans seeking careers in science, he went to Germany for graduate study. Owing to the work of figures such as Ferdinand Braun, Heinrich Hertz, Adolf Slaby, and Jonathan Zenneck, Germany was a center for the art and science of electromagnetic waves. Taylor's original goal was to study wireless with Braun at the University of Strasbourg. But after hearing that Braun kept most of his wireless research secret for commercial reasons, he headed for Göttingen. There he took courses in mathematical physics and applied electricity with Max Abraham, David Hilbert, and Hermann Voigt and wrote a PhD dissertation on the aluminum cell, an electrolytic Hertzian-wave detector. In 1909, Taylor returned home to become head of the physics department at the University of North Dakota in Grand Forks.[5]

Taylor's interest in wave propagation began in Grand Forks as he became a

3. For American science up to the early twentieth century, see Kevles, *The Physicists* (1997); Reingold, *Science, American Style* (1991).

4. Haring, *Ham Radio's Technical Culture* (2007).

5. For Taylor's early life and career, see Taylor, *Radio Reminiscences* (1960), 7–30. A brief biography is also available in Brittain, "Albert Hoyt Taylor" (1994), 958.

ham and set up his own radio station at the university.[6] When "working" other amateur radio stations in the Midwest, he recognized diurnal and weather-related variations of received signals' intensity. Monitoring their long-term tendencies and seeking explanations for the variations interested Taylor. He performed an experiment on wave propagation with Albert Blatterman, owner of an amateur station and professor of electrical engineering at Washington University in St Louis. The two men exchanged nighttime radio signals at wavelengths of 1,500, 850, and 500 meters. Their intensity data had fading in which the intensity extremes at 1,500 meters always corresponded to the extremes at 500 meters. Since 1,500 was an exact multiple of 500, it implied that the waveforms at both wavelengths were superpositions of rays with wavelength-independent paths. Thus Taylor and Blatterman argued that the received signals were the sum of reflected waves from clouds, a diurnally varying ionized layer, and the permanent Kennelly-Heaviside layer.[7] This experience showed Taylor the complexity of radio-wave propagation.

Taylor presented these experimental results in 1916 at the New York meeting of the Institute of Radio Engineers (IRE). There he met Louis Austin, W. H. G. Bullard, and Stanford Hooper from the U.S. Navy and thereby began a lifelong relationship with the navy. Right before the U.S. declared war against Germany in 1917, Taylor made a decision that profoundly affected his future—he left the University of North Dakota to join the military. He took a commission in the Naval Reserve to become a lieutenant. Throughout the war, he served at the navy's wireless stations in various places in the Midwest and the East: Great Lakes station in Michigan, Belmar station in New Jersey, and Bar Harbor station in Maine.[8]

Taylor was a keen experimenter. In addition to monitoring regular military communications, he often tried new ideas in antennae and signal reception, and several highly capable technicians, including Louis Gebhardt and Leo Young, helped him. Taylor recruited them through his personal network of radio amateurs; these men—talented telegraph operators, tinkerers, and inventors—had been radio amateurs before enlisting in the navy, and they continued their hobby. Taylor took them with him as a team wherever he transferred in the navy,[9] and later they would assist in the navy's short-wave propagation trials.

6. Taylor, *Radio Reminiscences* (1960), 34.

7. Taylor and Blatterman, "Variations in nocturnal transmission" (1916), 131–55.

8. Taylor, *Radio Reminiscences* (1960), 39–60.

9. David K. Allison, "An Interview with Dr. Louis August Gebhardt," 12 and 19 Sept. and 3 Oct. 1977, 7–8, Historical Archives, Naval Research Laboratory, Washington, DC.

After the war Taylor chose to stay in the U.S. Navy to explore wireless, gradually establishing a reputation as a radio experimenter. In 1918, he became head of the Experimental Section at the Naval Air Station in Hampton Roads, Virginia, with the primary responsibility of developing aviation radio. In 1919, his radio team formed a Naval Aircraft Radio Laboratory under the Bureau of Engineering and moved to the Naval Air Station in Anacostia (Washington, DC).[10]

The task of the Naval Aircraft Radio Laboratory was to build wireless communications systems for the new corps of naval airplanes. Taylor designed noise-free aircraft-mounted receivers and measured the errors of long-range direction finders. Together with Leo Young, he explored the possibility of radio detection of moving objects—a precursor of radar.[11] The major technical issue for the lab was high-frequency radio. Taylor, Gebhard, Young, and others developed short-wave wireless devices: double-stage superheterodyne receivers, continuous-wave generators with accurate frequencies, and transmitting tubes with low parasite capacitance.[12] In 1923, the navy integrated the Naval Aircraft Radio Laboratory into the Naval Research Laboratory.

Taylor resembled his predecessor Louis Austin in many ways. Both did graduate studies in physics in Germany, were famous for wireless research, and built their careers in the U.S. Navy. Yet they had significant differences. Austin was a product of the German Reichsanstalt. His preoccupation was accurate measurements of physical quantities. His relations with the National Bureau of Standards had been strong, but he did not have much contact with the navy's line officers. In contrast, Taylor was once an officer and a radio amateur. His ethos was trying new things and making them work, rather than scientific rigor. And he maintained personal contacts with servicemen and amateurs that were crucial for collecting data on short-wave propagation. Austin and Taylor represented different approaches to scientific research in the military before and after the Great War: contractual projects in the hands of independent scientists versus fully fledged development in the service of the military.

Resonating with Taylor's personal history, the establishment of the Naval Research Laboratory also reflected the fundamental features of American sci-

10. Taylor, *Radio Reminiscences* (1960), 61–71.

11. Howeth, *History of Communication-Electronics* (1963), 320–25.

12. Taylor, *Radio Reminiscences* (1960), 85–100. Some of these products, such as the crystal-controlled CW generator, appeared after the Naval Aircraft Radio Laboratory merged into the Naval Research Laboratory in 1923.

ence and technology about the time of World War I. As soon as war broke out in Europe in 1914, American industrialists and scientists foresaw its economic and military threats to their country. The United States must prepare, they believed. It must find ways to overcome the shortage of imported materials and goods resulting from the British fleets' and German submarines' interruption of transatlantic trade. More important, it must sharpen its arms to be ready to protect its citizens, even to enter the war. To these people, science and technology were crucial means to prepare the country. Although historians tend not to see World War I (unlike its sequel) as a scientists' war, U.S. organizational endeavors to facilitate war research and development were still significant. The government mobilized chemists to manufacture poison gas, replace German glass, and find alternatives to mineral imports. It asked physicists to improve airplanes, radio, submarine detectors, and various optical and acoustic sensors. It set up major national research establishments, such as the National Advisory Committee for Aeronautics, the National Research Council, and the Naval Research Laboratory.[13]

The U.S. government considered the idea of a Naval Research Laboratory before it entered the war. In 1915, Thomas Edison proposed in the *New York Times* that the navy should establish "a department of invention and development" to fulfill the requirements of modern warfare. Navy Secretary Josephus Daniels convened a consulting board with twenty-two members from academia and industry (including Edison) to prepare for setting up the facility. The board's idea, following Edison's thought, was to extrapolate the model of industrial laboratories to the military sector. The extrapolation did not go well at the beginning, however. The government commissioned the Naval Research Laboratory in 1917 in Bellevue (Washington, DC), but it remained just a name for years. Edison boycotted the board after it voted for a site in Bellevue instead of Sandy Hook, New Jersey, much closer to his West Orange Laboratory. Funding from Congress came belatedly and reluctantly. Most bureaus in the navy were unclear about how to use this lab-to-be, which had no building or personnel until 1923.[14]

Although Daniels's vision was to make NRL a research agency to serve

13. Kevles, *Physicists* (1997), 102–38.

14. Amato, *Pushing the Horizon* (1997), 16–29; Howeth, *History of Communication-Electronics* (1963), 326. For the early history of the U.S. Naval Research Laboratory, also see Allison, *New Eye for the Navy* (1981), chaps. 1–4; Hevly, "Basic research within a military context" (1987), chap. 1; Taylor, *The First Twenty-five Years* (1948).

the entire navy, it did not develop that way. Seven of the eight naval bureaus opposed the idea of an agency serving all the navy's research needs, and even the only supportive bureau, the Bureau of Engineering, had a selective view about its future, which was not to involve just any kind of naval research. Stanford Hooper, the head of the bureau's radio division, thought that a modern navy's most important need for research and development lay in sensor and information technologies, crucial components of the command-control-communication systems that preoccupied the U.S. Navy from the 1920s onwards.[15]

At the time, radio and underwater sounding of submarines were the primary naval information and sensor technologies at stake. Therefore NRL should be a venue to provide the research capacity for that, integrating the navy's existing research organizations on radio and underwater sounding. Under this rubric, the bureau combined four research units—three on radio and one on sound—into NRL. Thus NRL subsumed the Engineering Experiment Station under Harvey C. Hayes in Annapolis, Maryland; Taylor's Naval Aircraft Radio Laboratory in Anacostia; the Naval Radio Laboratory under Louis Austin at the National Bureau of Standards; and the Radio Test Shop that William Eaton directed at the Washington (DC) Navy Yard.[16]

The Naval Research Laboratory opened on 2 July 1923. Though a military establishment, NRL had mainly civilian members; only Director E. L. Bennett and Assistant Director Edgar Oberlin were naval officers. (The navy released Taylor, who became one of its civilian employees.)[17] At the start, the facility had two divisions—radio and sound. Taylor's Radio Division combined the Naval Aircraft Radio Laboratory, the Naval Radio Laboratory, and the Radio Test Shop. It had groups working on aircraft radio, direction finders, general research, precision measurement, receivers, and transmitters.[18] At first,

15. For the history of how the U.S. Navy came to terms with the command-control-communication systems in the early twentieth century, see Wolters, "Managing a sea of information" (2003).

16. Bureau of Engineering (memorandum), "Centralization of radio and sound research and development at the U.S. Naval Research Laboratory at Bellevue," 12 Feb. 1925, Historical Archives, NRL.

17. Allison, *New Eye for the Navy* (1981), 33–47. Bennett worked in the Navy Department; only Oberlin was in the laboratory to supervise its daily activities.

18. Taylor, *Radio Reminiscences* (1960), 14. The precision-measurement unit comprised people from the National Bureau of Standards, the receiver group from the Radio Test Shop, and other groups from the Radio Aircraft Station.

it focused on developing antennae, crystal frequency stabilizers, heterodyne receivers, and thermionic tubes—elements of short-wave radio. As Taylor recalled, "the most important service of the Radio Division in the early days was the 'selling' of the high-frequency program to the Navy and, indirectly, to the radio communication industry."[19]

Selling short-wave radio could not succeed just by promoting design, however. To make short-wave radio a practical communications system, the navy had to know the ranges that signals could reach at different frequencies, times of day, and seasons, information that the Kennelly-Heaviside-Watson theory and the Austin-Cohen formula failed to generate. Moreover, knowledge about how short waves propagated in space was relevant not only to assessing the feasibility of short-wave communications systems, but also to constructing operational guidelines for them. So NRL launched an experimental study on short-wave propagation.

### *Discovery of the Skip Zone*

Taylor and his team of technicians performed propagation experiments at short wavelengths as early as 1920 at the Naval Air Station in Anacostia. Like their work at various wireless stations during the war, these ex-amateurs conducted their activities in off-duty hours. When the station's high-power transmitter was not in official use at night, they turned up its frequencies and executed a variety of tests to see how far short waves could reach. Soon they found that the Anacostia station had a widespread circle of radio-amateur listeners whenever they operated at night. This was an asset for propagation experiments, since listeners' reports on signal reception were valuable evidence. To attract more listeners, Taylor et al. even started to broadcast programs in 1920. With codename NSF among radio amateurs, the station broadcast phonographic and live music, public health lectures, chess games, and a speech by President Warren Harding.[20]

Taylor's group also collaborated with radio-amateur organizations in wave-propagation experiments. In 1920–21, it participated in the ARRL-NBS fading tests using the transmitter at Anacostia.[21] After moving to Bellevue, it used NRL's radio station to participate in ARRL's fourth transatlantic tests

19. Ibid., 17.
20. Ibid., 91–93.
21. Dellinger et al., "A study of radio signal fading" (1923), 196–98.

in 1923.[22] In response to hobbyists' success in the experiments, tests followed everywhere to increase radio frequencies. Many American amateurs were competing to reach other stations with shorter and shorter wavelengths. NRL joined the movement by initiating long-range tests with amateurs.

The enthusiasm for short-wave radio was an outcome of practical concerns. In a letter to the chief of naval operations, Taylor listed the advantages of high-frequency wireless—great potential for long-range communications, much lower cost than low-frequency stations that required high-power generators and giant antennae, faster communications because of wider bandwidths, and more directive transmission. According to Taylor, high-frequency radio was so promising that British Marconi, France's Military Telegraph Services, General Electric, the Italian navy, the Norwegian government, Telefunken in Germany, and Westinghouse were all experimenting on waves between 75 meters and 100 meters.[23] But replacing low with high frequencies was not straightforward. Short waves had unusual propagation characteristics. At short wavelengths, propagating range depended on time of day, season, and weather. Such dependences varied with frequency, too. Different environmental conditions entailed different optimum frequencies for communications. To help select the best frequencies, empirical data on wave propagation were essential.

From the amateurs' two-way transatlantic test in 1923, radio technologists witnessed the remarkable ranges that waves at 3,000 kHz (wavelength 100 meters) could achieve at night. Many endeavored to, as an amateur stated, "see just how far down it [wavelength] was possible to go and still maintain two-way contact."[24] In November 1923, NRL started short-wave propagation tests with radio stations around the country. In December, Taylor sent a report to call the navy's attention to "the phenomenal results obtained with radio transmitters operating in the neighborhood of 3000 kcs [kHz]." The recent experimental findings had shown that "if the frequency is high enough, some apparently new phenomena of transmissions take place and the intensity of signals apparently does not in any way follow the well known Austin-Cohen transmission formula." For instance, engineers at Westinghouse discovered that transmission at 3,000 kHz between Pittsburgh and Cleveland

22. Taylor, *Radio Reminiscences*, 105.

23. Taylor to the chief of naval operation (letter), 21 Aug. 1924, folder "Radio-Atmospheric Effects," Historical Archives, NRL.

24. Reinartz, "A year's work" (1925), 1894.

was barely possible during the day but better after dark, whereas transmission at 3,500 kHz was better in daytime.[25] The observations showed the necessity of distinct propagation tests at day and night.

NRL continued short-wave propagation tests in 1924. It constructed four special transmitting sets to push frequency to 4,000 kHz (wavelength 75 meters). Daytime and nocturnal experiments followed—the former at the lab, the latter at Taylor's home. Radio stations all over the country collaborated: Detroit, Jacksonville, Minneapolis, and so on.[26] Among the participants were wireless amateurs from ARRL, including H. T. Dalrymple (Akron, Ohio), William Justice Lee (Orlando, Florida), and John Reinartz (Hartford, Connecticut).[27] These people, especially Reinartz—an officer in the Naval Reserve and a prominent amateur who invented the Reinartz tuning circuit—played a valuable role. In March, Taylor, Young, and Reinartz boosted frequency to 6,000 kHz (wavelength 50 meters). They found that waves higher than 3,000 kHz behaved quite differently from those between 1,500 kHz and 3,000 kHz (100–200 meters). While the 1,500–3,000 kHz waves had long communications ranges only at night (which was why amateurs' transatlantic experiments of 1920–23 all took place after dark), those higher than 3,000 kHz had considerable ranges in daytime.[28] Moreover, daytime signal strength *increased* with frequency between 3,000 kHz and 6,000 kHz. It seemed that increasing frequency further might generate even better conditions for long-range communications.

Throughout March 1924, Reinartz, Taylor, and Young measured signal intensity with decreasing wavelength. Reinartz's station had a variable-frequency transmitter. So the tests went one way from Hartford to Bellevue. They began with contact at moderate frequencies that worked well both ways. Then Reinartz slowly raised the frequency while Young monitored the variation of signal intensity and notified Reinartz when the signals ceased to reach him. Signal intensity usually went up with increasing frequency but *fell sharply*

25. Taylor to the Bureau of Engineering (letter), 17 Dec. 1923, Bureau of Ships Records (RG 19), Naval Research Laboratory (unclassified), box 83, folder "Transmitting and Receiving Systems—1923#1."

26. Taylor to the Bureau of Engineering (letter), 15 March 1924, ibid.

27. Reinartz was active in ARRL in organizing the long-range tests and developed the Reinartz tuning circuit in wide use by amateurs in the 1920s. He had come to know Taylor when he served in the U.S. Navy as a radioman during World War I. Taylor, *Radio Reminiscences* (1960), 100–115.

28. Reinartz, "A year's work" (1925), 1894; Taylor, *Radio Reminiscences*, 109.

*to zero* at a threshold frequency, which changed day by day.[29] By 26 March, Reinartz had kept a stable link at 44 meters and had occasionally connected with Bellevue at 23, 27, and 30 meters.[30] Frequencies between 5,880 kHz and 7,690 kHz (39–51 meters) seemed to be the optimum band for daytime communications. But the same frequencies that appeared satisfactory during the day were totally useless after dark: twenty minutes after sunset, the signals faded out within a few seconds to one percent of their daytime intensity. When frequency dropped to 2,500–4,000 kHz (75–120 meters), however, nighttime results were almost as good as daytime ones.[31] Why did the increasing signal intensity suddenly fall to zero at certain threshold frequencies? Why did the optimum frequencies for daytime communications become useless after dark? Why were waves shorter than 40 meters unable to get through steadily? A phenomenon shed some light on these questions.

At the end of March 1924, Dalrymple in Akron reported that he, farther from Reinartz than NRL, could receive the signals from Hartford at 22 meter wavelength when NRL received no signals during the tests at noon. William Justice Lee in Orlando, several hundred miles farther from Hartford than Bellevue, also reported reception of Reinartz's signals at night when Young and Taylor were unable to receive any message from Hartford. A third station, 9BRI, reported similar results later.[32]

In summer 1924, the Department of Commerce opened the bands of 4–5, 20–22, 40–43, and 75–80 meters to amateurs. Within a few months, many amateur stations "worked" Reinartz at the free wavelengths at about 21 meters and 40 meters and hence supplied additional empirical data. Reinartz and NRL's data indicated a zone of silence around a transmitter at wavelengths shorter than 40 meters. This "skip zone" could receive no signal except at very close distances; outside the zone, signals came through with clarity. The skip-zone effect occurred only when the frequency was higher than a certain threshold. This is why Young observed a sudden fall of signal strength to zero beyond some frequency—NRL fell into the skip zone of Reinartz's station. Furthermore, the threshold frequency and the skip distance seemed to change diurnally and with atmospheric conditions. This explained why waves under

29. Taylor, *Radio Reminiscences* (1960), 109.

30. Reinartz, "A year's work" (1925), 1894.

31. Taylor to the Bureau of Engineering (letter), 2 April 1924, Bureau of Ships Records (RG 19), Naval Research Laboratory (unclassified), box 83, folder "Transmitting and Receiving Systems—1923#1."

32. Reinartz, "A year's work" (1925), 1894; Taylor, *Radio Reminiscences* (1960), 109.

40 meters from Hartford sometimes reached Bellevue (when they either did not skip far enough over Bellevue or did not skip at all) but sometimes did not (when they skipped over Bellevue).

### *More Comprehensive Range Data at High Frequencies*

Waves shorter than 100 meters traveled considerably farther than those between 100 meters and 200 meters, and the skip-zone effect appeared at wavelengths shorter than 40 meters. The propagation of electromagnetic waves in the shorter-than-100-meter region had quite different characteristics from the longer waves. How to explain this novel behavior? How could it help in building high-frequency communications systems? Only more experimental data on wave propagation at higher frequencies could answer these questions.

The Naval Research Laboratory conducted more experiments with amateurs in 1924 and 1925.[33] Taylor and his colleagues reached 20 MHz (wavelength 15 meters) and extended the transmitter-receiver separation to 2,000 miles. When measuring signal strength at longer distances than the separations between NRL and most amateur stations in North America, they collaborated with government stations in U.S. territories overseas. For instance, they used the daily reports of Major J. O. Mauborgne, who took observations of the army's radio communications between New York and Panama at 16, 20.8, 32, and 41.7 meters.[34] These experiments sought to estimate the maximum ranges of radio signals at frequencies of 100–20,000 kHz. This estimate would quantify more precisely the transmission of short waves, which was useful for scientific studies of short-wave propagation and as a guide in "formulating policies looking forward to the possible wider adoption of high frequency communication in the Naval service."[35]

Taylor published the estimated frequency-dependent ranges in May 1925.[36] He obtained these figures with a 5 kilowatt antenna, average antenna installation, and connections between points on the same meridian (to sim-

33. By September 1924, 240 amateur stations cooperated with NRL on the short-wave propagation experiments. Taylor to the Bureau of Engineering (letter), 15 Sept. 1924, Bureau of Ships Records (RG 19), Naval Research Laboratory (unclassified), box 83, folder "Transmitting and Receiving Systems—1923#1."

34. Taylor, "An investigation" (1925), 681.

35. Taylor to the Bureau of Engineering (letter), 21 April 1925, Bureau of Ships Records (RG 19), Naval Research Laboratory (unclassified), box 83, folder "Transmitting and Receiving Systems—1923#1."

36. Taylor, "An investigation" (1925), 677–83.

plify the effect of different time zones); the numbers represented data from measurements in daytime, summer nighttime, and winter nighttime.[37] Taylor presented the results in a range chart that used a line to represent a valid range of communications at a frequency (figure 6.1). Note that Taylor, unlike Austin, did not present the data of signal intensity over distance. He gave the maximum range of communications, which was a measurable from technical practice, not a direct physical quantity. As he admitted, "The purpose is rather to serve as a practical guide to indicate what ranges may be covered at different frequencies and what ranges remain to be explored, and what we hope to get in the un-explored regions."[38]

Nevertheless, this chart presented significant scientific evidence. The average range data confirmed the early discoveries concerning short waves. Between 100 kHz and 500 kHz (600–3,000 meters), the waves behaved no differently from what observers had known in the 1910s: the daytime range decreased with frequency as the Austin-Cohen formula had predicted. The extreme nighttime range in winter was longer than that in summer, and both were considerably longer than the daytime range. Between 500 kHz and 2,000 kHz (150–600 meters), propagation was consistent with that in the amateur experiments in the late 1910s and early 1920s: the daytime range diminished further following the Austin-Cohen formula—it became only 125 miles at 2,000 kHz. The summer nighttime range was not much better; it rarely exceeded 500 miles. But the extreme winter nighttime range could be as long as 2,000 miles (which therefore made the transatlantic experiments successful), and communications could remain reliable up to 1,000 miles at 500 kHz and 1,000 kHz and up to 500 miles at 2,000 kHz.

Between 3 MHz and 6 MHz (50–100 meters), however, the trend became different. Contrary to Austin-Cohen, the daytime range began to *increase* with frequency. This made long-distance daytime communications at higher frequencies feasible—the daytime range was 750 miles at 4 MHz and 1,000 miles at 6 MHz. Moreover, the nighttime range (in both summer and winter) was much longer than that at lower frequencies. Winter-night signals at 3 MHz might reach 8,000 miles. At 6 MHz, records of communications across 10,000 miles existed. Between 7.5 MHz and 20 MHz (15–40 meters), the nighttime range remained extremely long, the daytime range increased with

37. The daytime data did not split into summer and winter sets because the seasonal difference for the daytime data was much less significant than that of the nighttime data. Taylor to Bureau of Engineering, 21 April 1925.

38. Taylor, "An investigation" (1925), 681.

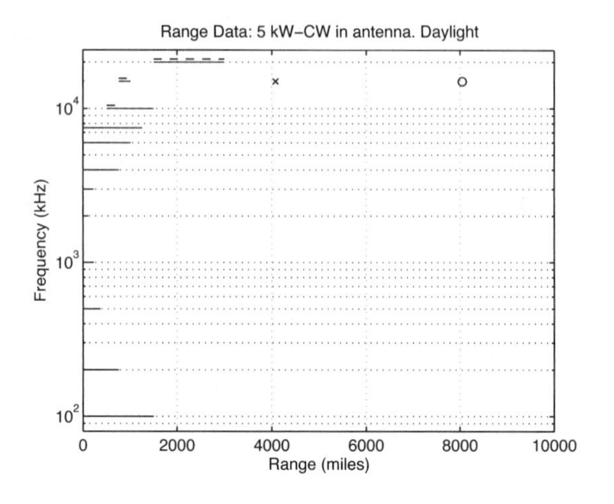


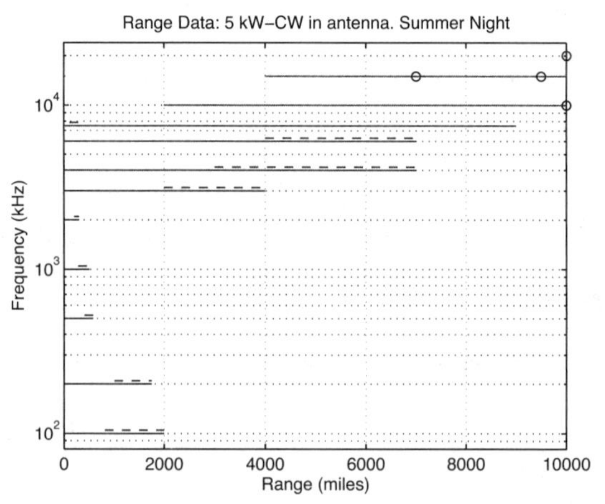


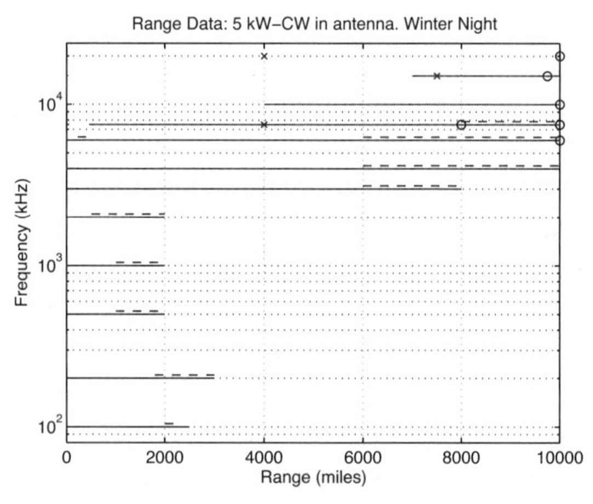


FIGURE 6.1. NRL range chart. Replotted from Taylor, "An investigation" (1925). (a) Daylight; (b) summer night; (c) winter night. The solid lines denote data; the dash lines above, the range uncertainty; x denotes the limit of exploration; and o, unexplored ranges. The discontinuity of lines refers to skip regions.

frequency, and the skip-zone effect appeared. Signal skipping began to show up both day and night at 7.5 MHz. Like the daytime range, the skip distance also increased with frequency.

Taylor's 1925 range chart provided crucial empirical data for understanding short-wave propagation. Like the Austin-Cohen formula, it reflected an epistemic transformation from technology-specific practical knowledge to general facts of nature. It summarized the important experimental results from radio amateurs, NRL, and other American military establishments. These results showed two characteristics surprising to the world before 1923: the daytime range of propagation first decreased but *then increased* with frequency, and its minimum occurred between 1.5 MHz and 2 MHz (150–200 meters); the skip-zone phenomenon appeared for waves over 7.5 MHz (under 40 meters), and skip distance increased with frequency. These features at short wavelengths challenged a decades-old belief that radio-wave propagation exhibited monotonic regularities (i.e., one quantity increased or decreased strictly with another) because of absorption, diffraction, and reflection.

Such empirical findings raised a number of questions. Why did the propagation range *not* vary monotonically with frequency? What was magic about the band of 1.5–2 MHz? Why did the skip-zone effect occur? Why did it occur only at frequencies above 7.5 MHz? Why did the skip distance increase with frequency? Why were the usable frequencies and the sizes of skip zones different between day and night, summer and winter?

One of these questions already had strong clues to an answer—the atmosphere. The diurnal and seasonal variations of usable frequencies and skip zones suggested that radio-wave propagation depended on the sun's position with respect to the earth, which was plausible if waves propagated via an atmospheric layer that resulted from solar radiation. This concept was not at all new: Kennelly and Heaviside's conjecture of a reflective upper sheath and Eccles's model of an ionic refractive layer seemed to refer to the same thing. Nevertheless, they could *not* refer to the same thing. Although the upper layer might account for the exotic short-wave phenomena, it could not do so in terms of the atmospheric reflection or refraction theories of the 1900s and 1910s. If Kennelly, Heaviside, and Watson were correct that the atmospheric layer was a homogeneous reflector, then the Austin-Cohen formula would have governed everything, skip zones would not have existed, and radio amateurs would have been unable to send 200 meter waves across the Atlantic. Eccles's model was more promising. But it still fell short of explaining why the propagation range was minimal at certain frequencies. To answer the questions that the recent discoveries raised at short wavelengths, radio scientists needed a new theory of wave propagation deriving from ionic refraction.

### MAGNETO-IONIC THEORY FOR SHORT WAVES

Following the discovery of several important short-wave anomalies in the early 1920s, physicists and engineers developed theories of wave propagation. These theories all came from William Henry Eccles's groundbreaking work on ionic refraction in 1912: they all shared the assumption that solar radiation created gaseous ions—free electrons or atoms with electric charges—in the upper atmosphere, where ion density increased with height and the upgoing wave-propagating direction bent downward because increasing ion density lowered refractive index and hence increased wave speed at higher altitudes.

But ionic refraction underwent reformulation and two revisions in the mid-1920s. First, Joseph Larmor at Cambridge restated Eccles's theory in a simple form that gained currency among scientific and engineering communities. Second, Edward Appleton at Cambridge's Cavendish Laboratory and Harold Nichols and John Schelleng at AT&T's research departments introduced the geomagnetic field's effect to account for the nonmonotonic variation of propagating range with frequency. Third and finally, Albert Hoyt Taylor and Edward Hulburt at NRL explained the skip-zone effect with the finite thickness of the ionized upper layer.

In all these developments, short radio waves provided fertile ground for the theory of ionic refraction. In the 1910s, the new theory was simply a hypothesis to explain how electromagnetic waves could bend in the atmosphere and why radio signals and noise changed over space and time. For several years, the idea experienced generic acceptance, but wave propagation researchers did not look further into it. The short-wave anomalies that surfaced in the early 1920s rekindled physicists and engineers' interest in further exploring the theoretical implications of ionic refraction. By mid-decade Larmor et al. had revitalized and enriched the theory with a more complicated physical structure—the geomagnetic effect on dispersion and polarization change—and an extended set of unexpected explananda that the theory could successfully account for, such as the skip zone and nonmonotonic variations of the propagating range. The magneto-ionic theory emerged as a more powerful, encompassing model and became the standard doctrine of radio-wave propagation.

## *Eccles's Ionic Refraction Theory*

Before delving into the work of Appleton, Hulburt, Larmor, Nichols, Schelleng, and Taylor, I outline the reasoning in William Henry Eccles's ionic-refraction theory (which I did not do in chapter 4). In his theory of 1912, a radio wave propagated along a downward curve because of refraction in an ionized atmosphere. Suppose the radio wave was a plane wave propagating along direction $z$, linearly polarized, with electric field $E$ along direction $x$ and magnetic field $H$ along direction $y$ and depending only on $z$. Then Maxwell's equations for the wave in an ionized medium were $\partial E/\partial z = -i\omega\mu_0 H$ (Faraday's law) and $\partial H/\partial z = -i\omega\varepsilon_0 E + J$ (Ampère's law), where $i=\sqrt{-1}$, $\mu_0$ was free-space permeability, $\varepsilon_0$ free-space dielectric constant, and $J$ the space current density resulting from ions' moving because of the wave's electric field. The ionic space current density $J$ was the central part of the theory. It was the sum of all motions of ions (electrons and positive ions) that fluctuated randomly according to statistical mechanics.

Unlike contemporary microphysicists such as Owen Richardson, who was developing a gas theory for metal conductivity, Eccles ignored the statistical nature of ions by assuming that at the macroscopic level it was adequate to consider the current resulting from ions' *average* motions. So $J$ was approximately the sum of individual ions' average motions, $J = Nev$ ($N$ was ion number density, $e$ an ion's charge, and $v$ an ion's mean velocity). And these ions were most likely electrons, as the positive ions were much heavier and thus had much smaller motions. The mean velocity was a function of Newton's second law of motion, $mdv/dt + gv = i\omega mv + gv = eE$, where $\omega = 2\pi f$ was angular frequency and the term $gv$ corresponded to friction ($g$ was the frictional coefficient). Substituting $J = Nev$ and Newton's law into Maxwell's equations led to the new wave equations $\partial E/\partial z = -i\omega\mu_0 H$ and $\partial H/\partial z = -i\omega\varepsilon_0[1+Ne^2/i\omega\varepsilon_0(im\omega+g)]E + J$. These new equations were equivalent to the wave equations for a material with free-space permeability $\mu_0$ and effective dielectric constant

$$\text{(6.1)} \quad \varepsilon_{eff} = \varepsilon_0\left[1+\frac{Ne^2}{i\omega\varepsilon_0(im\omega+g)}\right].$$

The wave equations' solution was $E = E_0\exp(-\kappa z)\exp(i\omega z/p)$ and $H = H_0\exp(-\kappa z)\exp(i\omega z/p)$, where the attenuation constant $\kappa$ and the phase velocity $p$ depended on the imaginary and real parts of the dielectric constant $\varepsilon_0[1+Ne^2/i\omega\varepsilon_0(im\omega+g)]$, respectively. Following Eccles's demonstration,

$p$ increased with ion density $N$, since a larger $N$ reduced the effective dielectric constant's real part. Thus upgoing waves gradually bent downward in an atmosphere whose ion density increased with height (because solar radiation, which was stronger at higher altitude, generated more ions at higher altitude). And this downwardly curving refraction looked like a reflection from the atmosphere. The friction from collisions of ions with uncharged gas particles attenuated wave intensity. It should have been larger at the lower atmosphere, because the gas density was higher at lower altitude. Eccles hypothesized this relation between attenuation and height to explain why wave intensity was stronger at night: only the less absorbent higher atmosphere had enough ions to bend waves after dark.

Eccles's theory seemed to provide a more plausible model of atmospheric reflection than the Kennelly-Heaviside layer. Unable to yield the Austin-Cohen formula, however, the theory remained on the sidelines until the 1920s, when it began to receive more attention.[39] As radio amateurs and engineers set up long-range links with shorter wavelengths, atmospheric reflection or refraction became more probably the key mechanism of propagation. In 1924, another Briton, Joseph Larmor, readdressed Eccles's theory.

### *Larmor's Ionic Refraction Theory*

A native of Magheragal, in the northern part of Ireland, Joseph Larmor studied at Queen's College, Belfast, and St John's College, Cambridge, where he became senior wrangler. After graduation, he taught physics at Queen's, returned as lecturer to St John's, and assumed the Lucasian professorship in 1903 after the death of George Stokes.[40]

Larmor started as a Maxwellian, a follower of Maxwell's approach in constructing a coherent electromagnetic theory. His major scientific feat was to explore a continuum-ether theory to interpret both Maxwell's electromagnetic formulations and the new physics of electrons and special relativity.[41] He also worked on geomagnetism and perturbed rotary motions of the earth.[42]

39. In March 1913, Jakob Salpeter at the University of Göttingen independently published on wave propagation in an ionized medium. Salpeter, "Das Reflexionsvermögen" (1914), 247–53. Salpeter's theory had two differences from Eccles's: he focused on laboratory-produced ionized gas instead of ionized atmosphere and concentrated more on the microscopic foundation of friction than on propagation phenomena in nature (see chapter 10 below).

40. Eddington, "Larmor" (1942–44), 197–98.

41. Ibid., 198–204, Warwick, *Masters of Theory* (2003), 357–98.

42. Eddington, "Larmor" (1942–44), 204.

Radio was not Larmor's favorite topic, but he monitored the development of wireless technology through his friends such as John Ambrose Fleming.[43] The discussions on long-distance radio-wave propagation were all too familiar to Larmor, as they had much in common with his work on waves and the transformation from Maxwellian physics to microphysics that he witnessed and in which he actively participated. When preparing his course notes on electromagnetism at Cambridge in 1924, Larmor developed and published a theory of ionic refraction, gaining the attention of people seeking a theory of wave propagation in an ionized atmosphere.[44]

Larmor believed that ions' influence on the atmosphere's electrical properties was dielectric—modifying the dielectric constant without incurring dissipation. When an oscillating electric field entered an ionized gas, it forced the ions into motion. If the gas was rarefied enough and the period of oscillation was long enough so that the ionic mean free path was much longer than the distance an externally driven ion traveled within an oscillating period, then an average ion did not collide with others within the period. This implied that ions were virtually free under the time scale of the applied electric field, and the collision-induced friction $gv$ vanished. Thus the ionized medium's effective dielectric constant became the following—cf., equation (6.1)—

$$(6.1') \quad \varepsilon_{eff} = \varepsilon_0 \left[1 - (Ne^2 / m\varepsilon_0 \omega^2)\right].$$

The wave's phase velocity was accordingly $p = c/(1-Ne^2/m\varepsilon_0\omega^2)^{1/2}$, where $c$ was the speed of light in free space. The phase velocity $p$ increased with ion number density $N$. So Larmor could use the same argument as Eccles did to explain why waves bent along the earth's curvature. In fact, he not only gave the qualitative explication, but he also evaluated quantitatively the extent to which ionized air bent a radio wave's propagating direction.[45]

43. Fleming to Larmor (letters), 1906–1925, Joseph Larmor Papers, Special Collection, St John's College Library.

44. Larmor, "Wireless electric rays" (1924), 1025–36. Larmor probably wrote the paper independently of Eccles. He stated: "A theory satisfying this criterion [the dielectric-type influence] was hammered out in class-lectures at Cambridge on electric waves last February, and has been, in fact, already expounded in answers in the Mathematical Tripos" ( ibid., 1027). He knew Eccles's 1912 paper (ibid., 1028), but learned about it perhaps only after developing his theory.

45. Larmor suggested that the curvature of rays propagating in an ionized medium was $d(p/c)/ds$, where $s$ was path length. When the rays were approximately horizontal, $s$ was almost equal to height $h$. The condition for long-range propagation was thus that $d(p/c)/dh$ approxi-

Larmor's theory was a special case of Eccles's with null friction.[46] In both theories, radio waves propagated along the earth's surface because of bending by refraction in an ionized atmosphere whose ion density increased with height. The explanation was convincing in the early 1920s when the Kennelly-Heaviside hypothesis appeared consistent with the Austin-Cohen formula and when findings by radio engineers and amateurs demonstrated the correlation between atmospheric conditions and fading, static, and other propagation phenomena.

Nevertheless, the Eccles-Larmor theory of ionic refraction had a problem. Equations (6.1) and (6.1′) indicated that refraction decreased monotonically with frequency: the larger $\omega$, the closer the refractive index $\varepsilon_{eff}/\varepsilon_0$ was to 1. Therefore rays deflected less with lower frequency and hence traveled farther. This result, however, contradicted NRL's range data from the mid-1920s, which exhibited a range minimum between 1.5 MHz (200 meters) and 2 MHz (150 meters). For frequencies higher than 2 MHz, measured propagating distance increased monotonically with frequency; but at less than 1.5 MHz, unlike in the theory, it *decreased* with frequency.

That the empirical range data had a minimum at a critical frequency suggested that radio-wave propagation underwent resonance. If that were true, then what would be the physical mechanism of the resonance? Eccles and Larmor's theory was unable to answer this question. The answer, from Americans Harold Nichols and John Schelleng and the Englishman Edward Appleton, lay in the effect of the geomagnetic field.

## *Effects of Geomagnetism: Nichols, Schelleng, and Appleton*

Two early developers of the theory of magnetized ionic refraction were Harold W. Nichols and John C. Schelleng at the engineering department of Western Electric, manufacturer for the American Telephone and Telegraph Company (AT&T). (I consider Edward V. Appleton of England below.) In the 1920s,

---

mated the earth's curvature. The only parameter in $p$ that varied with $h$ was ion density $N$. So $d(p/c)/dh = (\partial p/\partial N)(dN/dh)/c$. For given $m$, $e$, $\epsilon$, and $\omega$, one could calculate the gradient $dN/dh$ from the fact that $d(p/c)/dh$ was equal to the earth's curvature. Larmor found out that for the 100 meter wave the value of $N$ at the height of 10 km was 0.3 per cubic cm (for electrons) or 500 per cubic cm (for hydrogen ions) (ibid., 1031).

46. Eccles's results had more complicated frequency dependence because of the friction coefficient $g$. But after expanding over $g$, his $p/c$ was a function of the second and higher orders of $g$. This implied that when friction was small its effect on phase velocity was negligible. Thus Larmor's theory held not only for $g = 0$ but also for small $g$.

AT&T had built a nationwide telephonic network with a monopoly—the Bell System. The emergence of wireless technology allowed extension of the system without the constraint of wires. The Bell Companies had worked on wireless since the 1910s. They experimented on radiotelephony (50–100 KHz) in 1914 between Montauk, Long Island, and Wilmington, Delaware, and in 1915 between Arlington and stations in California, Hawaii, Panama, and Paris. During World War I, they developed naval and aviation radio for the U.S. military. After the war, they focused on transatlantic telephony. Their first transatlantic system still used frequencies below 100 kHz.[47] But amateurs' short-wave experiments inspired them to probe frequencies above 1.5 MHz.

Their teams investigating short-wave radio were the research branch of Western Electric's engineering department and AT&T's development and research group. As predecessors of the Bell Telephone Laboratories, they were among the earliest American industrial-research establishments to employ college graduates to conduct research not relating to immediate manufacturing needs.[48] In the 1910s, the Bell research groups hired C. R. Englund from the University of Chicago, R. A. Heising from the University of Wisconsin, and H. T. Fritts from the Royal Technical College of Denmark to develop high-power transmitters and measuring instruments. After AT&T built its megacycle transmitter in New Jersey in the mid-1920s, Bell researchers used it to perform experiments on long-distance wave propagation.[49]

Nichols and Schelleng were AT&T's first researchers to work on wave propagation. A graduate of the University of Chicago, Nichols entered Western Electric's engineering department in 1914. Schelleng taught physics at Cornell University before joining Western Electric in 1919.[50] When experimenting with propagation in 1924, the two men focused on the "selective effect"—the minimum range at the wavelength of about 200 meters. The effect had practical implications for AT&T's development of long-range wireless telephony, since it affected the telephone system's optimum radio frequencies. Nichols and Schelleng suggested that the action of the earth's magnetic field on electrons and positive ions in the atmosphere caused the selective effect; they published their theory in April 1925.[51]

47. Fagen, *Bell System* (1975), 368–74, 391–405.

48. For early American corporate research, see Reich, *American Industrial Research* (1985).

49. Fagen, *Bell System* (1975), 406–9, 912–18; Millman, *Bell System*, (1984), 194–96.

50. Fagen, *Bell System* (1985), 917.

51. Nichols and Schelleng, "Propagation" (1925), 215–34.

Physicists had been familiar with the external magnetic field's effect on electromagnetic waves. In the nineteenth century, Helmholtzians, Maxwellians, and Weberians sought support for their own theories in magneto-optics, which had close connections to the development of microphysics.[52] Paul Drude, Hendrik Antoon Lorentz, and Hermann Voigt had used microphysical arguments to deduce a theory of electromagnetic-wave propagation in a magnetized gas of electrons and ions.[53] Nichols and Schelleng's problem was similar to that in magneto-optics—the former dealt with radio waves in the geomagnetic field, and the latter with light in an artificial magnetic field. They followed Larmor's assumption that the upper atmosphere was rarefied gas consisting of free ions, and they borrowed Lorentz's approach: the equation of motion (in vector form) for an ion in such a medium was $m\,d\vec{v}/dt = i\omega m\vec{v} = e\vec{E} + e\vec{v}\times\vec{B}_g$, where $\vec{v}$ was ion speed, $\vec{E}$ the wave's electric field intensity, $\vec{B}_g$ the geomagnetic field's flux density, $\times$ vector cross product, $e$ ion charge, and $m$ mass. Unlike Larmor's equation of motion, this one had an additional term $e\vec{v}\times\vec{B}_g$ that corresponded to the Lorentz force that the geomagnetic field incurred on the moving ion. Suppose the geomagnetic field was uniform along the $z$ direction; viz., $\vec{B}_g = \hat{z}B_g$. Then one could solve the equation of motion to express $v_x$, $v_y$, and $v_z$ in terms of linear combinations of $E_x$, $E_y$, and $E_z$ parametrized by $B_g$, $m$, $e$, and $\omega$. One could similarly express the ionic drift current density $\vec{J}$ because of the relation $\vec{J} = Ne\vec{v}$. Plugging the expression for $\vec{J}$ into Ampère's law $\nabla\times\vec{H} = i\omega\varepsilon_0\vec{E} + \vec{J}$ for the wave's magnetic field intensity led to

(6.2)

$$\nabla\times\vec{H} = i\omega\varepsilon_0 \begin{pmatrix} 1+\dfrac{e^2N/m\varepsilon_0}{\omega_g^2-\omega^2} & -i\left(\dfrac{\omega_g}{\omega}\right)\dfrac{e^2N/m\varepsilon_0}{\omega_g^2-\omega^2} & 0 \\ i\left(\dfrac{\omega_g}{\omega}\right)\dfrac{e^2N/m\varepsilon_0}{\omega_g^2-\omega^2} & 1+\dfrac{e^2N/m\varepsilon_0}{\omega_g^2-\omega^2} & 0 \\ 0 & 0 & 1-\dfrac{e^2N/m\varepsilon_0}{\omega^2} \end{pmatrix} \begin{pmatrix} E_x \\ E_y \\ E_z \end{pmatrix},$$

where $\omega_g = eB_g/m$ was the "gyro" frequency (a term from optics, referring to the angular velocity, or angular frequency, of an ion's rotation in a magnetic field).

52. Buchwald, *From Maxwell to Microphysics* (1985).

53. For example, see Lorentz, *The Theory of Electrons* (1952), chap. 4.

Equation (6.2) indicated that the effective dielectric constant in a magnetized ionized medium was a tensor or matrix, not a scalar. In other words, the ionized gas was no longer isotropic when the geomagnetic field was present: waves propagating along different directions had different speeds. Moreover, some elements of this tensor went to infinity as the wave frequency $\omega$ equaled the gyro-frequency $\omega_g$, which was the rotational frequency of an ion driven by the geomagnetic field. This meant that resonance occurred at $\omega = \omega_g$, which prohibited waves from further propagation (the infinite dielectric constant retarded wave speed indefinitely).

Nichols and Schelleng took this to explain the propagating-range minimum. They assumed the value of the earth's magnetic field $B_g$ to be 0.5 gauss, leading to a gyro-frequency of 1.4 MHz (wavelength 214 meters) for electrons and 800 Hz (375 km) for hydrogen ions.[54] The frequency of electronic resonance was approximately equal to that of the empirical range minimum, so the observed selective effect was a result of electrons' magneto-ionic resonance.

Nichols and Schelleng went further to solve the wave equations with a tensor-like effective dielectric constant in equation (6.2). The solution was expressible in plane waves, as in Eccles's and Larmor's theories. Nonetheless, the wave-propagating velocity and polarization differed with direction, since the effective dielectric constant, being a tensor, was sensitive to orientation. Without having to tackle the more complicated general solution, Nichols and Schelleng (following Lorentz) gave solutions in three special cases. (i) When the wave's electric field was parallel to the geomagnetic field, the geomagnetic effect was totally absent. The solution for refractive index was $n^2 = 1-Ne^2/m\varepsilon_0\omega^2$, identical to Larmor's formula, and the wave polarization was linear (refractive index $n$ determined the wave's phase velocity $p$ as $p = c/n$, where $c$ was the speed of light in free space). (ii) When the wave propagated parallel to the geomagnetic field, the wave equations had two solutions for refractive index:

$$(6.3)\quad n^2 = 1 - \frac{Ne^2}{m\varepsilon_0\omega^2\left(1 \mp \omega_g/\omega\right)}.$$

These solutions corresponded to waves with two distinct phase velocities; they had right-handed and left-handed circular polarization, respectively. A linearly polarized wave propagating along the geomagnetic field thus decomposed into two circularly polarized waves with different phase velocities,

54. Nichols and Schelleng, "Propagation" (1925), 218.

which engendered a phase difference between the two waves after a distance, which led again to a linear polarization, as the two waves recombined, but rotated away from the original polarization (the well-known "Faraday rotation" in optics). (iii) When both the electric field and the wave-propagating direction were normal to the geomagnetic field, there was a solution

$$(6.4)\quad n^2 = 1 - \frac{Ne^2}{m\varepsilon_0\omega^2\left[1-\left(\omega_g^2/\omega^2\right)\left(1-Ne^2/m\varepsilon_0\omega^2\right)^{-1}\right]}$$

with an elliptical polarization.

Nichols and Schelleng's conclusion was much more complex than Eccles's and Larmor's. Two waves with different velocities and polarizations yielded a Faraday rotation when the propagating direction aligned with the geomagnetic field—case (ii). A double refraction—waves with distinct propagating directions, polarizations, and velocities—occurred when waves propagated normally to the geomagnetic field—cases (i) and (iii). The ionized atmosphere amid terrestrial magnetism resembled an optically exotic crystal! Also, these waves' different ray-path curvatures led to different propagating ranges.[55] A wave in a magnetized ionized gas did not have a single refracting ray path; it split into multiple components with different speeds, polarizations, and ranges.

Nichols and Schelleng were not the only researchers to consider the geomagnetic effect on radio-wave propagation. In November 1924, Edward Victor Appleton at the Cavendish Laboratory proposed a similar idea in a conference in South Kensington, London (more about Appleton in chapter 7).[56] His presentation discussed the diurnal, seasonal, and random variations of short-wave intensity and their deviations from the Duddell-Tissot and Austin-Cohen formulae. He believed that ionic refraction bent radio waves along the earth's surface. In the ionic refraction theory, whether waves could travel far without dissipation depended on whether they met Larmor's collision-free condition (ionic mean free path much longer than wavelength).

But the collision-free status for electrons differed from that for molecular ions. To estimate which frequencies could accommodate nonabsorptive long-range transmission, one had to determine whether electrons or molecular ions were deflecting radio waves. Appleton suggested an empirical way to do so

55. Ibid., 224–26.

56. Appleton, "Geophysical influences" (1924–25), 16D–22D.

using the geomagnetic effect. He found the same results as Nichols and Schelleng's for waves propagating along the geomagnetic field: two components with different phase velocities and opposite circular polarizations. One of the phase velocities went to zero—a resonance occurred—when $\omega = \omega_g = eB_g/m$ (equation (6.3)). Appleton took the value of the geomagnetic field $B_g$ to be 0.18 gauss and obtained a gyro frequency $\omega_g$ = 520 kHz (wavelength 580 m) for electrons and 300 Hz (1,000 km) for hydrogen ions. Therefore "peculiar things might be expected to happen" at about 580 meters if electrons were the dominant carriers in atmospheric refraction.

Appleton and the team of Nichols and Schelleng developed the theory of magnetized ionic refraction (later the magneto-ionic theory) independent of each other.[57] They estimated the value of the gyro wavelength differently—Nichols and Schelleng had 214 meters, and Appleton 580 meters—because Appleton factored in only the horizontal component of the geomagnetic field.[58] After this correction, however, both theories explained a puzzling phenomenon in the short-wave range data—the minimum of measured propagating distance at the wavelength of about 200 meters—in terms of the resonance of electrons' gyro-magnetic motions. Within a year, Albert Hoyt Taylor and Edward Hulburt at NRL would develop the magneto-ionic theory further to explain another puzzling short-wave effect: the existence of the skip zone.

## *Explaining the Skip Zone: Taylor and Hulburt*

American radio amateur John Reinartz made the first attempt to explain the skip effect. After his tests with NRL in 1923 that led to discovery of the phenomenon, he continued to "work" other amateurs on waves shorter than 50 meters and tried to make sense of his observations. In April 1925, he published an article in *QST* to explain his findings with a hypothesis.[59] He focused on three questions: why was the skip effect possible, why did the range of wireless signals vary diurnally, and why did it differ with frequency?

His answer: the shape and height of the Kennelly-Heaviside layer. First, he assumed that the layer's most effective angle of reflection was 45°. The sky-

57. Nichols and Schelleng mentioned ("Propagation" (1925), 218) that "on March 7, after this paper had been written, the February 15 issue of the Proceedings of the Physical Society of London arrived in New York."

58. Ibid.

59. Reinartz, "The reflection of short waves" (1925), 9–12.

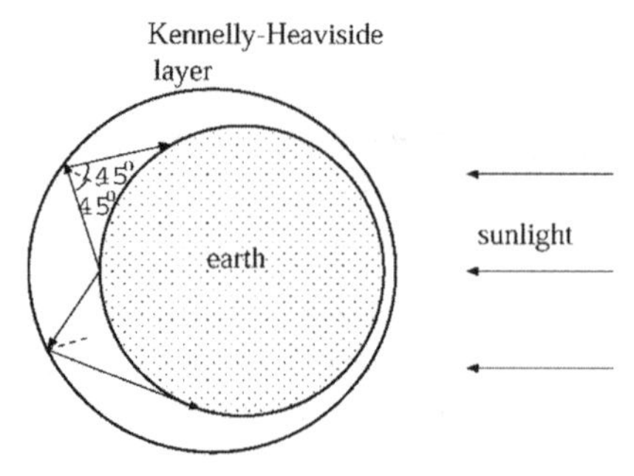


FIGURE 6.2. Reinartz's reflection model for skip distance. Reconstructed from Reinartz, "The reflection of short waves," *QST* (April 1925), figure 4, courtesy of *QST*.

wave intensity diminished as the receiver displaced from the reach of 45° rays from the transmitter via the Kennelly-Heaviside layer. A skip zone was the region between the transmitter and the point where the 45° rays reached the ground (figure 6.2). Moreover, Reinartz assumed the layer to be an oval, not a sphere, stretching away from the sun because of the nonuniform sunlight at different parts of the earth. The asymmetric shape explained why the eastward skip distances were longer at dawn while the westward skip distances were longer at dusk: the reflected rays reached farther in those cases, since the reflecting layer was higher. Finally, he assumed that the shorter the wavelength, the deeper the waves penetrated into the layer. This explained why propagating range dropped with increasing wavelengths as they were shorter than 100 meters.

But Reinartz's theory had two shortcomings. That shorter waves penetrated deeper into the upper atmosphere contradicted a common belief that shorter waves dissipated more in air. Also, why did the upper layer have optimum reflection at 45°? Reinartz was unable to reconcile his model with these concerns.[60] In the meantime, Taylor and E. O. Hulburt at NRL developed a theory of skip distance citing ionic refraction that gained wider acceptance.

Hulburt was another NRL researcher of the World War I generation. Born

60. The *QST* technical editor raised these points (ibid., 12). He tried to solve the second problem by treating the Kennelly-Heaviside layer as a diffuse reflector. At a small incident angle, the energy scattering from the layer spread widely. At a larger incident angle, it concentrated more within a small bundle of directions and therefore reflected better. Meanwhile, at a larger incident angle, the ray path traversing the low-altitude region was longer, so the ray encountered more obstructions on the terrain. A compromise was the median between 0° and 90°. This account was no less hypothetical than Reinartz's.

in Vermillion, South Dakota, and growing up in Baltimore, Edward Olson Hulburt received a PhD in physics from Johns Hopkins University. His graduate training was in physical optics; he wrote a dissertation on experiments and theories of metallic reflectivity at ultraviolet band. He became a physics instructor at Case Western Reserve University after graduation. As the United States entered the war in 1917, the Army Signal Corps drafted him and dispatched him to its Radio Laboratory in Paris, where he worked under Edwin Armstrong and Gustav Ferrié. This experience opened a pathway for him to radio. After the war, he taught at Johns Hopkins and the University of Iowa and eventually joined the Naval Research Laboratory in 1924.[61]

The laboratory he saw was humble, with fewer than two dozen researchers and staff members. Taylor and Hayes had recruited him, and his task was to use his expertise on physical optics to set up a heat and light division that Taylor and Hayes had been advocating. The new unit had only him; he went "to the Laboratory and with empty rooms. [Hulburt] ordered much equipment but nothing was going to come for weeks or months."[62] He had to work temporarily with Taylor on radio.

Taylor had just discovered the skip zone from his experiments on propagation of waves shorter than 50 meters. He took the puzzling range data to Hulburt and challenged him to explain them. Hulburt quickly recognized "those gaps [the skip zones] as phenomena of total internal reflection of electromagnetic waves."[63] Hulburt the physical optician knew that light rays moving from an optically dense medium (with higher refractive index) to an optically loose medium (with lower refractive index) underwent a total reflection when the incident angle exceeded a critical value. Replacing light rays with radio waves, the optically loose medium with the ionized layer, and the optically dense medium with the air below, one could see that total internal reflection implied that beyond a critical distance radio waves experienced complete reflection from the sky, while within it part of the wave energy penetrated the sky. Hulburt saw a skip-zone theory from the model of total internal reflection.

It took a year for Taylor and Hulburt to develop the theory, and they published their work in February 1926.[64] They explained the skip zone by first considering reflection of radio waves at the boundary between free space

61. David K. Allison, "An Interview with Dr. Edward Olson Hulburt," 22 Aug. and 8 Sept. 1977, v–vi, Historical Archives, NRL.

62. Reinartz, "The reflection of short waves" (1925), 14.

63. Ibid., 15.

64. Taylor and Hulburt, "Propagation" (1926), 189–215.

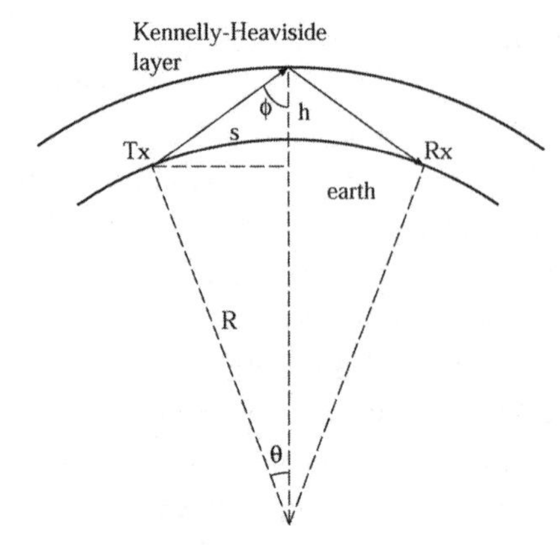


FIGURE 6.3. Taylor and Hulburt's reflection model for the skip zone. Reconstructed from Taylor and Hulburt, "Propagation" (1926), figure 4.

and the ionic-gas layer. Since the upper ionized medium's refractive index was smaller than free space's, the upper layer completely reflected the upgoing rays if the propagating direction's angle of elevation was smaller than the critical angle for total reflection. If not, then part of the wave energy disappeared in the sky. The critical angle $\phi$ for total reflection corresponded to a horizontal distance $2s$ (figure 6.3). All the transmitting rays arrived when the transmitter-receiver separation exceeded $2s$ because of total reflection, whereas only part of the rays arrived when the separation was less than $2s$, since the rest of the rays never returned to the earth. This distance $2s$ was therefore the skip distance.

The refractive index of the ionized layer was necessary for calculating the skip distance $2s$. Taylor and Hulburt had four choices: equations (6.1′), (6.3—two cases), and (6.4). Their choice depended on the wave-propagating direction and polarization. For waves shorter than 50 meters, however, they showed that the values of the four refractive indices were close to one another,[65] so they could choose any of the four. They selected equation (6.3) with the upper (minus) sign. They established a relation $n = \sin\phi$ between the critical angle $\phi$ and the layer's refractive index $n$ from Snell's law and obtained a relation $s = h\tan\phi$ from the geometry in figure 6.3 ($h$ was the layer's height). Substituting these two relations into equation (6.3) with the upper sign, they obtained a relation between the skip distance $2s$ and wavelength $\lambda$, with two parameters $h$ and $Ne^2/m\varepsilon_0$. They determined the values of $e$ and $m$ from the

65. Ibid., 197.

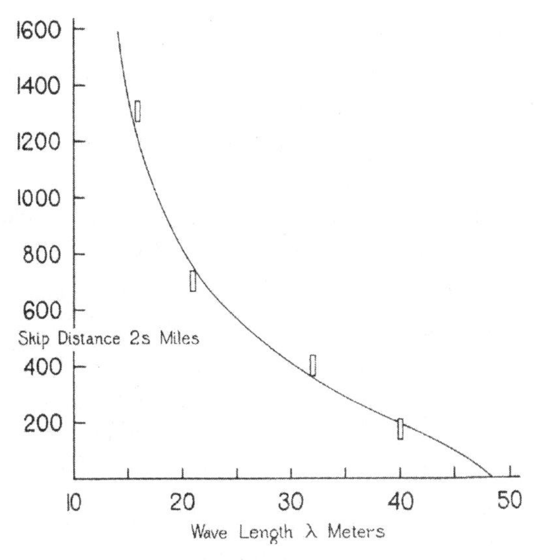


FIGURE 6.4. Skip distances: experimental data (bars) vs. theoretical prediction (curve). Taylor and Hulburt, "Propagation" (1926), figure 1.

assumption that the principal charged particles in the ionized layer were electrons and adjusted the number density $N$ and the layer's height $h$ to match the calculated values of skip distance $2s$ with the data from the NRL-ARRL experiments at wavelengths of 16 meters and 40 meters.[66] Then they plotted the theoretical and experimental values of skip distance (figure 6.4) The theoretical predictions agreed well with the experimental data at wavelengths 16, 21, 32, and 40 meters.

According to figure 6.4, the skip distance decreased with wavelength and totally vanished when wavelength was longer than 48 meters. This explained why Taylor and Reinartz observed the skip-zone phenomenon only for waves shorter than 50 meters. Also, the skip distance did not increase indefinitely with frequency. When wavelength was shorter than 14 meters, the critical angle was so close to 90° that the layer could not totally reflect even the horizontal rays. It was difficult to receive radio waves shorter than 14 meters at *all* distances.

Nevertheless, the reflection model had a limitation. Although total reflection ceased at small incident angles, partial reflection prevailed at all angles, implying that some radio energy still reflected into the range within the skip distance. If so, then a skip zone would not be completely silent. But radio ex-

66. Ibid., 199.

perimenters' experiences indicated that skip zones were entirely beyond the reach of signals. Thus the reflection theory needed modification. To do that, Taylor and Hulburt pursued a refraction theory in which the atmosphere was a heterogeneous medium with its electron density increasing continuously with height. Thus the ionized layer was no longer a homogeneous body. The electron density, one of the crucial characteristics of the ionized layer, was a continuous function of height instead of a fixed number.

Following Eccles and Larmor, Taylor and Hulburt derived the equation for the curving trajectory of a refracting ray in the ionized layer with a given electron density profile. From geometry and Snell's law, they obtained the trajectory's tangent $dy/dx$ at height $y$ and horizontal distance $x$ as a function of the ray's incident angle $\phi$ to the layer and the refractive index $n$ at height $y$. They could use this relation to solve the ray's trajectory for a given variation of $n$ with $y$. Equations (6.1′), (6.3), and (6.4) suggested that $n$ varied with $y$ only through ion density $N$.

Taylor and Hulburt considered four simple variations. First, they assumed that electron number density increased linearly with height ($N = \beta y$). In this case, the solution for the ray's trajectory was a downwardly bending parabola. The ray reached a maximum height $h$ and returned to the earth at a distance $2s$ from its origin, where $2s = 4h\tan\phi$. This result did not explain the skip-zone phenomenon, since a nondirectional transmitting antenna on the ground radiated rays in all possible directions from $\phi = 0°$ to $\phi = 90°$, which, according to the solution, corresponded to horizontal ranges from zero to infinity. Thus Taylor and Hulburt added an assumption: *the ionized layer had a finite thickness* $h_c$. Above $h_c$, the relation $N = \beta y$ did not hold, and the electron number density $N$ increased less rapidly with height. Suppose the initial incident angle yielding a ray path with the maximum height $h_c$ was $\phi_c$ and the path's horizontal range $2s_c$. Then any ray with an initial angle $\phi$ smaller than $\phi_c$ escaped indefinitely to the upper sky without returning to the ground, for the refractive power of the atmosphere above $h_c$ was not enough to bend it down. That is, a receiver less distant than $2s_c$ from the transmitter could not receive any ray. This explained the skip-zone phenomenon (figure 6.5).

The skip distance in this model was $2s_c = 4h_c\tan\phi_c$, where the critical angle $\phi_c$ followed Snell's law $n_c = \sin\phi_c$ ($n_c$ was the refractive index of the ionized layer at height $h_c$). This expression was in the same analytical form as that in Taylor and Hulburt's total-reflection model. Thus by choosing proper values for the layer's height and the rate of ion-density increase with height, they could produce a skip-distance curve identical to that in figure 6.4. The

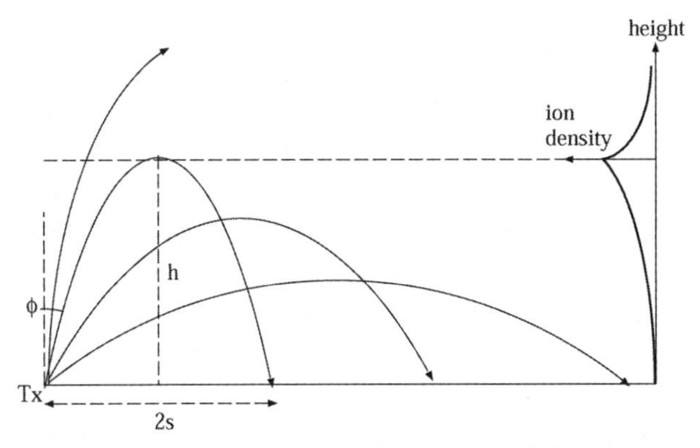


FIGURE 6.5. Taylor and Hulburt's refraction model for skip zones. Reconstructed from Taylor and Hulburt, "Propagation" (1926), figure 6.

refraction theory could also match the empirical data, and it could predict a real skipping not explainable by reflection.

Taylor and Hulburt considered three more cases, in which electron density grew with the square, the exponential, and the square root of height. Except for the last case, all the skip-distance formulae had the same functional form as the linear case. So one could obtain the same numerical results for skip distance by selecting proper values for the height of the ionized layer and the electron density at a given height.[67]

Taylor and Hulburt's contribution to studies of radio-wave propagation was their use of the ionized layer to explain not only long-range propagation (as many researchers did) but also the new skip effect. In contrast to Reinartz's ad hoc assumption of the angular dependence of the layer's reflectivity, they showed that the skip effect followed from the layer's finite thickness. In addition, their theory could generate quantitative predictions in good agreement with the short-wave experimental data. This agreement did not rely on a specific assumption about ion-density profile, wave propagating direction, or polarization. A number of possibilities, if not all, led to the same conclusion. By explaining the new 200 meter range minimum and skip zones, Nichols, Schelleng, Taylor, and Hulburt entrenched Eccles and Larmor's model of the ionized layer within a new theoretical structure of magneto-ionic refraction.

From Heaviside, Kennelly, and Eccles's models in the 1910s to the theory

67. Appleton also independently derived a similar theory of skip zones in 1926 after Taylor and Hulburt. See Appleton "On the diurnal variation" (1925–27), 155–61.

of ionic refraction in the early 1920s, the doctrine of atmospheric reflection/refraction underwent a major transformation. Above all, the ionic refraction theory could explain not only long-distance radio-wave propagation and atmospheric noise (as the earlier models had done) but also new propagation phenomena at short wavelengths—range minimum and the skip distance. More important than expanding the domain of explananda, however, the new theories began to exploit the explanatory power of the mathematical structures and quantitative arguments relating to the models. Thus Appleton, Hulburt, Nichols, Schelleng, and Taylor converted the set of questions that radio physicists had aimed to answer from why (e.g., why could radio waves propagate along the earth's curvature?) to how much (e.g., how much was the maximum range of propagation at a given wavelength?). In addressing these distinct sets of questions, the theories changed from answers to specific why-questions to devices to systematically generate answers to a whole realm of quantitative questions. The validity of the theories' quantitative reasoning derived from two aspects—their predictions agreed with experimental data, and such an agreement was substantially independent of the assumption about the ion-density profile.

Yet the ionic refraction theory of the early 1920s did not just offer quantitative explanations. More important, they opened up avenues to help convert radio into a major tool for probing the atmosphere: they recommended entities to measure. Larmor's revision of Eccles's theory stated explicitly that ionic refraction was the accumulative outcome of the entire ion-density profile over height. Thus radio waves propagating from the ground over the ionized layer to the ground should carry information about the ion-density profile. Moreover, Taylor and Hulburt's ionic refraction model of the skip zone suggested that the zone was a crucial geophysical index of the state of the ionized layer, such as its thickness and maximum electron density. Finally, Appleton, Nichol, and Schelleng's magneto-ionic theory indicated that a propagating radio wave under the influence of the geomagnetic field usually had two components with distinct speeds, intensities, and polarizations. Measuring the *difference* between these two components could often produce valuable information about the ionized layer and the earth's magnetic field.

From the U.S. Navy's discovery of the skip zone and frequency-dependent range minimum to the Anglo-American physicists' theories of ionic refraction, the studies of short waves had introduced within the years 1920–25 a rather different grasp of the interactions between radio and the earth. In such an understanding, the upper atmospheric layer deflected radio waves, as the

Kennelly-Heaviside model had stipulated before 1920. But unlike its predecessor, which equated the layer with a sharp conducting boundary, the new theory now modeled the layer with a mass of ionized gas of extended thickness and nonconstant electron density. The mechanism that formed the layer seemed more concrete and plausible—ionization by solar radiation. The layer could account for considerably more propagation phenomena, especially at short wavelengths. And radio seemed to turn into a promising tool to probe its characteristics. All these developments suggested that the ionized layer was becoming more and more *real*.

But was the ionized layer *real*? The primary questions were at least three in number. Was there more "direct" empirical evidence for its existence? Could scientists establish evidence for it independent of the presupposed physical and mathematical structures of the magnetized ionic refraction theory? Was it a real entity or merely a hypothesis to explain wave propagation? These fundamental questions bothered radio researchers increasingly as the magneto-ionic theory was emerging in the mid-1920s.

CHAPTER 7

# British Radio Research and the Moments of Discovery

On 10 December 1947, Edward Victor Appleton received the Nobel Prize for physics in Stockholm from King Gustav V of Sweden "for his investigations of the physics of the upper atmosphere." Two days later, the laureate delivered his Nobel Lecture. Titled "The ionosphere," the speech provided a historical review (peaking at his own work) of events leading to the discovery of this ionized upper region, whose name British scientist Robert Watson Watt had coined in 1926.[1]

According to Appleton, there were two lines of evidence before the mid-1920s suggesting that the high atmosphere might be electrically conducting. One came from nineteenth-century geophysicist Balfour Stewart, who hypothesized the electro-active layer as part of a gigantic planetary dynamo causing periodic variation of the earth's magnetic field. The other, "from the study of the long-distance propagation of radio waves," debuted with Marconi's transatlantic trial in 1901. The second line also included amateurs' finding of waves shorter than 200 meters in the early 1920s, which Appleton did not mention, although Professor E. Hulthén of the Nobel Committee for Physics did so in his presentation speech for the award.[2] The idea of a large

1. This term became widespread after 1932, when British radio researchers such as Appleton and Watson Watt put it in the titles of papers that they published; Gillmor, "The history of the term 'ionosphere'" (1976), 347–48.

2. Appleton, "The ionosphere" (1964), 79–80; E. Hulthén, "Presentation speech for Edward Victor Appleton" (10 Dec. 1947), in Nobel Prize Website: www.nobelprize.org/nobel_prizes/physics/laureates/1947/press.html (last accessed on 3 December 2012).

body of free charged particles somewhere in the upper sky seemed more and more credible.

Despite increasing empirical support from radio science and geomagnetism, Appleton continued, many scientists did not accept the ionosphere as "real." Some thought it just a clever hypothesis, regardless of how useful for explaining facts or organizing knowledge. It was not because they refused the existing evidence, but because they demanded extra evidence of a different kind. Before the mid-1920s, Appleton told his audience, "the Kennelly-Heaviside theory did not gain universal acceptance, for *direct evidence* of the existence of the conducting layer was lacking."[3]

## *Direct Evidence, Sounding-Echo Experiments, Operational Realism*

What would constitute "direct" evidence for the ionosphere? What would allow assertion of the reality of such a scientific object? Setting aside philosophical qualms about whether we can ever have direct evidence for anything at all, the answer is straightforward today: we can send a rocket, a satellite, or a space shuttle to the upper sky to measure the electron and ion densities there. If we find electron or ion concentration at high altitude much greater than that in ordinary air, then we can be confident about the reality of the ionized region in the upper sky. This is exactly what geophysicists have been doing since the 1950s—they have launched spacecraft carrying Langmuir probes, mass spectrometers, and other plasma-measuring instruments to measure the ionosphere in situ.[4] Yet this approach was unavailable in the 1920s: flying to the upper atmosphere was out of the question; unmanned meteorological balloons could go up perhaps 20 kilometers.[5] Most estimates placed the upper ionized layer (if it existed) at at least 80 kilometers up. If physically reaching the upper sky to measure its electrical properties was the only permissible way for scientists to secure direct evidence for the ionosphere, then they had to wait for more than two decades.

But they did not wait that long. Well before rockets and satellites, two groups of British scientists in the mid-1920s performed some radio tests that gained

3. Appleton, "The ionosphere" (1964), 80 (my emphasis).

4. Bowhill, "Space techniques" (1974), 2238–40.

5. Hulburt, "Early theory of the ionosphere" (1974), 2138. European meteorologists had pioneered high-altitude balloons in measuring the atmosphere since the late nineteenth century. Among them was the Frenchman Léon Philippe Teisserenc de Bort, who discovered the stratosphere. See Wenstrom, "Milestones in meteorology" (1940), 230; Fonton, "De Bort" (2004).

wide acceptance as establishing the ionosphere's reality. Reginald Smith-Rose and R. H. Barfield at the National Physical Laboratory in Teddington investigated the errors in wireless direction finders, while Appleton and Miles Barnett first at the Cavendish Laboratory and then at King's College, London worked on artificial fading, man-made undulation of received signals due to a continuous change of radio frequency. Both sets of trials looked like other radio propagation experiments: the researchers sent signals from transmitters and collected them at receivers. So why were they more direct than, for example, those of ARRL, Austin, Eccles, Marconi, or Taylor? Why did they seem to carry more ontological weight to contemporaries?

To some historians, these questions were social rather than epistemic; and scientists really did not have legitimate reasons for confidence in the value of those experiments. C. Stewart Gillmor has asserted that the radio tools that early researchers used and the associated cultural ideas overstated the reality of the ionospheric layer; for those who did not employ the radio method, the notion was far from obvious.[6] Recently, Aitor Anduaga has further expanded Gillmor's thesis. He has claimed that "the vertical structure of the atmosphere and its partition in layers was a *conceptual representation*, rather than the result of a cascade of discoveries." In fact, "there was nothing uncontestable or unproblematic in the series of historical declarations on those 'discoveries.'" Instead of a scientific object that emerged through careful analysis of experimental results, the reflective ionosphere layer was actually a "*reification* of metaphors" that flowed from "a strong tendency to ascribe material existence to theoretical entities." Why did scientists come up with this realist interpretation of the upper atmosphere? Because, Anduaga has argued, an "aura" in the broader social and material context encouraged such an interpretation. In brief, Anduaga explains the accepted reality of the ionosphere in terms of a zeitgeist emerging from the social, political, cultural, and economical ambience.[7]

While Anduaga's account may explain popular dissemination of the idea of the ionosphere, it is much less clear whether his perspective illuminates the *beginning* of that process. Before short-wave communication, people were no less enthusiastic about wireless, and the hypothesis of the reflective layer was widely available. Why did that atmosphere not lead to a realist interpretation of the upper layer? Why did scientists and engineers value so highly Appleton and Barnett's and Breit and Tuve's experiments? Why did they believe

6. Gillmor, "Threshold to space" (1981), 102–5.

7. Anduaga, *Wireless and Empire* (2009), 274–328.

that the new radio method—unlike geomagnetic observations and the old propagation tests—generated more direct evidence? The zeitgeist does not address the *epistemic process* whereby scientists decided on the upper layer's existence.

In contrast with Anduaga's sociocultural account, I argue that three epistemic reasons explain the emphasis on those radio experiments of the mid-1920s. First, in their experiments on direction finding and artificial fading, Appleton, Barfield, Barnett, and Smith-Rose did nothing but demonstrate the reality of "sky waves"—that radio waves from the transmitter reached the receiver via the upper sky—from measured data. Unlike Marconi and amateurs' transoceanic success, the Austin-Cohen formula, the skip zone, and the short-wave range profiles, those British researchers' results did not rely on any specific wave-propagation theory to become evidence for the upper layer. They needed no knowledge about the ionosphere's structure or the diffractive or refractive behavior of electromagnetic waves. Sky waves entailed the wave-deflecting layer from elementary geometrical optics (what else could be responsible for that effect?). In those mid-1920s' experiments, the quest for the evidence of the ionosphere was therefore translated into the search for evidence of sky waves. Here "direct" meant free from the mediation of wave propagation theory—except the most obvious one, such as basic ray tracing.

Second, it was possible for the Britons to retrieve directly certain crucial properties of the scientific object. The designs for both Barfield and Smith-Rose's direction-finding experiment and Appleton and Barnett's fading experiment allowed the scientists to determine, in a straightforward manner, the height of the upper ionized layer—measuring the sky wave's incident angle in the former case and the path difference between sky wave and ground wave in the latter. By contrast, neither the long-wave propagation trials in the 1910s nor the short-wave range tests in the early 1920s allowed easy calculation of the layer's height. One could do that only by evoking Watson's atmospheric reflection theory or the magneto-ionic refraction theory; and even so, the layer's height lay deep within the theories' complex mathematical structures and was not easily measurable.

Yet there was another sense of "directness" in Appleton and Barnett's experimental practice. Unlike the earlier wireless tests, these two Cambridge researchers exercised more active control of the experimental—particularly the signal-transmitting—conditions. Before the mid-1920s, scientists and engineers used existing commercial, military, or amateur wireless communications facilities to perform experiments. They might have some control of the sending power and frequency but had to stay with the ordinary signal patterns of telegraphic communications—namely, dots and dashes. By contrast, Appleton and Barnett designed *waveforms* exclusively for the experimental purpose. Specifically, they changed the frequency of a continuous-wave transmitter and monitored the variation of signal intensity at the receiver. This marked the first appearance in the history of radio of the modern "sounding-echo" method—sending electromagnetic waves to the sky and observing the signal change at the returns.

The traditional, pre-1920s propagation experiments and the sounding-echo experiments had quite different rationales. While the former were intended to uncover the relationship between propagating wave intensity and distance (and accordingly derived the existence of the upper layer), the latter was simply to examine whether the upward waves bounced back from the sky or not. To identify the echoes, the experimenters adjusted the sending waveforms and followed the *immediate* changes at the incoming signals. Moreover, this echo-sounding scheme conformed to the two criteria of directness that I mentioned above. The immediate changes in the transmitting signals would both be independent of any specific wave propagation theory and give straightforward, easy-to-recognize measures of the layer's height. Thus the active manipulability of the sounding-echo method allowed more direct experimental probing of the ionosphere. It is therefore not a surprise that Hulthén highlighted Appleton's "frequency variation method" as his first step on the road towards the Nobel Prize.[8]

The scientific acclaim for those mid-1920s British wireless experiments (especially Appleton and Barnett's) revealed certain critical issues about scientific realism that philosophers Ian Hacking and Nancy Cartwright have emphasized. Cartwright has maintained that causing tangible effects that are adjustable has been an important attribute of scientific objects. Similarly, Hacking has argued that intervening has been more crucial to modern science than representing or explaining. To him, scientists accept an entity to be real when they can manipulate it, tinker with it, and use it as a tool to make further investigations.[9]

Why did scientists of the 1920s believe that Appleton and Barnett's experiments provided direct evidence for the ionosphere? The reason was clear from the perspective of Hacking's operational realism, if we broaden the meaning of "operation" from Hacking's intervention and manipulation to

8. Hulthén, "Presentation speech" (1947).

9. Cartwright, *How the Laws of Physics Lie* (1983); Hacking, *Representing and Intervening* (1983).

active mediation in general. Scientists of the 1920s removed doubts about the ionosphere's reality not because they could manipulate or control the ionosphere (they could not), but because they could find a way to actively probe it. And actively probing the ionosphere involved manipulating some other scientific object (radio waves) to perturb the ionosphere's condition and monitoring the consequent effects.

### *A British System of Radio Ionospheric Research*

Hacking's operational realism reminds us of the significance of pragmatic contexts beneath any ontological claim about a scientific object. The ionosphere was no exception. Barfield and Smith-Rose's direction-finding tests and Appleton and Barnett's fading experiments in the mid-1920s were products of the British government's new infrastructure for radio research. Since World War I, the British state and society had been rethinking the old laissez-faire attitude towards scientific research and technological development. Many people believed that the state should help sponsor, guide, and shape the country's science and technology, or else it would lag behind Germany and the United States. During the war the government more actively supervised and funded research, development, and manufacturing. On the one hand, this system foreshadowed a military-industrial-technocratic complex—historian David Edgerton's "warfare state."[10] On the other hand, it gave birth to an umbrella civilian agency—the Department of Scientific and Industrial Research (DSIR)—to coordinate and sponsor most scientific and engineering investigations in the country. Wireless was on the agenda of DSIR officials for its preeminent practical implications. Soon after peace came, the department established a Radio Research Board.

The Radio Research Board rapidly became a key institution in radio science. It integrated Britain's existing wireless investigations in industry and academia; it incorporated the government's National Physical Laboratory into radio studies; it built its own research stations; and it helped create a crucial network of propagation research that eventually spread into British dominions overseas. Anduaga has charted interwar Britain's political and institutional structures for radio ionospheric work, with the board playing a central part.[11] With its sponsorship, both Barfield and Smith-Rose's work on direction finding and Appleton and Barnett's on fading epitomized the new

10. Edgerton, *Warfare State* (2006).

11. Anduaga, *Wireless and Empire* (2009), 1–120.

direction in research on radio propagation that the board helped initiate. Emphasis shifted: from analyzing the behavior of radio waves to radio probing of the ionosphere's physical properties, and from long-distance propagation measurements at mobile platforms (such as ships) or widespread commercial, military, or amateur stations to shorter-range propagation experiments at much sparser research stations with better instruments. Radio ionospheric studies were beginning to have real laboratories.

## DIRECTION FINDING, WAVE POLARIZATION, AND THE IONOSPHERE

An evidential search for the ionosphere grew from the engineering work on direction finding. Since the beginning of wireless, people had thought of using electromagnetic waves in positioning as well as in communications. Both applications built on the same radio technology. While communications delivered messages with electromagnetic waves, positioning determined the sources of radiation by identifying the propagating directions of the transmitting electromagnetic waves. As early as the 1890s, inventors and engineers had found that certain antenna setups could receive (and send, too) particularly strong radio waves in specific directions. This characteristic encouraged experiments with wireless direction finding: Marconi developed the tilting "wave antenna" with an aerial rod slanting or lying horizontal to the opposite of the wave-propagating direction (chapter 2). A. Blondel in France, Ferdinand Braun and Jonathan Zenneck in Germany, S. G. Brown in Britain, and J. Stone Stone and Lee de Forest in the United States explored the multiple-antenna arrangements that synthesized signals arriving at distinct, virtually identical, and equally spaced antennae (the predecessors of the modern antenna arrays).[12] But neither the wave antenna nor the array was as popular as another kind of early wireless direction finding—the rotating loop antenna, or the "frame aerial," as contemporaries termed it.

### *Loop Direction Finders and Their Problems*

Loop was the oldest type of receiving antenna. In Hertz's wireless experiments in 1886–88, the "Nebenkreis" that he used to detect the remote wave effect was a looped metal wire interrupted by a spark-inducing gap. The loop

12. Zenneck, *Wireless Telegraphy* (1915), 340–46; Keen, *Direction and Position Finding* (1922), 4–8.

antenna detected the direction of an incoming wave via rotation until its energy was at a minimum. Researchers had early on figured out that the loop's orientation affected the arriving signal's intensity. Hertz himself showed that a loop antenna induced a strong spark at a certain orientation but failed to induce any spark at all when the orientation changed. In 1905–6, Henry Round at the British Marconi Company performed a series of experiments, turning this effect into a direction-finding device.[13]

The basic principle of the loop-antenna method (as well as the wave-antenna method) was to determine a radio wave's propagating direction from its polarization, i.e., the direction of its magnetic (or electric) field. According to the Maxwellian theory of electromagnetism, a wave's magnetic (or electric) field in free space was normal to its propagating direction, and the simplest way to detect the magnetic field was to use a loop. In the loop, from Faraday's law, the wave's time-variant magnetic field induced an electromotive force proportional to the time derivative of the magnetic flux, which was the product of the field's time derivative, the loop area, and the cosine of the angle between the field and the loop's normal direction. Because of this angular dependence, no electromotive force emerged when the magnetic field was parallel to the loop. Thus, when a receiving loop directly faced a transmitting antenna (so that the loop's normal aligned with the opposite of the wave-propagating direction), the detected signal intensity was null. One could therefore find the transmitter's direction by rotating the vertical receiving loop around the vertical axis until the detected signal strength was minimum.

The rotating loop antenna was a simple and convenient design for wireless direction finding. Technologists immediately grasped its potential in practical applications. One was navigation. Suppose a ship in mid-Atlantic could determine the direction of radio waves coming from a maritime station whose location it knew. Then the crew would gain invaluable information in positioning the ship. To the armed forces, direction finders were an even more useful tool for gathering intelligence. Accurate direction-finding technology could help locate an enemy's airplanes, ground troops, radio stations, and submarines whenever these units communicated wirelessly. In response to transportation and military needs, researchers had developed various techniques in the 1900s and 1910s to improve the performance of rotating-loop direction finders. Above all, multiple frames replaced the single loop, and the frame area enlarged to strengthen incoming signals, for the induced electro-

13. Keen, *Direction and Position Finding* (1922), 8; Round, "Direction and position finding" (1920), 224–27.

motive force grew with both the number and the area of the loops. Frame antennae several feet long and wide became common. In addition, to make the signal minimum more discernible, engineers such as the Royal Navy Captain J. Robinson attached an auxiliary loop orthogonal to the main loop and rotated the antenna set altogether.[14]

Antenna setups such as these indeed gave more sensitive signal output, yet rotating a ten-foot-wide frame aerial was a clumsy and time-consuming operation, especially for military applications. A way to overcome this disadvantage was to fix the antennae. In 1908, two Italians, E. Bellini and A. Tosi, figured out how to do so.[15] The Bellini-Tosi system had two mutually perpendicular vertical loop antennae. Incoming signals on the two loops were proportional to the cosine and sine of the angle between the wave-propagating direction and the first loop's normal, respectively. The signals went into a "goniometer," consisting of two field coils and a rotating search coil. Each field coil ran parallel to a loop antenna and coupled its signal. So the total magnetic field from the field coils was parallel to the wave propagating direction. The coil-induced magnetic field rotated the search coil until its axis aligned with the field's direction, which was the same as the wave's (figure 7.1) In this manner, the Bellini-Tosi system reproduced a radio wave's magnetic polarization in a smaller coil without having to rotate the big antenna loops.[16]

Another major breakthrough came with refinement of vacuum tubes, a direct consequence of World War I. The war generated large demand for wireless communications sets, and the new thermionic tube turned out to be a powerful electronic device to amplify wireless signals. After August 1914, the British, French, German, and American governments invested heavily in research and development of thermionic tubes. Their endeavors turned the primitive Fleming valves and de Forest audions into high-quality vacuum tubes. This benefited not only wireless communications but also direction finding, for the high-gain tube amplifiers could now magnify very weak enemy radio waves from afar and reduce the need for large frame aerials.[17]

With all these improvements, loop-antenna direction finders became a prevailing military technology during World War I. The French Army and the

14. Keen, *Direction and Position Finding* (1922), 45–46.

15. Bellini and Tosi, "A directive system of wireless telegraphy" (1907), 771–75, (1908), 348–51.

16. On the principles of the single-loop and Bellini-Tosi systems, see Smith-Rose, *Direction-Finding* (1923), 1–9; Zenneck, *Wireless Telegraphy* (1915), 347–52; Eccles, *Wireless Telegraphy and Telephony* (1918), 445–48; Keen, *Direction and Position Finding* (1922), 49–57.

17. Smith-Rose, *Direction-Finding* (1927), 1.

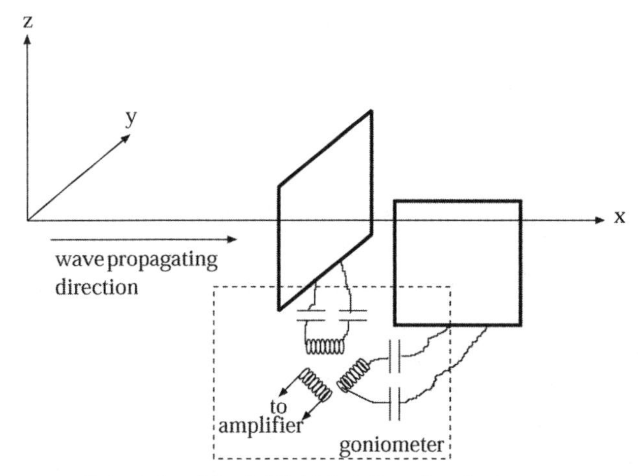


FIGURE 7.1. Principle of the Bellini-Tosi direction finder.

U.S. Navy adopted the rotating frame antennae as their standard direction finders. These instruments contained compact revolving loops and high-gain electronic amplifiers using vacuum tubes manufactured by France's Military Telegraph Services. After the United States entered the war in 1917, France's Military Telegraph Services and the U.S. Signal Corps cosponsored a laboratory in Paris to develop wireless sets, including direction finders for detecting incoming enemy airplanes from their engines' radio noise.[18]

The United Kingdom was even more active in developing and using wireless direction finders, especially the Bellini-Tosi sets. Early in the war, the Royal Navy commissioned William Duddell of the Royal Society to design direction finders that detected long-wave wireless signals from ships, and later it transferred this work to the National Physical Laboratory. On the ground, Round and his team at the Marconi Company assisted British military intelligence in setting up direction finding stations behind the Western Front to pinpoint transmitters in Germany. Later, Round transferred to the Admiralty in London, where he created networks of wireless direction finding facilities ("B" and "Y" stations) on British coasts to detect signals from intruding aircraft, ships, and submarines. "Room 40" in the Admiralty, which processed and analyzed information from these stations, became legendary for its ability to locate German U-boats and zeppelins; and its famous activities resumed in World War II. Direction finders found uses away from homeland defense and

18. Keen, *Direction and Position Finding* (1922), 8–9; Lessing, *Armstrong* (1956), chaps. 7–8.

the Western Front: the Royal Engineers deployed quite a number of stations in the Balkans, the Middle East, and North Africa to monitor wireless communications by Austria-Hungary and the Ottoman Empire.[19]

The Great War turned wireless direction finding from a rudimentary idea into a practical technology. Nevertheless, this technology did not always work. In field operations, Bellini-Tosi, frame aerial, and similar systems all experienced considerable problems. In the 1900s, engineers had noted that direction finders using the principle of magnetic-field polarization had random errors as large as 40°. Twenty years of improvements helped suppress many of the random errors in the daytime, but not at night. During the war, military operators and technicians found an odd night effect in all loop-based direction finders: a nicely working instrument began to turn erratic at sunset and fluctuated persistently throughout the night.[20] Perfecting antenna loops, goniometers, rotating mechanisms, and tube amplifiers did not eliminate the errors, which seemed to be the result of a physical process that designers ignored. In the later 1910s, T. L. Eckersley, working at the Royal Engineers, argued that sky waves reflecting from the upper atmosphere caused these errors—direction finders' night effect seemed to indicate the existence of a wave-deflecting layer.

## *Thomas Eckersley's Work on Polarization*

Thomas Lydwell Eckersley was an engineer and physicist who built his entire career on radio, but seemed to experience estrangement and contradiction. He preferred theory to experiment but ended up working in a company. He was neither an abstract-science academician nor a pragmatic inventor-entrepreneur, although he received education in both areas. He was a pioneer in radio-wave propagation in the upper atmosphere (he initiated the research on atmospheric whistlers; see chapter 9). Yet his theoretical work was difficult to follow, and, probably for the same reason, few geophysicists and radio scientists genuinely acknowledged his contribution, even though he became a fellow of the Royal Society.

A native of London, Eckersley studied engineering at University College, London. But he found the practical aspects of engineering unattractive and

19. Hartcup, *War of Invention* (1988), 123–27.

20. Smith-Rose, *Direction Finding* (1927), 1–2. Reginald Fessenden made the earliest observations on this characteristic of direction finding errors between 1901 and 1907.

achieved only a second-class degree. After leaving the college, he joined the National Physical Laboratory. There he worked on the behavior of iron under the influence of alternating magnetic fields, which led him first to magnetic detectors for radio waves and then to radio in general. In 1910, he entered Trinity College, Cambridge, to study for the Mathematical Tripos. After his exams, he worked briefly at the Cavendish Laboratory, but he loved theoretical rather than experimental inquiries, and his research led nowhere.

Eckersley left Cambridge and headed east. He took a position as an inspector with the Egyptian Government Survey to make astronomical observations. As the First World War began, he enlisted in the Royal Engineers in the eastern Mediterranean. The battles in the Balkans escalated when Greece in September 1915 declared war against the Central Powers and invited Allied forces to land in Salonika, which became a stronghold of British troops. In Egypt and Salonika, Eckersley worked out the arriving directions of radio waves from the enemy's transmitting stations using data from cities under British control on the Mediterranean.[21]

While on duty, Eckersley noted that the daytime errors of good single-loop or Bellini-Tosi direction finders for wavelengths of 300–5,000 meters and distances of 300–500 miles were consistently within 2°, but their nighttime errors were much more serious, fluctuating, and unpredictable.[22] He proposed a hypothesis: the direction-finding errors happened because loop antennae at night received not only direct waves from transmitters, but also sky waves via the Kennelly-Heaviside layer, which were presumably stronger at night. The device's designers assumed that radio waves propagated *directly* from the transmitting station to the direction-finding station. Since both facilities were on the ground and their separation was much shorter than the earth's radius, this direct radiation should have been "ground waves" traveling on the horizontal plane, and the upright loop should have measured correctly the ground waves' magnetic polarization. Yet reflection from the Kennelly-Heaviside layer invalidated this assumption. With the addition of nonhorizontal sky waves, the loop measured not the direct waves' magnetic polarization, but a composite effect of both ground and sky waves.

Eckersley deduced quantitative predictions for the detected composite polarizations and verified them with wireless experiments in Cairo, Damascus, and Salonika. These Mediterranean cities turned out to be more suitable

21. Ratcliffe, "Eckersley" (1959), 69–70.

22. Eckersley, "The effect" (1921), 60–62.

for such experiments than England or France, for the lower latitudes made direction-finding fluctuations at night more conspicuous. In 1916, Eckersley submitted to the War Office "the first scientific discussion of this subject."[23]

After the war, the military released Eckersley, who returned to England. He joined British Marconi and, among other things, published his declassified wartime discovery in 1921.[24] The paper recounted his work in 1915–16 by considering a transmitter and a single-vertical-loop direction finder on the flat earth with high conductivity (both soil and seawater are conductive at radio frequencies). The direct wave from the transmitter propagated horizontally along direction $x$, its electric field tilted slightly from direction $z$ (vertical) to $x$ (direct-wave propagating), and its magnetic field's direction ($y$) was normal to the $x$-$z$ plane, according to Zenneck's surface-wave theory (chapter 2). The indirect wave was the superposition of a downward wave from the upper sky and its reflection from the ground; both propagating directions were (Eckersley assumed) on the $x$-$z$ plane.[25] The indirect-wave polarization could be any linear combination of "transverse-electric" (when the electric field was along direction $y$ and the magnetic field on the $x$-$z$ plane) and "transverse-magnetic" (when the opposite was the case). Since the transverse-magnetic indirect wave had the same magnetic polarization as the direct wave, it did not affect direction finding. So Eckersley considered the presence only of the direct wave and the transverse-electric sky wave in which the $x$ and $z$ components of the overall magnetic field came exclusively from the indirect wave, and the $y$ component from the direct wave, $H_x = H_x^{\text{indirect}}$, $H_z = H_z^{\text{indirect}}$, and $H_y = H_y^{\text{direct}}$ (figure 7.2).

Eckersley designed three experiments to disentangle the mixed polarization of the sky and direct waves, basing them all on the designs for direction finders. The first was the ordinary direction-finding measurement using a single vertical loop antenna (see figure 7.2a and b). The error $\theta$—the angular deviation of the measured direction from the transmitter's true direction—resulted from the nonzero magnetic field $H_x$. From the principle of single-loop direction finding, $\tan\theta = (\partial H_x/\partial t)/(\partial H_y/\partial t)$. Eckersley's second experiment was similar to the Bellini-Tosi setup except that he turned a vertical loop hori-

23. Ratcliffe, "Eckersley" (1959), 70. Eckersley also found that surface waves propagating from land to sea or vice versa experienced refraction. He expanded this discovery into a theory in 1920. Eckersley, "Refraction of electric waves" (1920), 421–28.

24. Eckersley, "The effect" (1921), 60–65, 231–48.

25. Eckersley later confirmed experimentally the assumption that sky waves did not deviate from the plane of direct-wave propagation.

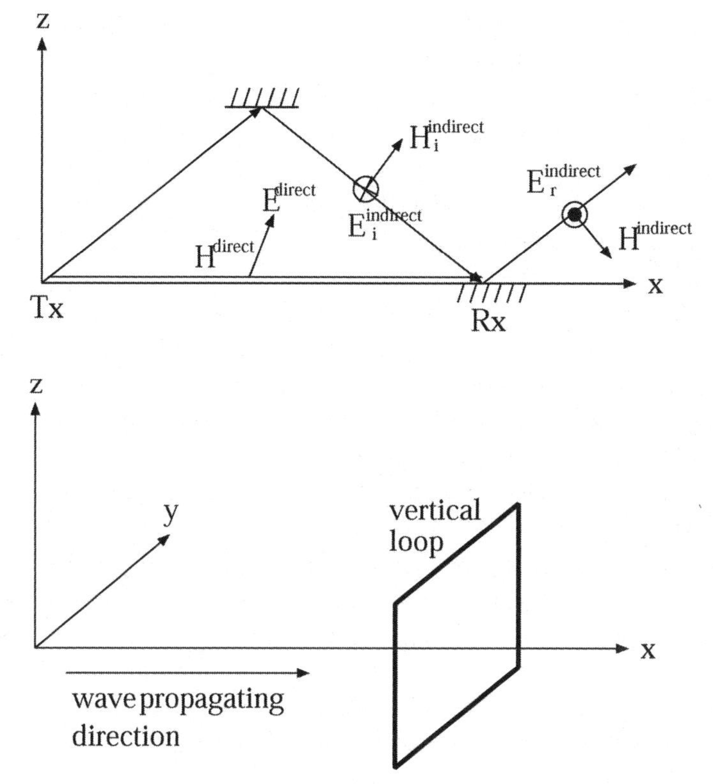


FIGURE 7.2. Configuration of Thomas L. Eckersley's direction-finding experiments.

zontal. He placed a reversing switch in the horizontal loop to flip the polarity of its electromotive force. As the polarity flipped back and forth, the goniometer's rotating coil pointed to two different directions. The angle $\theta_1$ between these two directions followed $\tan(\theta_1/2) = [(\partial H_z/\partial t)/(\partial H_y/\partial t)] \times$(an antenna-dependent constant). Eckersley's third experiment was direction finding with a pair of mutually perpendicular horizontal rods. Since the antenna was a straight aerial rather than a closed loop, the measurable was the electric field rather than the magnetic. The angle $\theta_2$ between the measured direction and the transmitter's true direction therefore followed $\tan\theta_2 = E_y/E_x$.

Eckersley argued that he could obtain evidence for the hypothesis that sky waves caused direction-finding errors by comparing data from the three experiments. First, he showed that $\tan(\theta_1/2)/\tan\theta = [(\partial H_z/\partial t)/(\partial H_x/\partial t)] \times$ (an antenna-dependent constant). The factor $(\partial H_z/\partial t)/(\partial H_x/\partial t)$ was a function of the indirect wave, since the direct wave did not have $H_x$ and $H_z$ components. In

Eckersley's model, in which an indirect wave was the sum of a sky wave and its reflection from the ground, this factor was independent of the sky-wave intensity at the ground level; it depended only on the ground's reflectivity, which resulted from the ground's conductivity and the dielectric constant at radio frequencies.

This implied that the ratio $\tan(\theta_1/2)/\tan\theta$ of the measured data from the first and second experiments should be a constant if Eckersley's theory was right. With a similar but more complicated analysis, he showed that the ratio $\tan(\theta_2)/\tan\theta$ of the data from the first and third experiments should also be a constant. His wartime experimental data from Cairo, Damascus, and Salonika did show a strong linear correlation between the direction errors $\tan\theta$ and $\tan(\theta_1/2)$ and a reasonable linear correlation between the direction errors $\tan\theta$ and $\tan\theta_2$. Therefore the sky waves were quite likely the source of direction-finding errors. The flip side of the coin was that a reflective Kennelly-Heaviside layer was real.[26]

Although Eckersley's experimental design and data interpretation grew from several premises regarding the polarizations of electromagnetic waves, those premises were actually independent of the theories of radio-wave propagation he meant to check. His premises did not stipulate whether radio waves traversed space through diffraction, reflection, or refraction. More important, they remained agnostic about the existence of the Kennelly-Heaviside layer. Eckersley assumed only that (i) in addition to direct ground waves, certain sky waves were also present at direction finders; and (ii) polarizations of the sky waves followed basic Maxwellian optics—the wave-propagating direction, magnetic field, and electric field were mutually orthogonal, the polarization of the sum of a sky wave and its reflection from the ground were analyzable if one decomposed the waves into the transverse-electric and transverse-magnetic elements, and so on. This theoretical independence would remain a central feature of the forthcoming polarization experiments in Britain and Australia.

Eckersley did not conclusively prove the ionized layer's existence, however. He was indeed one step ahead of Eccles, Watson, and other pre-1920 propagation researchers, for he did not merely explain his wartime observations on direction-finding errors with a model of atmospheric reflection. He designed three experimental setups using direction finders and with them investigated the polarization of incoming waves. His experiments did show some regularity that plausibly implied the reality of the upper ionized layer. But something was missing. His results did not impress contemporaries as direct evidence

26. Ibid., 236, 240.

for the ontological claim. The problem lay in his choice of measurables. The approximate constancy of $\tan(\theta_1/2)/\tan\theta$ and $\tan\theta_2/\tan\theta$ at best showed the empirical consistency of the sky-wave model.

Such results could not offer insights critical to grappling with the reality of the ionized layer: researchers could not use his experimental data to determine a sky wave's incident angle and intensity, crucial details for estimating the layer's height and reflectivity. Not that a sky wave's incident angle and intensity were entirely absent from his mathematical structure; but he buried them under the mathematical formulae for $\tan(\theta_1/2)/\tan\theta$ and $\tan\theta_2/\tan\theta$. Obtaining their values would necessitate solving complicated equations involving ground conductivity and the dielectric constant. Determining these values was, if not totally implausible, not straightforward. Without empirical information about the layer's height and reflectivity, its existence remained questionable.

Yet Eckersley opened a new field of inquiry that focused on wave polarization. His work inspired two lines of thinking—technologically, it might help make direction finders more accurate by coping with sky waves; scientifically, it showed how measuring polarization might lead to experimental evidence for the upper ionized layer. Britain's National Physical Laboratory and its Radio Research Board pursued both lines of research.

### *National Physical Laboratory, Radio Research Board, and Smith-Rose*

In the United Kingdom, the National Physical Laboratory (NPL) had been a center for research and development in wireless direction finding since World War I. Its history traced back to the Victorian era—it was a product of industrial standardization. The expansion of British industry and manufacturing after the Industrial Revolution created demand for precise determination of physical units. After Germany established the Physikalisch-Technische Reichsanstalt in 1883, British scientists and engineers urged their government to build a national laboratory, and it created NPL in 1900. The Royal Society initially ran NPL, with the Board of Trade secretary sitting on its general board. NPL started in Teddington, a London suburb, with two departments: engineering and physics. During World War I, it split part of physics to form an electricity department to measure electrical units, calibrate devices, standardize photometry, and undertake radio research.[27]

Science and technology became a national project as the war broke out.

27. Pyatt, *The National Physical Laboratory* (1983), 14–96.

Many observers worried that Britain's industry had been inferior to Germany's and wanted the government to play a more active role in applied and industrial research. In 1915, David Lloyd George's cabinet appointed a committee of scientists and industrialists to advise it about a possible ministry. The cabinet set up the Department of Scientific and Industrial Research (DSIR) in 1916.[28]

DSIR was an all-encompassing institution. It aimed to encourage, sponsor, manage, and coordinate applied and industrial research. It took over NPL from the Royal Society, set up research grants, helped industrial sectors to form research associations, created its own research branches, and distributed scientific information. While civil servants and entrepreneurs preferred research by private organizations, technocrats and scientists favored more investment in DSIR's own laboratories. As the latter view gradually won out, DSIR grew. After the war, its responsibility expanded from industrial to all applied research (including the military). Four coordinating research boards—for chemistry, engineering, physics, and radio—emerged in 1920 under the watchful eye of DSIR Assistant Secretary Henry Tizard.[29]

The Radio Research Board originated from a decision by the Imperial Communications Committee, which looked at policies regarding overland telegraphy, submarine cables, visual signaling, and wireless telegraphy. In 1919, its wireless subcommittee recommended that DSIR establish a research board on wireless telegraphy. This new body consisted of representatives from the existing branches concerning wireless technology, including the Admiralty, the Air Ministry, the General Post Office, and the War Office.[30] It started work in 1920, with a chair (Henry Jackson), representatives from the four establishments, and three civilian technical experts.[31]

The Radio Research Board was to sponsor, coordinate, and discuss research into practical problems involving radio. To members from the armed forces and the Post Office who had worked on long-range communications, the crucial problems related to the characteristics of radio channels (i.e., the behavior of wave propagation and the properties of interference and noise), not to individual transmitting or receiving technologies. The board's three

28. Melville, *Department* (1962), 23–28; Varcoe, *Organizing for Science in Britain* (1974), 9–16; Edwards, *Co-operative Industrial Research* (1949), 32–40; Hull, "War of words" (1999), 461–81.

29. Melville, *Department* (1962), 15–34.

30. Conclusion of 10th meeting of Imperial Communications Committee, 28 Oct. 1919, DSIR 11/19, Public Record Office.

31. Minute, 1920, AVIA 8/14, Public Record Office.

earliest subcommittees—Atmospherics, Directional Wireless, and Propagation of Waves—reflected this emphasis. Atmospherics sponsored Robert Watson Watt at the Meteorology Office to investigate radio methods of locating thunderstorms and the nature of atmospheric noise.[32] Directional Wireless supported research on direction finding at the Wireless Division of NPL's electricity department, which installed devices around the British Isles to measure signals from high-power stations on the Continent,[33] and a young man, Reginald Smith-Rose, took responsibility for this project. Propagation of Waves supported Edward Appleton's work at the Cavendish Laboratory on fading and signal-intensity measurements.

Reginald Leslie Smith-Rose spent his adult life workng for the national scientific establishment. He would later become NPL's acting director and head two international bodies: the International Radio Consultative Committee (French acronym CCIR) and the International Union of Radio Science (French acronym URSI). A native of London, he studied physics at Imperial College, London, where he obtained his AB (1914) and later his PhD (1923) and ScD (1926). After working for four years at Siemens Brothers in Woolwich as an assistant engineer, he joined NPL's electricity department to perform radio research, and there he stayed.[34]

Smith-Rose's first project at NPL was direction finding. The Royal Navy had given NPL experience with wireless direction finding early in the war. Modeling itself on the Admiralty's B and Y stations, NPL after the war set up networks of direction-finding stands, experimented with accurate positioning, and observed the long-term effects of environmental conditions on positioning. The project began with its installation of experimental sets in a few places. It built ten observing stations around the British Isles, including one at NPL, in 1921. These direction finders were to detect high-power commercial spark stations (wavelengths 2,000–9,000 meters) on the Continent. The experimenters worked on a regular program, with daily observations that took three to four hours. Measurements began in 1921 and continued until 1924.[35] Like the yearlong fading and propagation measurements by NBS and NRL in the United States, this project was accumulating large amounts of data.[36]

32. Subcommittee B on Atmospherics (Radio Research Board), "Program and estimates for a general investigation of atmospherics," Sub-Committee B Paper No. 9, R.R. Board Paper No. 18, DSIR 36/4478, Public Record Office.

33. Smith-Rose, *Direction Finding* (1927), 3.

34. C. W. Oatley, "Smith-Rose" (1986), 787–88.

35. Smith-Rose, *Direction Finding* (1927), 3–4.

36. For a detailed discussion of the results, see ibid., 1–36.

### *Smith-Rose and Barfield's Experiments*

After the project ended, Smith-Rose focused on Eckersley's pursuit of direct evidence for sky waves from direction-finding experiments and polarization measurements. His experience had taught him how to measure polarization accurately, and he tackled the problem that Eckersley had left. He and his assistant, R. H. Barfield, began their board experiments on sky waves in early 1924. As Eckersley had shown, quests for "direct" evidence were about choosing the appropriate quantities to measure. To avoid his pitfall, Smith-Rose and Barfield wanted to be able to determine a sky wave's angle of incidence, so they designed their experiment differently. Instead of adapting the Bellini-Tosi system, they used a rotating straight aerial and a rotating loop to measure polarizations.

They made three assumptions. First, only indirect waves were present in the receiving antenna when the transmitter and receiver were far apart, since direct waves (Zenneck's surface waves) attenuated considerably beyond several hundred miles. Second, a downward sky wave propagated on the plane spanning the vertical $z$ direction and the horizontal $x$ direction connecting transmitter and receiver, and its propagating direction vis-à-vis the vertical $z$ was $\theta$; viz., sky waves did not have lateral deviation. (Eckersley had already given some empirical support for this assumption.) Third, the receiving antenna could not separate the sky wave and its reflection from the ground wave; both were plane waves whose polarizations were linear combinations of the transverse-electric and transverse-magnetic components (figure 7.3).

Under these assumptions, one could determine a sky wave's incident angle $\theta$ by measuring two polarimetric quantities: $\tan\beta \equiv E_x/E_z$ and $\tan\delta \equiv H_z/(H_x^2+H_y^2)^{1/2}$ ($E_x$, $E_y$, $E_z$ and $H_x$, $H_y$, $H_z$ were the overall electric and magnetic fields along directions $x$, $y$, and $z$, respectively); $\tan\beta$ measured the tilt of the electric field from the vertical axis, and $\tan\delta$ the deviation of the magnetic field from the horizontal plane. Experimentally, one measured $\tan\beta$ by rotating a vertical rod antenna around the $y$-axis until the incoming signal's strength was minimum; the angle between the rod and the $z$-axis was $90°-\beta$. One measured $\tan\delta$ by (i) rotating a vertical loop antenna around the $z$-axis until the received signal's strength was maximum, at which point the loop's normal direction was $y'$ and the horizontal direction perpendicular to both $z$ and $y'$ was $x'$; and (ii) rotating the loop around the $x'$-axis until the incoming signal's strength was minimum. At this position, the angle between the loop's normal direction and the $y'$-axis was $90°-\delta$ (see figure 7.3).

Theoretically, the ground earth at the receiver's location determined the

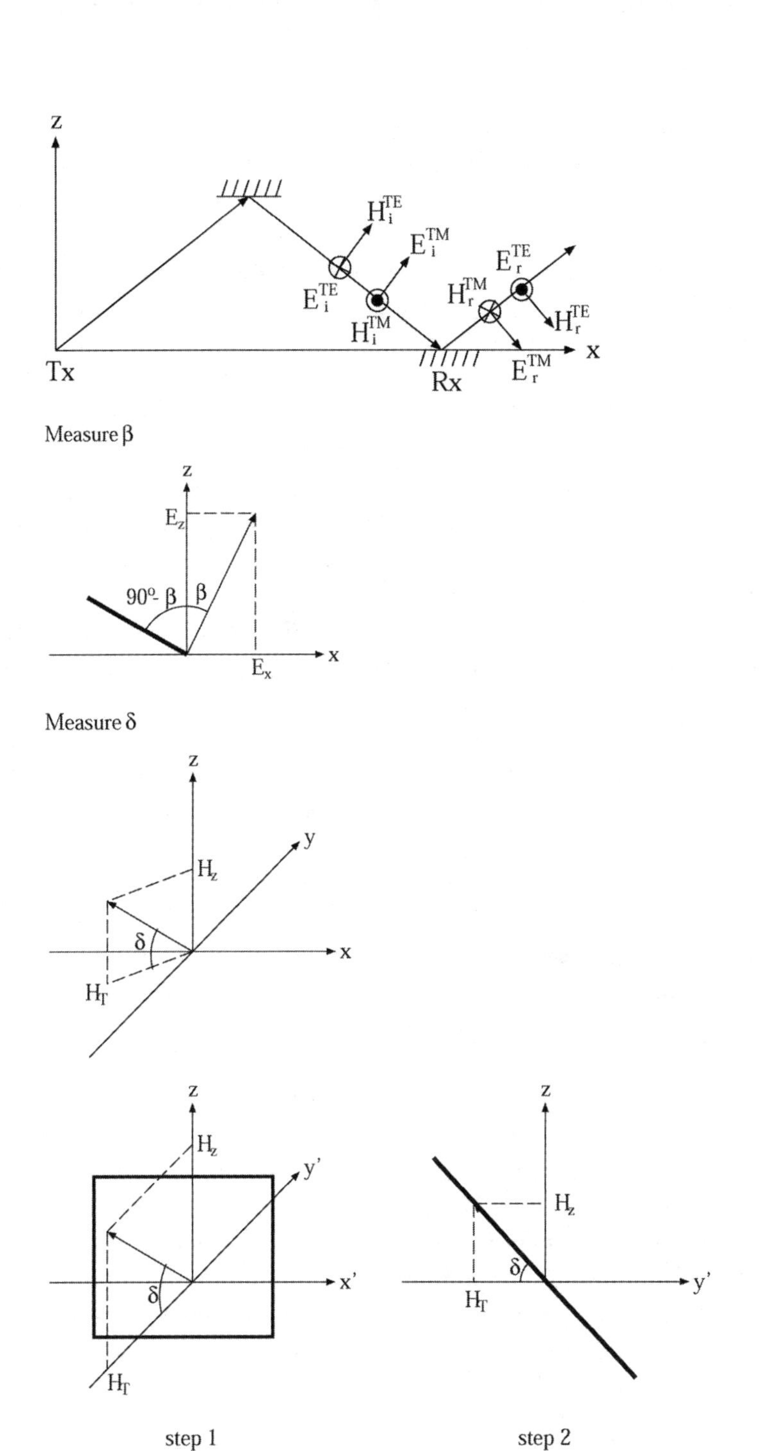


FIGURE 7.3. Configuration of Smith-Rose and Barfield's polarimetric experiments.

angles $\beta$ and $\delta$. The optics of plane-wave reflection showed that a transverse-magnetic incident sky wave (along with its reflection from the ground) yielded $\delta = 0$ and $\tan\beta \;\; (\omega\varepsilon/2\sigma)^{1/2}/\sin\theta$ ($\sigma$ was the ground conductivity, $\varepsilon$ the ground dielectric constant, and $\omega = 2\pi f$ the wave's angular frequency). Similarly, a transverse-electric sky wave (along with its reflection from the ground) yielded $\beta = 0$ and $\tan\delta \;\; (\omega\varepsilon/2\sigma)^{1/2}\sin\theta$. For any mixed polarization, the transverse-magnetic component did not contribute to the value of $\delta$, nor the transverse-electric component to $\beta$. So one could calculate a sky wave's incident angle $\theta$ from the measured $\beta$ and the relation $\tan\beta \;\; (\omega\varepsilon/2\sigma)^{1/2}/\sin\theta$ *or* from the measured $\delta$ and the relation $\tan\delta \;\; (\omega\varepsilon/2\sigma)^{1/2}\sin\theta$. In other words, each of Smith-Rose and Barfield's experimental setups automatically disentangled the transverse-electric component from the transverse-magnetic component and measured the sky wave's incident angle, the most significant quantity for determining the height of the upper reflective layer.[37]

One problem remained for Smith-Rose and Barfield: to deduce the value of $\theta$ from the above argument required the value of $\sigma/\omega\varepsilon$. They designed an experiment using Zenneck's surface-wave theory to measure this quantity near the receiver's location. From the theory, a finite ground conductivity tilted a surface wave's electric field from the vertical direction, and the tilting angle $\beta_0$ obeyed $\tan\beta_0 \cong (\omega\varepsilon/2\sigma)^{1/2}$. One could measure the angle $\beta_0$ by the same method as that for determining $\beta$. The only difference was that in the $\beta$-measurements the transmitting station was hundreds of miles away from the receiver, while in the $\beta_0$-measurements it was only a few miles away.

Smith-Rose and Barfield performed the experiment in 1924. They measured the soil conductivities (to determine $\beta_0$) in Teddington and Slough, where the receiving sets were. They also measured $\beta$ and $\delta$ for waves from stations in Königswusterhausen, Leafield, Nantes, Nauen, Ongar, Paris, Teddington, and Tours, ranging from 10 miles to 600 miles away from the receiving sites. But the experimental design spoiled the results. At the frequency band 44–670 kHz, the soil conductivity was quite high ($7\times10^7$ electrostatic units). Such a factor made the values of $(\omega\varepsilon/2\sigma)^{1/2}$ and hence $\beta_0$ extremely small—the measured values of $\beta_0$ were all within 2°. Since $\beta_0$ was extremely small, $\beta$ and $\delta$ were tiny as well—the measured values of $\beta$ did not exceed 3°, and those of $\delta$ were all within 1°. The quantity in question—$\sin\theta$—which equaled either $\tan\beta_0/\tan\beta$ or $\tan\delta/\tan\beta_0$, was therefore a small number divided by another small number. Consequently, the data errors of $\sin\theta$ were very large.

37. Smith-Rose and Barfield, "On the determination" (1925), 592.

This was exactly what Smith-Rose and Barfield observed. The measured values of $\tan\beta_0$, $\tan\beta$, and $\tan\delta$ were so tiny that no definite conclusion was possible about the values of $\theta$.[38] Smith-Rose and Barfield did not make any mistake in the theoretical argument underlying their experimental design. But the quantities they chose to measure were pragmatically hard to retrieve with adequate precision. Whereas Eckersley's choice of measurables in his polarization experiment made it difficult in principle to obtain direct evidence for sky waves (the incident angle), their choice was theoretically sound but practically problematic. Knowing this limitation, they proposed three possible methods of improvement—first, to increase the accuracy of the measuring devices or select other measuring quantities; second, to carry out the experiment at shorter wavelengths—$\tan\beta_0 = (\omega\varepsilon/2\sigma)^{1/2}$ increased with frequency $\omega$; and third, to find a different site with less ground conductivity. But experimentation was about competing with time. Because of the wartime development of wireless technology, the state's strategic stress on radio, and the findings from recent short-wave experiments, the mid-1920s saw a craving for exploration of the upper atmospheric layer. Smith-Rose and Barfield at NPL were not the only team to work on this project. Before they improved the polarization experiments, Edward Appleton, working for the same Radio Research Board, obtained evidence for sky waves' existence from another approach.

## FREQUENCY-CHANGE EXPERIMENT AND DISCOVERING THE IONOSPHERE

While measuring polarization with direction finders was the first breakthrough in propagation experiments before the mid-1920s, the fading-induced frequency-change method launched a new regime for radio ionospheric research. Scientists started to perform "active" experimentation with the ionosphere only after introduction of the frequency-change method, which Edward Appleton invented.

### *Edward Victor Appleton*

Edward Victor Appleton was a founder of ionosphere studies. His career focused on using radio experimentation to explain the interactions of the upper atmosphere, the geomagnetic field, and solar radiation and the diverse physical characteristics of the ionosphere. He drew together geophysics, me-

38. Ibid., 597–99.

teorology, radio science, and wireless engineering and made possible global short-wave communications, radar, and space programs. Before switching to administrative work in 1939 (first as secretary of DSIR and then as principal and vice-chancellor of the University of Edinburgh), he and his coworkers had made England a center of ionospheric research. Today, Britain's national establishment for space and high-energy physics in Didcot is named the Rutherford-Appleton Laboratory to memorialize his contribution. He and Ernest Rutherford represented two traditions of modern British experimental physics: one looking inward to the constitution of microscopic particles, the other looking outward to the structure of nature. Acquiring direct evidence for the ionosphere started Appleton's career on the latter track.

A native of Bradford in Yorkshire, Appleton went to St John's College, Cambridge, to prepare for the Tripos. He studied geology, mineralogy, and physics but was not particularly keen about mathematics; so he chose the newer Natural Sciences Tripos.[39] He gained first class in part 1 of the Tripos in 1913 and part 2 (Physics) in 1914 and began to work for the crystallographer William Lawrence Bragg, Jr., at the Cavendish Laboratory on the structures of metallic crystals. He left Cambridge in 1915 to join the Royal Engineers, where he specialized in signal duties, especially wireless communications. His war experience turned out to be invaluable. As he recalled later:[40]

> During my R.E. career, I was able to put a few problems "on the shelf" in my mind, to be attacked when the war was over. Two of these occupied my interest when I got back to Cambridge, the theory of the thermoionic valve and the theory of long-distance radio propagation.

Both problems shaped his forthcoming research.

Appleton returned to the Cavendish after the war. Most researchers there were pursuing molecular structures and subatomic particles, topics close to the hearts of directors Ernest Rutherford and John Joseph Thomson. But Appleton was keen to study radio physics. Fortunately, although Rutherford wanted him to do research on nuclear physics, he still said, "Go ahead, I will

39. Since the 1880s, the Natural Science Tripos had rivaled the Mathematical Tripos in attracting students of science. In the early twentieth century, students at the Cavendish Laboratory read for the Natural Science Tripos. See Warwick, *Masters of Theory* (2003), 218, 497.

40. Notes for an autobiographical interview that never took place; cited in Ratcliffe, "Appleton" (1966), 3.

back you."[41] The first problem that interested Appleton was the nonlinear characteristics of electronic tubes, several of which he brought with him from the British army. He noted their complex voltage-current relationship and worked with the Dutch engineer Balthasar van der Pol on the electrical properties of tube oscillators.[42] Another problem that attracted him was the nature of atmospheric noise. Under the influence of C. T. R. Wilson, who had been conducting cloud-chamber experiments to mimic thunderstorms at the Cavendish,[43] Appleton attempted to connect atmospherics with thunderstorms and other electrically excited weather processes.[44]

### *The Frequency-Change Experiment*

In the early 1920s, Appleton extended his research from atmospherics to fading. American radio amateurs with ARRL and NBS had performed some large-scale fading experiments. Fearing that fading would hamper the emerging short-wave communications, the Radio Research Board in Britain sponsored Appleton to conduct fading research. A Cambridge student, Miles Barnett from New Zealand, assisted him. Later another Cambridge pupil, John Ratcliffe, who had studied chemistry, geology, and physics with C. T. R. Wilson and Larmor, joined the team.[45]

The introduction of radio broadcasting and the creation of the British Broadcasting Company (BBC) in the early 1920s greatly assisted Appleton's investigation—"powerful continuous-wave senders became generally available for the first time."[46] In 1923, Appleton and his assistants at the Cavendish's receiving station monitored the diurnal variations of radio signals from the BBC's transmitters in London and Bournemouth. Signal intensity remained constant in the daytime but varied significantly at night. For a nearby transmitting station (London), daytime signals were reasonably strong and nighttime signals had about the same mean values but with more or fewer periodic variations. For a distant station (Bournemouth), signals in the day were

41. Appleton, *Science and the Nation* (1957), 39.

42. Ratcliffe, "Appleton" (1966), 9. Appleton and van der Pol were lifelong friends. After the latter went back to Holland they corresponded about nonlinear electronic circuits and other research topics. See Stumpers, "Appleton and van der Pol" (1975), 344–56.

43. Galison and Assmus, "Artificial clouds" (1989), 225–74.

44. Ratcliffe, "Appleton" (1966), 9–10.

45. Budden, "Ratcliffe" (1988), 671–73.

46. Appleton, "The ionosphere" (1964), 80.

weak and those at night had much higher mean values but with much larger, much less periodic, and less rapid variations.[47]

To explain their observations, Appleton and Barnett hypothesized fading as a result of interference between direct ground waves and indirect sky waves: ground waves propagated along the earth's surface while the ionized layer deflected sky waves. Stronger solar rays during the day made the ionized layer thicker and closer to the ground, so sky waves deflected at lower heights. The higher concentration of gas molecules at lower heights caused deflecting waves more frictional loss because of more frequent collisions between ions and molecules. Thus sky waves were weak in the day, and incoming signals were mostly from invariable ground waves. In contrast, sky waves were strong at night. Incoming signals were the superposition of both sky and ground waves at short transmitter-receiver distances, and sky waves dominated at long distances. Since a sky wave's phase, amplitude, and polarization changed continually because of varying atmospheric conditions, signals at night fluctuated considerably. At a long transmitter-receiver distance, the mean values of nighttime signals were much stronger than those of daytime signals, since sky waves (unlike ground waves) did not attenuate quickly over distance. Appleton and Barnett accounted for the existence of sky waves in terms of the magneto-ionic theory, in which fluctuations of the ionized layer's height and ion density and of the geomagnetic field caused a sky wave's phase, amplitude, and polarization to vary.[48]

Neither Appleton and Barnett's experimental results nor their hypothesis were new. De Forest, Fuller, and Pierce in the 1910s and ARRL and NBS in the early 1920s all made sense of fading along the same lines. But the Cambridge researchers proposed to find direct evidence for the sky-wave hypothesis. Since fading supposedly resulted from the change of sky waves with respect to ground waves, actively producing *artificial* fading in a predictable manner would verify this proposition. The Cambridge duo came up with the following idea (figure 7.4). The incoming signal intensity at a point on the earth was the superposition of a ground wave and a sky wave. These waves had the same frequency yet different phases, for their path lengths differed. If the sky-ray path was $a$, the ground-ray path was $a'$, and the radio wavelength was $\lambda$, then the phase difference between sky and ground wave was $2\pi(a-a')/\lambda$. From

47. Appleton and Barnett, "Local reflection" (1925), 333. Unlike American amateurs' fading tests, which recorded the results with Eccles's audibility scales, Appleton and Barnett used galvanometers to measure signal intensity.

48. Appleton, "Geophysical influences" (1924–25), 17D–21D.

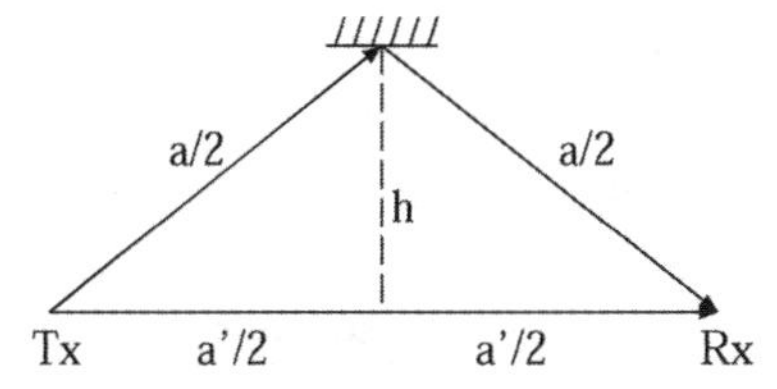


FIGURE 7.4. Appleton and Barnett's geometric model for sky waves.

the fundamental properties of waves, the superposition of the two waves had the greatest amplitude when $(a-a')/\lambda$ was an integer and the minimum when $(a-a')/\lambda$ was an integer plus a half. *Changing continuously* the radio wavelength from $\lambda$ to $\lambda'$ would alter the two waves' phase difference continuously as well, and the amplitude of the superposed wave would change accordingly from maximum to minimum to maximum and so on. The number $\mathcal{N}_m$ of the observed signal maxima should be

$$(7.1) \quad \mathcal{N}_m = (a-a')\left(\frac{1}{\lambda}-\frac{1}{\lambda'}\right) \cong \frac{\lambda-\lambda'}{\lambda^2}(a-a') = \frac{\delta\lambda}{\lambda^2}(a-a')\cdot$$

The approximation held, since the wavelength change $\delta\lambda$ was much smaller than the wavelength proper $\lambda$ or $\lambda'$. Within the short period of wavelength change, the ionized layer's height (and hence the path-length difference) was invariant. So equation (7.1) implied that the number of observed signal maxima $\mathcal{N}_m$ was proportional to the wavelength change $\delta\lambda$ if the received wave was indeed the superposition of two waves with different path lengths.

Appleton and Barnett entertained the idea of frequency alteration in late 1924, after the former left Cambridge for a professorship at King's College, London. At the end of the year, the BBC gave them permission to conduct the experiment after broadcasting hours at its sending station in Bournemouth. The optimum transmitter-receiver distance was 160 kilometers (100 miles), where sky and ground waves had approximately the same intensity and would create the most conspicuous interfering effects. The Cavendish Laboratory was not at this distance from the Bournemouth station. So the scientists moved the receiving station to Oxford's Electrical Laboratory, 140 kilometers from Bournemouth.

Their primary technical issue was that the transmitter had to be able to send signals with a continuous and uniform wavelength change of 5–10 meters while keeping a constant antenna current. A. G. D. West of the BBC helped construct such equipment. Also, the receiver should retain a flat tuning within this wavelength interval in order not to modulate the measured intensity. Appleton and Barnett used a four-stage high-frequency electronic-tube amplifier with tuned tube anode coupling and an Einthoven galvanometer that had served them in the Cambridge-London fading experiments.[49] The period of frequency change was 10–30 seconds, so the wavelength should alter slowly enough that the time constant of the receiver's valve amplifier would not smooth out the fringes.[50] This receiving circuit accommodated a quadratic relationship between the galvanometer current $I$ and the received electromotive force $v$, which the researchers represented by $I = a\nu + b\nu^2$. When the incoming wave was a superposition of two waves with different phases—$\nu = \kappa[E\sin(\omega t) + e\sin(\omega t + \phi)]$—the time-averaged current reading in the galvanometer would be the time average of $b\kappa^2[E\sin(\omega t) + e\sin(\omega t + \phi)]^2$, which was $(b\kappa^2/2)[E^2 + e^2 + 2eE\cos\phi]$. A wavelength change that altered the phase difference $\phi$ would modulate the galvanometer reading.

Appleton and Barnett performed the experiment between Bournemouth and Oxford on 11 December 1924 and 17 February 1925. Their measurements were encouraging: number of fringes $N_m$ was indeed proportional to wavelength change $\delta\lambda$. For wavelengths between 385 meters and 392 meters, the average number of fringes was 4.5, leading to the ratio $N_m/\delta\lambda = 0.64$; for those between 385 meters and 395 meters, it was 7.0, leading to $N_m/\delta\lambda = 0.64$ as well. The experimenters' two-wave hypothesis was thus consistent with the experimental data. Moreover, they used measured $N_m$ to evaluate height $h$ of the reflective layer from equation (7.1) and the geometry $h = (a'^2 - a^2)^{1/2}/2$ (see figure 7.4). In February 1925, they reported to *Nature* their results and an estimate of 85 kilometers for $h$. Months later, they slightly revised the estimate to 80–90 kilometers. This, they believed, was by far the most concrete evidence for the Kennelly-Heaviside layer.[51]

Appleton and Barnett's frequency-altering tests on 11 December 1924 and 17 February 1925 later gained the reputation as the crucial experiment towards discovery of the ionosphere. Thanks to these tests, many people credited Appleton with discovering the ionosphere, which was definitely a major

49. Appleton and Barnett, "On some direct evidence" (1925), 627–28.

50. Ratcliffe, "Some memories of Sir Edward Appleton: the earliest ionosphere experiments," *R.S.R.S. Newsletter*, 49 (May 1965), box "Radio Research Board Committees, 1920–40s," Historical Records, Rutherford Appleton Laboratory (Space Science Department), Didcot, England.

51. Appleton and Barnett, "Local reflection" (1925), 333–34, and "On some direct evidence" (1925), 629.

contributing factor to his winning the Nobel Prize.[52] Because experiments on direction finding, fading, propagating ranges, and skip distances had been under way, overemphasizing the role of the 1924–25 frequency-change experiment may be problematic. But these tests were indeed distinct in two ways: first, crucial information about the ionized layer—its height—came directly from the measured data; second, evidence for sky waves came through *active* manipulation, not just *passive* observation, of radio signals.

Smith-Rose and Barfield at NPL—Appleton and Barnett's closest competitors—also aimed to prove the reality of the ionosphere by measuring its height, which was at least theoretically plausible if not practicable. In contrast to the NPL researchers, who measured but did not tinker with the transmitting radio waves, however, the Cambridge scientists performed an intervening experiment by modulating the waves with frequency change and observing the consequences. Such active manipulations of artificial fading characterized Appleton and Barnett's experimental efforts and made their evidence more direct support to contemporaries for the ionized layer's reality.

Nevertheless, their work had a problem—the empirical evidence showed only that the incoming signals consisted of two waves with a phase difference; it did not show that one of the two waves propagated downward from the upper sky. They could both have been ground waves traversing different horizontal paths! To determine the wave-propagating directions required measurements of polarization.

After February 1924, Appleton and Barnett were busy producing more data from the frequency-change method and finding a way to measure polarization free from Smith-Rose and Barfield's problem with ground conductivity. Their solution: compare signals from a vertical rod antenna with those from a vertical loop parallel to the plane of propagation (figure 7.5). Suppose that a ground wave and a sky wave propagated between transmitter and receiver, the ground was nearly perfectly conductive, and the sky wave disappeared during the day. The vertical rod received the ground wave's vertical electric field $E_0\sin(\omega t)$ in the daytime. At night, it received the vertical electric field of the ground wave, the sky wave's transverse-magnetic component, and its reflection from the ground, $E_0\sin(\omega t) + 2E_1\sin\theta\sin(\omega t+\phi)$ ($\theta$ denoted the sky wave's incident angle, and $\phi$ the phase difference between the sky wave and the ground wave). The indirect-wave component caused fluctuations. So the nighttime reading of galvanometer current deviated from the

52. Ratcliffe, "Appleton" (1966), 10; Appleton, "The ionosphere" (1964), 79–86.

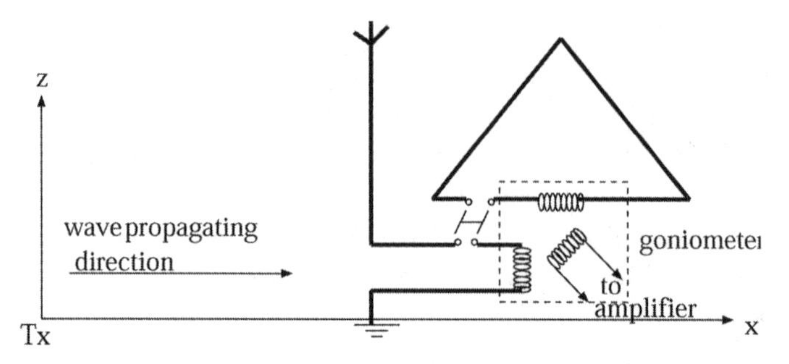


FIGURE 7.5. Configuration of Appleton and Barnett's polarimetric experiment.

daytime reading, and the time average of such a deviation was proportional to $E_0E_1\sin\theta\cos\phi$.

Similarly, the vertical loop received the ground wave's vertical magnetic field $H_0\sin(\omega t)$ in the daytime. At night, it received the vertical magnetic field of the ground wave, the sky wave's transverse-magnetic component, and its reflection from the ground, $H_0\sin(\omega t)+2H_1\sin(\omega t+\phi)$. The time average of its nighttime current reading's deviation from its daytime counterpart was proportional to $H_0H_1\cos\phi$. Therefore the ratio of the loop's average nighttime current deviation from its daytime value to the same factors for the rod was $K = (H_0H_1\cos\phi)/(E_0E_1\sin\theta\cos\phi) = \csc\theta(H_0H_1/E_0E_1) = \csc\theta$ (when electric field was in electrostatic unit and magnetic field in electromagnetic unit so that $E_0=H_0$ and $E_1=H_1$). Thus nighttime fading because of sky waves made signal intensity vary more in the vertical loop than in the vertical rod, since $\csc\theta > 1$ for all values of $\theta$. In other words, the sky almost certainly was the source of fluctuations if the measured ratio $K$ was larger than 1.

By working exclusively on the transverse magnetic components of electromagnetic waves, Appleton and Barnett found a way to measure polarization that avoided Smith-Rose and Barfield's problem: large ground conductivity was no longer a practical problem in the rod-loop scheme. Moreover, the measure of the ratio $K$ could directly determine a sky wave's incident angle θ, which could also help in estimating the height of the ionized layer.

Appleton and Barnett implemented this idea and measured signals sent from the BBC's London station to the Cavendish. Most of the time they found a ratio $K$ of the loop variation to the rod variation larger than 1—most frequently 2.85, corresponding to $\theta = 21°$. Then they measured signals from Birmingham, 140 kilometers from Cambridge. $K$ was still larger than 1, and the angle θ from measured $K$ was 42°, consistent with the fact that the sky wave's incident angle increased with separation between transmitter and receiver.[53]

## *Smith-Rose and Barfield Try Again*

As Appleton and Barnett were busy with their sky-wave experiments, Smith-Rose and Barfield made progress in 1925, too. Their preoccupation after the unsuccessful attempt in 1924 was to solve the problem of small $(\omega\varepsilon/\sigma)^{1/2}$ (because of high ground conductivity) in measuring polarization. To do so, they increased the frequencies. They installed a medium-wave (300–500 meters) transmitting station in Bournemouth and accordingly upgraded the receiving set at the Radio Research Board's Slough station, where they had been conducting direction-finding experiments. At the higher frequencies, $(\omega\varepsilon/\sigma)^{1/2}$ increased by 2–6 times, reducing the requirement for receiving sensitivity.[54]

Yet they needed to make more fundamental changes to obtain more precise results. They aimed to demonstrate that sky waves dominated long-range signals and to determine the waves' incident angles from polarization measurements. They relied on three fundamental polarimetric relations. They had found in 1924 that $\tan\beta \cong (\omega\varepsilon/2\sigma)^{1/2}/\sin\theta$ and $\tan\delta \cong (\omega\varepsilon/2\sigma)^{1/2}\sin\theta$. They had measured $\tan\beta = E_x/E_z$ by a rotating rod and $\tan\delta = H_z/(H_x^2 + H_y^2)^{1/2}$ by a rotating loop. In addition, they also found that the ratio of the signal in a vertical loop on the plane of propagation to the signal in a vertical rod was $R = H_y/E_z = 1/\sin\theta$.[55]

All three formulae presupposed the absence of ground waves, which was true for the researchers' 1924 experiment between long-wave transmitters in Continental Europe and Slough but untrue for their new experiment between the short-wave sender in Bournemouth and Slough, which were too close to each other. Thus they revised these formulae by incorporating the ground-wave components. The new formulae, along with a few facts about ground waves that they deduced from Zenneck's surface-wave theory, formed a set of mathematical relations that they could use to determine a sky wave's incident angle θ from any one of the three groups of measurable quantities: (i) $(\omega\varepsilon/\sigma)^{1/2}$, $\tan\beta$, $E_z$, and $E_z^{\text{ground}}$; (ii) $(\omega\varepsilon_0/\sigma)^{1/2}$ and $\tan\delta$; and (iii) $R$, $E_z$, and

53. For Appleton and Barnett's loop-aerial scheme and the experimental results, see Appleton and Barnett, "On some direct evidence" (1925), 630–34.

54. Smith-Rose and Barfield, "An investigation" (1926), 581.

55. Ibid., 590.

$E_z^{\text{ground}}$. $E_z$ was the vertical component of the total electric field measurable by a vertical rod at night, and $E_z^{\text{ground}}$ was the vertical component of the ground wave's electric field, which one could estimate from the same measurement during the day.[56]

Smith-Rose and Barfield's three groups of mathematical relations corresponded to three ways of measuring a sky wave's incident angle. The first used instruments to measure wave-propagating direction with a rotating rod and wave intensity with a vertical rod. The second employed devices to measure direction with a rotating rod and direction with a rotating loop. The third used a direction-measuring instrument with a vertical loop, a vertical rod, and an instrument that measured intensity with a vertical rod. With these methods, the *much shorter transmitter-receiver distance* than Smith-Rose and Barfield's previous ones strengthened signals and thus reduced error.

However, direct ground waves no longer disappeared at the receiver, which made it more difficult to use measured data to determine $\theta$, for the direct-wave effect required separate measurement. Moreover, methods (i) and (ii) still needed the value of $(\omega\varepsilon/\sigma)^{1/2}$—a function of ground conductivity $\sigma$—to determine $\theta$. So these two methods encountered the same problem as before: obtaining $\theta$ by dividing a small measurable with another small measurable.

Nevertheless Smith-Rose and Barfield's three new methods could work together to fight these problems. Method (ii) was independent of the direct-wave effect, because both ground waves and sky waves were on the same plane of polarization. Meanwhile, method (iii) did not depend at all on the value of $(\omega\varepsilon/\sigma)^{1/2}$, because it (like Appleton and Barnett's polarimetric scheme) involved the ratio of the measured results from a loop to those from a rod. Taking such a ratio cancelled the effect of $(\omega\varepsilon/\sigma)^{1/2}$. Therefore Smith-Rose and Barfield could explore the individual advantages of the three methods and crosscheck their results with one another.

Applying the three methods, the two men performed measurements between Slough and Bournemouth from May to July 1925, plotted the measured waveforms of $\tan\beta$, $\tan\delta$, and $R$, and used sampled values of these waveforms to evaluate the corresponding $\theta$. The results, though fluctuating, showed

56. Whether $E_z$ and $E_z^{\text{ground}}$ were in phase with each other affected the estimate's accuracy. In general, they were not in phase, and accurate evaluation of their phase difference was impossible. But one could estimate the maximum and minimum of $\theta$ without knowing the phase. From such estimates, Smith-Rose and Barfield found that the errors of $\theta$ due to this phase effect were not significant (ibid., 609).

clearly that most θ values were between 13° and 34°, meaning that nighttime waves were indeed from elevated incidence angles. This formed another piece of empirical evidence for sky waves. Smith-Rose and Barfield took the smallest angle $\theta = 13°$ from the measurements to estimate the maximum height of the wave-deflecting layer in the sky between Slough and Bournemouth. The value was about 88 kilometers, consistent with Appleton and Barnett's estimate. Finding encouragement in this result, they did more measuring between July and December in Birmingham, London, and Newcastle[57] and reported in December 1925.

## *Improving Direction Finding as a By-product*

Seeking evidence for sky waves could help improve direction-finding technology, which interested the NPL and the Radio Research Board. Under pressure from these organizations, Smith-Rose and Barfield quickly found their scientific studies useful to direction-finding engineering. Their results indicated that any direction finder made of rotating loops to detect the polarizations of incoming waves had errors because of sky waves with elevated incident directions. But a sky wave rarely deviated laterally from the direct ground wave's plane of propagation; viz., the horizontal component of a sky wave's propagating direction aligned with the ground wave's. In addition, a wave's phase was a function solely of its propagating direction, not its polarization. So a direction finder that detected the horizontal components of incoming waves' phases could eliminate the errors resulting from sky wave's polarizations.

British inventor F. Adcock had patented a phase-detecting direction finder in 1919.[58] His system had two mutually orthogonal pairs of identical vertical straightrod antennae connecting to a goniometer and to the ground (figure 7.6). In each pair, the distance between the two rods created a phase difference between the wave-induced signals on the rods. When a rod was much smaller than the wavelength, the difference of the electromotive forces on a

57. Smith-Rose and Barfield, "Further measurements on wireless waves received from the upper atmosphere," Radio Research Board, Propagation of Waves Committee, S.C.A. Paper 129, box "Radio Research Committees, 1920–40s," Historical Records, Rutherford Appleton Laboratory.

58. Adcock, "Improvement in means for determining the direction of a distant source of electromagnetic radiation," British Patent 130490/1919; ibid.

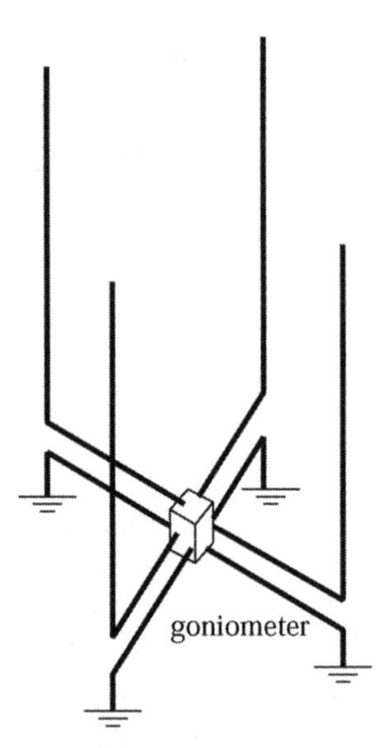


FIGURE 7.6. The Adcock system.

pair of rods approximated the phase difference, which was proportional to $\sin\theta\cos\psi$ (where $\psi$ was the angle between the horizontal component of the wave's propagating direction and the horizontal line connecting the two rods). Similarly, the difference on the other pair of rods (orthogonal to the original pair) was proportional to $\sin\theta\sin\psi$. Consequently, the goniometer responding to these two differences pointed to the direction determined exclusively by $\psi$, not the angle of elevation $\theta$. Since $\psi$ represented the horizontal component of the wave-propagating direction, which was identical for both direct and indirect waves, it measured the true direction of the transmitting station.[59]

Nevertheless, the Adcock system did not perform as perfectly as the theory predicted; it failed to replace loop-type direction finders in the early 1920s. The major problem: it was highly sensitive to the symmetric condition between the two antenna pairs. A slight difference of electrical properties because of stray capacitance between the baselines and the ground earth drove the system considerably away from balance and caused serious errors. To maintain electrical symmetry, Smith-Rose and Barfield lifted the goniometer above the ground so that each vertical rod was more like a complete Hertzian dipole with two branches. In the new design, signals were registered in a more symmetrical manner as they were measured from the midpoints than from the

59. Smith-Rose and Barfield, "The cause and elimination" (1926), 833.

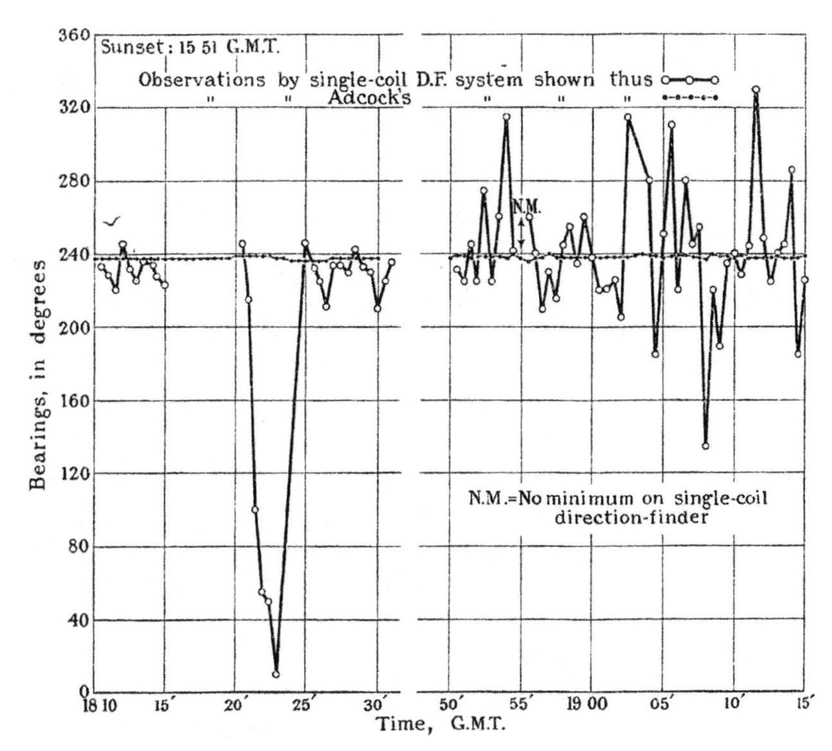


FIGURE 7.7. Direction-finding errors of the revised Adcock vs. the single-coil system; bearing on Bournemouth, 10 December 1925 (wavelength 386 meters). Smith-Rose and Barfield, "The cause and elimination" (1926), figure 7.

endpoints of the rods. Also, the effect of stray capacitance was much smaller, since the horizontal part was now well above ground.[60]

In late 1925, the two men tested the revised Adcock system in Slough. The results were exceptionally good. For signals from Birmingham (wavelength 475 meters), Bournemouth (386 meters), Cardiff (353 meters), London (365 meters), and Newcastle (404 meters), it performed much better than the single-coil system (figure 7.7) The new design controlled the direction errors within 1°. This was barely imaginable when Eckersley studied direction finders in World War I. Smith-Rose and Barfield's investigation of sky waves thus led to a breakthrough in direction-finding technology. From a different perspective, their success also proved that the major source of errors in loop-type direction finders was changes of polarization because of sky waves'

60. Ibid., 833–34.

elevated incident angles. When they used a phase instead of a polarimetric detector, they could eliminate such errors. This provided additional evidence for sky waves. Once again, scientific discovery intertwined with technological development.

Between 1923 and 1925, Appleton, Barnett, Smith-Rose, and Barfield launched the evidential search that "proved" the reality of the upper ionized layer. To the scientific community, the conjecture dating back to Stewart, Schuster, Kennelly, and Heaviside eventually established itself firmly. Breaking from tradition, these British researchers did not build their empirical inquiries on sophisticated theories of geomagnetic dynamics or radio-wave propagation. Instead, all they did was seek radio waves from the sky. To achieve this goal, Smith-Rose and Barfield measured polarization while leaving transmitting waves intact. Appleton and Barnett, in contrast, actively manipulated transmitting waves by varying continuously their frequencies and drew their empirical conclusion from observing the resulting pattern of incoming waveforms. Both approaches were able to measure the height of the ionosphere, which made its existence even more plausible. But Appleton and Barnett's method opened up a new possibility of using radio as an *active probe* to explore the ionosphere's mysteries. Radio scientists could now design and control transmitting waves, send them into the upper atmosphere, wait for their return, and use them to uncover the ionosphere's properties. This was similar to what mariners had been practicing as "echo sounding": sending sound underwater and determining the ocean's depth from the returning echoes. The experimental means that retrieved direct empirical support for the hypothetical ionosphere was becoming an experimental tool to explore the unknown characteristics of the "real" ionosphere.

Through the mid-twentieth century, radio sounding would constitute the central means to experiment with the upper atmosphere. Nevertheless, the most popular method of radio sounding was not Appleton and Barnett's frequency modulation. It was a scheme that sent radio *pulses* rather than continuous waves and detected their echoes from the ionosphere. The pulse-echo method, soon dominating radio sensing and radar technology, came from the Americans Gregory Breit and Merle Tuve at the Department of Terrestrial Magnetism, Carnegie Institution of Washington. Their effort was an attempt parallel to the Britons' to seek direct evidence for the ionosphere.

CHAPTER 8

# Pulse Echo, CIW, and Radio Probing of the Ionosphere

At the same time as Appleton and Barnett were working on the frequency-change experiment and Smith-Rose and Barfield were performing the polarimetric test, two U.S. researchers were also using radio to demonstrate the existence of the upper reflective layer. A few months after the Britons' announcements of discovery in December 1924, Gregory Breit and Merle Tuve at the Carnegie Institution of Washington (CIW) reported the third set of "direct" evidence for the ionosphere. Like the Britons, Breit and Tuve sought to measure the upper layer's height. And similar to Appleton and Barnett, they designed waveforms, sent them up, and retrieved data from the returning waves. But their scheme did not rely on frequency variation; rather, they transmitted sharp, periodically repetitive pulses.

Breit and Tuve's pulse-echo and Appleton and Barnett's frequency-change schemes went beyond soliciting evidence for the ionosphere. By the late 1920s, they became the major experimental means to explore the ionosphere's structure. Dissemination and appropriation of the two methods during this period shaped ionosphere studies for decades to follow. The American and British scientific institutions able to use such means became the centers of radio ionospheric research. In the United States, scientists and engineers under the CIW's influence preferred the pulse-echo approach for its direct visualization of the measured geophysical properties. In Britain, experts following Appleton's research program favored frequency change for its superb data-reduction theory and less complex equipment.

Yet both sounding-echo methods showed promise in identifying multiple reflections of radio waves from the ionosphere because of its heterogeneous layout. Frequency change bore the first fruit. With the apparatus that he invented with Barnett, Appleton and his protégés at London and Cambridge found in 1927 that the ionosphere actually had an additional layer to Kennelly-Heaviside. Calling it the "F layer," the Britons showed that the ionosphere had a more sophisticated structure than researchers originally thought, and radio seemed to be the only means to tease it out.

This chapter concerns the major development of radio ionospheric research immediately after Appleton and Barnett's crucial experiment. It focuses on Breit and Tuve's pulse-echo experiment around 1925, which led to further evidence for the ionosphere, and Appleton's discovery of the F layer using the frequency-change instrument. Both historical episodes played a crucial part in the epistemic transition from radio-wave propagation studies to radio probing of the ionosphere.

## PULSE-ECHO EXPERIMENTS IN THE UNITED STATES

The American efforts to prove the ionosphere's existence related closely to the nation's scientific interests in radio. While British Marconi, the Cavendish Laboratory, King's College, London, the National Physical Laboratory (NPL), and the Radio Research Board's station at Slough formed a virtual consortium to study radio-wave propagation during the 1920s, scientists and engineers were doing the same in and around the U.S. capital—what historian Bruce Hevly has called the "Washington network." This community included people from the CIW, the National Bureau of Standards (NBS—e.g., Austin and Dellinger), and the Naval Research Laboratory (NRL—e.g., Taylor and Hulburt).[1] But whereas NBS and NRL conducted radio research to improve wireless technology, CIW came to it via scholarly interest in atmospheric electricity and geomagnetism, a tradition tracing back to U.S. preoccupation with Humboldt-type geoscience in the nineteenth century. Two young physicists—Gregory Breit and Merle Tuve—connected the two areas by developing a radio pulse-echo method to experiment with the ionosphere.

### *Carnegie Institution of Washington, Gregory Breit, Merle Tuve*

The CIW exemplified the roles of philanthropic foundations in American scientific research before World War II. Today, the legacies of the Carnegies,

1. For the Washington network, see Hevly, "Building a Washington network" (1994), 143–48.

the Guggenheims, and the Rockefellers in U.S. academic establishments are still visible. The influence of this private patronage of science reached its peak in the early twentieth century. In 1901, American steel baron Andrew Carnegie donated ten million dollars to establish a research organization in the nation's capital. The Carnegie Institution of Washington had a president, a secretary, and a board of trustees. Its heads had a strong background in geoscience: the second president, Robert Woodward, had been in the U.S. Coast and Geodesic Survey; the first secretary, Charles Doolittle Walcott, director of the U.S. Geological Survey; and Carnegie a mining engineer. Under their influence, CIW's research directions mixed American pragmatism favoring experiments over theories with learned societies' natural-history tradition, which emphasized field observations. CIW had active programs of planetary science from its beginning. In 1903, it founded a Department of Terrestrial Magnetism (DTM).[2]

Terrestrial magnetism—the study of the earth's magnetic field—had been a subject in natural philosophy since Antiquity. To nineteenth-century scholars such as Gauss, Humboldt, and Sabine, the most critical problem in terrestrial magnetism was the lack of global data. From then on, the major activities of terrestrial magnetism consisted of transnational large-scale campaigns for geomagnetic measurements. The methods of geomagnetic research were therefore consistent with the Humboldtian tradition of expeditions and large-scale field science.

The DTM inherited this tradition. Louis Bauer, its advocate and first director, was well aware of this trend. In his view, DTM's preoccupation should be global magnetic surveys. He organized expeditions to measure variations of the geomagnetic field in numerous regions and lobbied for the building of CIW's famous all-wood sailing vessel—the *Carnegie*—for magnetic measurements. At its pinnacle, DTM was performing magnetic surveys in Africa, Australia, South America, the Atlantic, and the Pacific.[3] Its prosperity dwindled after the mid-1910s. The First World War and the shifting priorities of CIW leaders were the immediate reasons. The outbreak of war ended international collaboration and made global magnetic surveys much more dangerous. After the war, CIW's new president, John Merriam, thought that DTM should

2. For the CIW's early history, see Kohler, *Partners in Science* (1991); Reingold, "Carnegie Institution of Washington" (1979), 313–41; Servos, "Geophysical Laboratory of the Carnegie Institution of Washington," (1983), 147–85; Yochelson, "Andrew Carnegie and Charles Doolittle Walcott" (1994), 1–19; and Yoder, "Geophysical laboratory" (1994), 21–28.

3. Good, "Vision of a global physics" (1994), 29–36.

replace such endeavors with theoretical analysis, cooperative studies of the earth's crust and atmosphere with other departments, and laboratory or intervening experiments. Bauer's influence shrank, and DTM's assistant director, John Adam Fleming, took more responsibility.[4]

DTM diverted its focus away from magnetic surveys because of geoscientists' increasing interest in connecting geomagnetism with phenomena above the ground (atmospheric science), in addition to those below it (geology). Such a connection had existed: Stewart and Schuster's conducting-layer hypothesis preceding Kennelly and Heaviside marked the relevance of the upper atmosphere in geomagnetism. As the idea of the ionosphere increasingly prevailed, scientists began to contemplate a sophisticated natural system above the earth, in which solar radiation, molecules of the upper atmosphere, and the geomagnetic field interacted with one another. The timely rise of shortwave radio in the 1920s seemed to provide a promising tool for intervening experiments to unveil the ionospheric interactions. And the British researchers were already on the way.

Under such circumstances, Merriam and Fleming believed that DTM should study not only geomagnetism but also atmospheric electricity and the ionosphere via wireless. To promote such research, they employed not magneticians, geologists, and meteorologists as in the past, but young physicists with backgrounds in radio science and technology. Enter Gregory Breit and Merle Tuve.

Gregory Breit became one of the earliest U.S. theoretical nuclear physicists, but he started with ionospheric studies. Born in Nikolayev, Russia, he immigrated to Baltimore in 1915 and studied physics at Johns Hopkins University, where he trained as a radio scientist. After his BA in 1918 came his two theses: on transmission of electromagnetic waves in wireless telegraphy (AM, 1920) under Edward Oberlin Hulburt and on the distributive capacity of inductive coils (PhD, 1921) under department head Joseph Sweetman Ames. While in Baltimore, he was already in touch with the Washington network of atmospheric researchers and, through Ames,[5] worked at NBS. After graduation he spent two years in Leiden on a National Research Council fellowship studying with Paul Ehrenfest, taught in the physics department at the University of Minnesota, and joined DTM in 1924 as its mathematical physicist.[6]

Breit's task at DTM was to investigate the electrical properties of the upper

4. Cornell, "Tuve" (1986), 146.

5. Gillmor, "The big story" (1994), 136.

6. Cornell, "Tuve" (1986), 97; Hull, "Breit" (1998), 27–56.

atmosphere. He had several ideas in mind, such as designing field experiments, reproducing and investigating atmospheric ionization in the laboratory, and studying propagation of radio waves. DTM's staff could assist him with none of these. So he sought an outside collaborating experimenter, his former Minnesota student Merle Anthony Tuve.

Like Breit, Tuve earned fame not for radio but for his pioneering work on high-energy particles. A native of Canton, South Dakota, Tuve had been a wireless amateur since youth. Together with his playmate Ernest O. Lawrence (the inventor of the cyclotron and founder of the Lawrence Berkeley Laboratory), Tuve tinkered with wireless sets and read hobbyist magazines. Their experience with amateur radio profoundly influenced their future careers.[7] Tuve entered the University of Minnesota to study electrical engineering (BS in 1922) but later changed to physics (AM in 1923). There he trained as an experimental physicist. A few faculty members tutored him closely: Breit, his adviser, John G. Frayne, and an Englishman, William Swann, a former DTM researcher whose idea of echo-pulse would become the prototype of the Breit-Tuve experiment. For his master's thesis, Tuve investigated the effect of bombarding sodium and cadmium with positive ions, learning vacuum-tube techniques and properties of ionized gases in the process. Afterwards, he became an instructor at Princeton University, where he intended to pursue a PhD. But Breit strongly recommended that he attend Johns Hopkins, and in May 1924 Tuve became a doctoral student there under the supervision of the department head, Ames. And Breit immediately brought him into DTM's project on radio sounding of the ionosphere.[8]

### *Pulsed Radio Sounding of the Ionosphere*

DTM's radio-sounding project started in July 1924. On 25 July, Breit told Tuve that "we shall try to reflect Hertzian waves from the Heaviside layer sending in Washington and receiving in Baltimore or in the opposite direction."[9] In the summer and the fall, Breit was arranging a proposal for this project, which was to find direct evidence of the Kennelly-Heaviside layer. Breit's original idea was similar to Appleton's: phase differences between sky and ground waves caused fading with changing frequencies, and it was possible to detect pola-

7. Cornell, "Tuve" (1986), 15–26.

8. Carey, "Tuve" (1999), 46–48; Cornell, "Tuve" (1986), 15–26, 68–90, 101–25.

9. Breit to Tuve (letter), 25 July 1924, box 401, Merle Tuve Papers, Library of Congress, Washington, DC, cited from Gillmor, "The big story" (1994), 136.

rimetric differences between sky and ground waves by altering antenna directions.[10] But he planned to construct a large parabolic reflector to focus the transmitting waves within a narrow beam, from which he could obtain directly a sky wave's elevation angle and the layer's height. This parabolic reflector, wood covered with copper-mesh window screen, would be 13 feet deep and 30 feet at its widest. The wavelengths, at which the reflector-generated beamwidths were less than 20°, were several meters. Breit formally proposed the project in November. In the meantime, he traveled on the east coast seeking cooperation from universities and companies. Between fall 1924 and spring 1925, he tested frequency change (without the parabolic reflector) between CIW and ARRL and Cornell, Johns Hopkins, and Wesleyan universities.[11]

The project did not go well, however. Bauer reduced the budget for the large parabolic reflector, forcing Breit to replace it with a simpler design. The preliminary frequency-change tests yielded no positive results. And Tuve thought the entire parabolic-reflector scheme impractical. The 30-foot reflector was to produce narrow beams at wavelengths of several meters. But from his previous experience with radio, Tuve predicted that these ultra-short waves could not propagate from Washington to Baltimore. Proper wavelengths should be at least longer than 50 meters. A parabolic reflector for 50 meter (or longer) waves was simply too huge to construct.

Instead, Tuve suggested using "interrupted continuous waves," essentially pulsed waves, to detect the Kennelly-Heaviside layer. His idea was to interrupt radio transmission periodically to yield pulse-modulated continuous waves. As a train of pulsed waves went to the sky, the ionosphere would reflect a train of pulsed echoes to the ground (if the layer existed). The time difference between a transmitting pulse and its echo pulse could indicate the layer's height. Tuve had learned the pulse-echo method from his Minnesota professors John Frayne and William Swann, who entertained the idea in 1921.[12] Their scheme integrated the transmitting and the receiving antennae (with a single antenna that switched between the two functions) to deliver radio waves vertically to the sky. Since transmitter and receiver were at the same location and hence no ground wave was present, only sky waves arrived. Frayne and Swann's experiments failed because of a technical problem with antenna multiplexing:

10. Breit, "Report on the trip taken to New York City, Schenectady, and Boston in connection with proposed Heaviside layer experiments," 5 March 1925, box 11, folder "Heaviside Layer," Merle Tuve Papers, Library of Congress.

11. Cornell, "Tuve" (1986), 153–54; Gillmor, "The big story" (1994), 137–38.

12. Swann, "The penetrating radiation" (1921), 65–73.

transmitting waves leaked into the receiver and overloaded it too much for it to detect any echo.[13] Tuve and Breit separated the transmitting and receiving locations: if the ionized layer existed, there should be two trains of pulses at the receiver—one from ground waves and the other from sky waves. The layer's height would be a function of the time difference between a sky-wave pulse and a ground-wave pulse.

The pulse-echo idea became final in December 1924, and in January 1925 Breit and Tuve published a note describing their idea.[14] They estimated the period of a pulse train to discriminate a sky-wave return from a ground-wave return. From the similar geometry as Appleton and Barnett's experiment (see figure 7.4), the fact that the speed of light was roughly 186,000 miles (300,000 km) per second, and the estimate that the height of the conducting layer was roughly 80 miles (128 km), the time lag $t$ between a sky-wave pulse and a ground-wave pulse would be 1/1,162, 1/1,801, 1/2,806, and 1/4,923 second for transmitter-receiver separations of 0, 80, 160, and 320 miles (0, 128, 256, and 512 km), respectively. Also, the duration of a pulse should not be much shorter than half of the pulse-train period because of a constraint in the electronic circuit. So the pulse-train period $T$ had to be approximately $2t$ to make ground-wave pulses not overlap with sky-wave pulses (figure 8.1): $T$ should be 1/581, 1/900, 1/1,403, and 1/2,461 second for transmitter-receiver separations of 0, 80, 160, and 320 miles, respectively.

In the following months, Breit traveled to New York City, Schenectady, and Boston to seek collaborators. F. H. Kroger of the Radio Corporation of America (RCA) was willing to send signals from RCA's station in Tuckerton, New Jersey, during the experiment.[15] DTM's receiver in Washington, DC, was a superheterodyne set with a four-stage voltage amplifier and a power amplifier. Its output went to an oscillograph for a photograph. The transmitter in Tuckerton had a wavelength of 70 meters. Continuous waves came from a master oscillator and a two-stage tube amplifier. To modulate continuous waves into pulses, the experimenters coupled a high-tension (500V) low-frequency (500–600 Hz) oscillator to the grid of the first amplifying tube via inductive coils. The sinusoidal signal from the low-frequency oscillator modulated the grid bias of the radio-frequency amplifier: when the grid bias was less than a threshold depending on the direct-current (DC) level of the grid voltage, the amplifier was turned off, which interrupted the transmitting wave. The

13. Gillmor, "The big story" (1994), 135.

14. Tuve and Breit, "Note on a radio method" (1925), 15–16.

15. Cornell, "Tuve" (1986), 157.

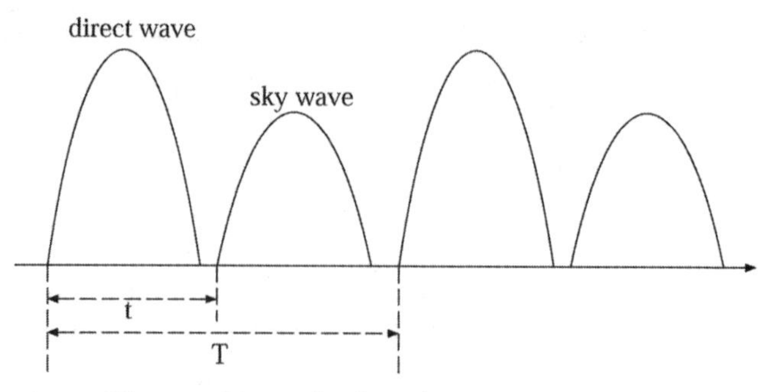


FIGURE 8.1. The waveforms of radio pulses.

period of interruption was 1/600–1/500 second (figure 8.2).[16] Changing the DC level of its grid voltage adjusted the amplifier's "on" time (the duration of a pulse). In this experiment, it was 1/5–1/3 of the period of interruption. The modulation scheme could not generate narrower pulses, since a further, higher threshold for reducing the "on" time would reduce too much the amplifier's effective grid bias voltage.

Preliminary tests between Washington and Tuckerton took place in April 1925, and a two-week experiment in late May and early June.[17] Although some incoming waveforms did look like the superposition of two trains of pulses, the phenomenon was actually a consequence of an unstable transmitter. The fluctuations of transmitting frequencies over time generated temporal variations at the output of the narrow-band superheterodyne receiver that resembled superposition of two trains of pulses.[18] This error reminded Breit and Tuve that the pulse-echo method depended on the precision of the transmitting and receiving waveforms.

That meant they had to use transmitters with more precise and stable radio frequencies. So they looked for other stations on the east coast. NBS, Westinghouse's station KDKA in Pittsburgh, and a few amateurs in the Washington, DC, area were willing to participate. Breit arranged tests with them in the summer. But these stations were not much better than Tuckerton at keeping frequencies stable. The crystal-controlled transmitter from Albert Hoyt Taylor's team at NRL appeared to be the only solution.

16. Breit and Tuve, "A test" (1926), 554–60; Dahl and Gebhardt, "Measurements" (1928), 290–92.

17. Cornell, "Tuve" (1986), 158.

18. J. A. Fleming to F. H. Kroger (letter), 6 July 1925, and Fleming to M. G. Grabau (letter), 24 June 1925, box 11, folder "Heaviside Layer," Merle Tuve Papers.

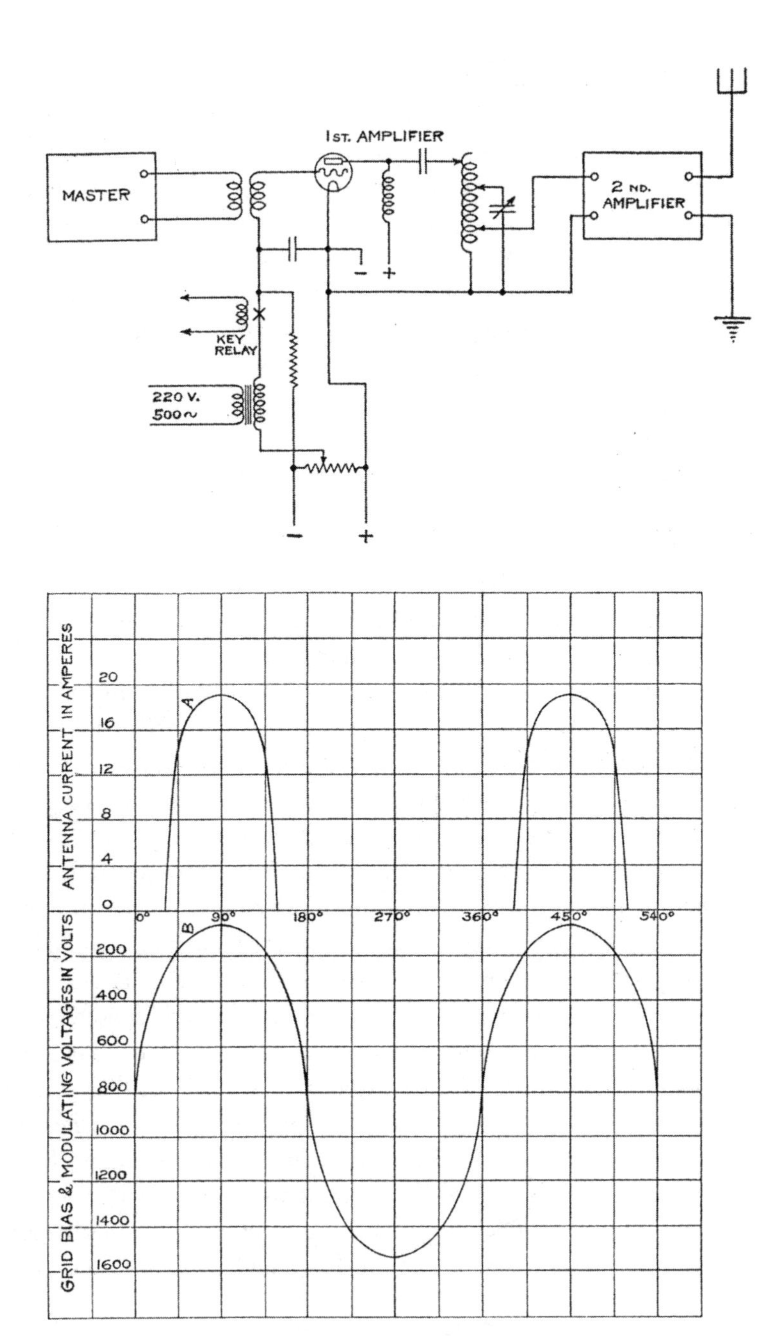


FIGURE 8.2. Pulse modulation of waveforms. Dahl and Gebhardt, "Measurements" (1928), figures 1-2.

In June 1925, Breit obtained permission from the navy and began the experimental setup with Taylor and NRL technicians Louis Gebhardt and Leo Young.[19] They sent signals from their transmitter at NRL to the receiver at DTM, eight miles away. They chose wavelengths of 71.3, 41.7, and 20 meters, and the period of pulse trains was 1/500 second. The experiment took place on 28 July and 6 August. Throughout, researchers monitored transmitting waveforms for stability. The results were positive. Double humps (not resulting from variations of transmitting waves) were clearly identifiable on some occasions.[20] The time lag between two trains of humps was about 1/1,700 second, corresponding to a 50-mile (80 km) layer height. There were a few triple humps, too (figure 8.3).[21]

Seven months after Appleton and Barnett's experiment, Breit and Tuve added another strong piece of evidence for the existence of the ionosphere: now the layer was reflecting radio pulses, and the heights from the two experiments were close to each other. Like Appleton and Barnett, Breit and Tuve demonstrated the usefulness of "sounding," actively manipulating the transmitted waveforms, in probing the upper atmosphere.

Breit and Tuve's discovery was not the conclusion of a search for evidence, but the beginning of an experimental program. Throughout September, they continued to improve the instrument and to make more measurements between DTM and NRL. In November, they also began experiments with the radio station at NBS at wavelengths of 50 and 75 meters. They found that the double humps did not always appear, and the time and intensity differences between the second (sky-wave) humps and the first (ground-wave) humps changed over time. Like the Britons, they thought that the layer's continually changing height and reflectivity caused these fluctuations. They could use the repetitive measurements to monitor variations in atmospheric conditions.

Like Appleton's frequency-change arrangement, therefore, the pulse-echo experiments shifted from seeking direct evidence for a hypothetical entity to monitoring the physical conditions of a real entity. These systematic measurements offered a chance to examine the atmosphere's effect on fading. Scientists found that the intensity of the second (sky-wave) humps, like fading, fluctuated considerably. Changing interference patterns between sky and ground waves as the result of path-difference variations could not be the cause, since

19. E. G. Oberlin to L. A. Bauer (letter), 30 June 1925, ibid.

20. Fleming to Grabau (letter), 10 Aug. 1925, and Bauer to O. B. Blackwell (letter), 14 Aug. 1925, ibid.

21. Breit and Tuve, "A radio method" (1925), 357.

was an improvement. However next day it
performed very poorly.

July 28, 1925

#18 NKF sending test at 71.3 meters

with A.C. on the plate. Observed double humps
visually, at times separated. used same element
as in #13. Film shows double hump very
plainly. Taken at 10:35. Test cont from 10:30
to about 10:50. At times a third hump seemed
visible

#19 NKF sending test at 71.3 meters

with A.C. on the plate. Three humps visible.
Their relative intensities varied during the
test very markedly, one of them appearing to
travel. Distance of "on" time traversed in perhaps
10 sec.. All three humps very visible. and
Film shows them. Fading marked. (3:10)

---

Simultaneous observations on NKF.

August 4, 1925

Tuve at Bellevue

Breit at Department

Steady deflection at Bellevue

Day cloudy

t

and occasionally

FIGURE 8.3. Tuve's research notebook, 28 July–7 August 1925. Box 11, folder 7, "Research Notebook: Heaviside Layer Oscillograms, 1925–30," in Ionospheric Section Records, 1927–59, courtesy of Carnegie Institution of Washington, Department of Terrestrial Magnetism Archives.

August 5. Source of trouble investigated at Bellevue.

Found that condenser in detector at Bellevue was across B battery and output of detector. This condenser being 1 μf and equivalent to about 30000 Ω gave

Eliminating condenser gave a small second hump

Putting more negative C battery on 2nd and 3d tube of transmitter ~~gave a~~ eliminated second hump. Last tube bias — 1100 v.
2nd tube bias — 110 v.

In previous tests Last tube — 800 v.
2nd tube — 80 v.

August 6

Tuve at Bellevue
Breit at Department

Tuve observed single humps with a slight broadening at the beginning (even though large amplitudes were used). Breit observed

FIGURE 8.3. Continued

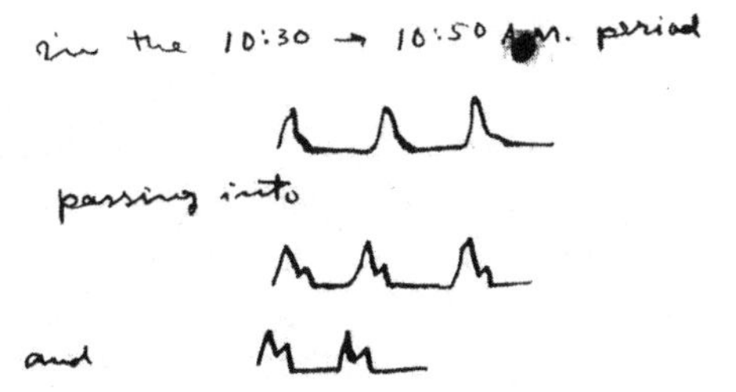
in the 10:30 → 10:50 A.M. period

passing into

and

and back 3 times.
The amplitude was kept low usually

In the 1:45 period the double hump was present all the time. At very low amplitudes it gave an appearance

This may be simply the "hiding" of the hump in the general thickness.

August 7, 1925

Visual observations on NKF at 1:45 P.M. to 2 P.M.
observers: Tuve and Breit.
λ = 71.3 meters
Received for about 7 minutes with single humps then for about 30 seconds a second hump appeared

→ t

disappeared and appeared no more.
Rain followed test

FIGURE 8.3. Continued

sky and ground waves, now in the form of pulses, completely separated from each other in time and were unable to interfere with each other. The cause had to be changes in the conducting layer's reflectivity, not its height. So Breit and Tuve argued that fluctuations in the upper layer's reflectivity rather than in its height more often caused the shifts.[22]

22. Breit and Tuve, "A test" (1926), 563–68.

## *The Meaning of Height*

The pulse-echo experiments marked the success of DTM's change from agency for magnetic surveys to laboratory for field experiments in atmospheric physics. Following the Britons, the team at the Carnegie Institution developed another useful radio method to probe the upper atmosphere. Breit and Tuve fully explored what they had achieved. DTM hired Tuve in August 1925 to work full-time on this project, and he turned it into his doctoral dissertation.[23] Breit took the opportunity to promote DTM's fame and his own. At the annual CIW exhibition on 11–14 December 1925, he demonstrated the double-hump experiment to visitors.[24] In front of an audience from all over the world, the magicians at CIW visualized the Kennelly-Heaviside layer in the splitting humps on oscillograms.

To any spectator, indeed, the visual feature of the pulse-echo setup created a compelling impression of the ionosphere's reality. Simple as it was, the Appleton-Barnett frequency-change method still required a trained expert to deduce the layer's existence from the number of continuous-wave peaks on an oscillogram. By contrast, Breit and Tuve's instrument presented a straightforward connection between experimental results and a scientific object: if, on an oscillogram, a radio echo pulse followed from a pulse traveling upward, then what could be the explanation except that there was really a reflective surface in the sky? Moreover, the way to determine the upper layer's height—a critical issue that made radio-sounding experiments stronger confirmation of the layer's existence than other propagation experiments—was more direct in the pulse-echo approach than in frequency change. While the former simply identified the time difference between the first and second humps with the path length of wave propagation (from which the height derived via elementary geometry), the latter had to deduce it from the number of continuous-wave peaks in terms of equation (7.1). The difference was especially obvious when there were two or more reflective ray paths: the former method could still easily handle different ray paths separately, since they corresponded to distinct nonoverlapping pulses, but the latter had difficulty counting the number of peaks, for the waveform on the oscillogram was now a superposition of multiple harmonics. Briefly, the pulse-echo instrument was a much more powerful method to measure the ionosphere's height.

Nevertheless, "height" was a tricky attribute of the ionosphere. Although

23. Cornell, "Tuve" (1986), 161.

24. Gillmor, "The big story" (1994), 139.

the operating principle of radio sounding *seemed to* treat the upper layer as a reflective surface, the reality was more complex. One simple fact: the "height" from radio sounding varied a lot. Breit and Tuve knew this fact well. From July to September 1925, their height measurements between DTM and NRL ranged from 55 miles (88 km) to 132 miles (211 km). The numbers sometimes changed with radio wavelength—the wavelengths of 71.3, 41.7, and 20 meters all gave different heights. For waves of 20 meters and shorter, reflection did not happen at all. Sometimes, the numbers also changed with the time of measurement, which was in the late morning (10:30 or 11:30 a.m.) or early afternoon (1:30 or 3:30 p.m.). Often, the "heights" just fluctuated without an apparent correlation.[25]

All this became more meaningful to Breit and Tuve as they delved into the ionic refraction theory of wave propagation. The theoretical work of Eccles, Hulburt, Larmor, Nichols, Schelleng, and Taylor had already indicated that the ionosphere was not the sharp reflective surface that Kennelly and Heaviside had assumed. Rather, it was an ionic medium whose thickness, depending on the strength of solar radiation, could increase and whose ion or electron density changed with height. Such a medium refracted the paths of radio waves and made the rays look like they were undergoing reflection. The time difference between a sky-wave pulse and a ground-wave pulse, Breit and Tuve claimed, did not directly measure the true path length, or equivalently, the height of the layer. It was a result of accumulative wave-propagating action in the ionosphere. (In optics, it was the "optical" path length.) The time of a sky-wave pulse's movement from ground to ground was thus an integral of the ratio of the differential true path length to the pulse's velocity, and the integral carried over the wave's entire path of refraction

$$(8.1)\quad \Delta t = \int_{path} \frac{ds}{p'} = \int_0^L \frac{n\,dx}{p'\sin\theta_0}.$$

In this integral, $n$ was the ionic medium's refractive index, $p' = c/[d(nf)/df]$ the pulse's "group" velocity, $ds$ the differential true path length, $dx$ the corresponding horizontal length, $L$ the distance between the path's initial point (the transmitter's location) and final point (the receiver's location) on the ground, and $\theta_0$ the path's initial angle vis-à-vis the vertical.

Breit and Tuve wanted to deduce from equation (8.1) the correct relationship between the measured time difference of pulses and the "true height"

25. Breit and Tuve, "A test" (1926), 567–68.

of the ionosphere. Key to this deduction was the expression for the refractive index $n$ of the ionosphere. As the ionic refraction theorists had shown, $n$ depended on the radio frequency, electron number density (cf., equation (6.1′) from Eccles and Larmor), and geomagnetic field (cf., equations (6.3) and (6.4) from Nichols, Schelleng, and Appleton). For simplicity, Breit and Tuve considered the cases that Nichols and Schelleng had examined: the wave-propagating direction was along or perpendicular to the geomagnetic field. The magneto-ionic theory yielded an immediate empirical implication for equation (8.1)—an incident sky-wave pulse geomagnetically split into two components with different velocities, corresponding to two separate pulses on the oscillogram. This explained why "triple humps" sometimes appeared: one hump associated with the ground wave, and two with magnetically dispersing sky waves.

But Breit and Tuve's more important result concerned the absence of the geomagnetic effect, which, according to Nichols and Schelleng, occurred when both the wave-propagating direction and the polarization were normal to the geomagnetic field. Under this condition, Breit and Tuve applied Larmor's refractive-index formula—i.e., equation (6.1′)—into equation (8.1) and obtained

$$(8.2)\quad \Delta t = \frac{L}{c \sin\theta_0}.$$

Equation (8.2) implied that the measured time lag for propagation along a refractive curve equaled the time that propagation would take in a vacuum along a wedged rectilinear path formed by the tangents of the curved path at the latter's initial and ending points (figure 8.4). In other words, the time delay of a wave refracting in the ionic layer equaled that of a wave reflecting from a sharp boundary. Ionic refraction and reflection were effectively identical in this case.[26]

Breit and Tuve's theoretical examination revealed that the pulse-echo experiments measured not the ionosphere's "true" height, but its "effective" or "virtual" height, at a certain radio frequency. Yet the ionic-refraction theory indicated that the optical path length of the wave refracting through the ionosphere equaled the true path length of the same wave reflecting at a sharp boundary whose height the sky wave's direction determined. This "equivalence theorem" therefore guaranteed that the pulse-echo method,

26. Ibid., 560–75.

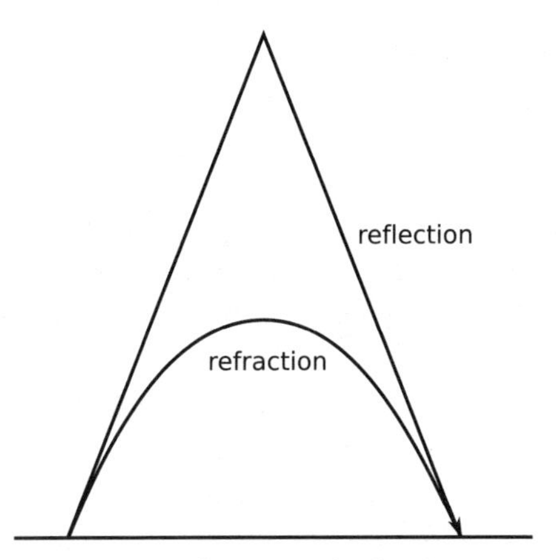


FIGURE 8.4. Equivalence of reflection and refraction.

the frequency-change method, and the polarization method should produce identical "heights." The effective height did not correspond to the actual locale at which a radio wave turned around, but the two should be reasonably close to each other, based on the estimates of the ionosphere's maximum electron number density. In this sense, the measured effective height should not deviate too much from the ionosphere's actual height.

This was good. Nevertheless, the large variation of the measured heights remained a strange fact, if not a puzzle, regardless of the relationship between the effective and the true heights. Breit and Tuve measured a wide range of heights, from 55 miles (88 km) to 132 miles (211 km)—the minimum was close to Appleton and Barnett's initial measurement, while the maximum was more than twice the minimum. It might not be unreasonable that the ionosphere's nature caused such considerable variation, but what dynamics and structure made it like that? What, indeed, was the ionosphere? From the experimental perspective, how could the effective height measurements help tackle these questions? What could DTM's results tell about the ionosphere, other than its height?

Breit and Tuve did not have answers. Nor did they seem particularly eager to pursue atmospheric science with radio sounding, as they were now shifting their attention to (much more visible) quantum physics. Their lack of interest was partly responsible for CIW's stagnancy in ionospheric research after the initial success. DTM's pulse-echo experiments after 1926 turned into routine practice. The focus became accumulation of data and perfection of experimental devices. Radio sounding became a program of making regular measurements and publishing experimental data, activities not dissimilar to magnetic surveys.

After 1926, engineers gradually displaced scientists in radio work at DTM; improving instruments was the main task. A Norwegian technician, Odd Dahl, joined CIW in 1927 to perform the regular radio sounding between DTM and NRL.[27] Dahl conducted experiments with NRL's Gebhardt on the unusual disturbances of the ionosphere's effective height and their possible connection with sunspots[28] and improved the Tuve radio sounder's modulating scheme from sinusoidal oscillator to multivibrator for the generation of narrow pulses.[29] The multivibrator, one of the earliest digital circuits, emerged in the late 1910s in England and France. It was an oscillator consisting of two symmetric amplifying tubes with their output (plates) connected to each other's input (grids). In contrast to sinusoidal oscillators, a multivibrator generated rectangular pulses; experimenters could adjust both the period of oscillation and the duration of pulses by varying the resistors and capacitors of the tubes' plate and grid circuits. A transmitter could produce flat rectangular pulses with narrow pulse widths using a multivibrator and hence could enhance the waveform's resolution. Tuve and Dahl applied the multivibrating radio sounder in the measurements in December 1927 and January 1928.[30]

The drive to perfect instruments did not reorient the entire program, however. DTM's ionospheric research waned after 1928. Breit turned to theoretical nuclear physics; he left CIW for New York University in 1929 and never returned to planetary science. Tuve stayed at DTM, but he also switched to particle physics. In 1929, he redirected his DTM team from radio sounding to building a high-energy electron accelerator, sharing the interest of his friend Lawrence. DTM's leading role in ionospheric studies diminished until a revival in the 1930s, when its employees and some technicians at NBS turned the pulse-echo setup into the automatic sweeping-frequency ionospheric sounder.[31]

## THE IONOSPHERE HAS A STRUCTURE

What Breit and Tuve left behind Appleton retained. While Smith-Rose and Barfield were busy improving direction finders and the DTM researchers

27. Cornell, "Tuve" (1986), 166–67.

28. Dahl and Gebhardt, "Measurements" (1928), 290–96.

29. Tuve and Dahl, "A transmitter modulating device" (1928), 794–98.

30. Breit, Tuve, and Dahl, "Effective heights" (1928), 1236–39.

31. Hull, "Breit" (1998), 31–32; Cornell, "Tuve" (1986), 176–97; Gillmor, "The big story" (1986), 140. Also see Yeang, "From mechanical objectivity to instrumentalizing theory" (2012).

were still developing the pulse-echo setup, Appleton was beginning to build a new research program growing from his and Barnett's discovery in early 1924. As soon as the frequency-altering and polarization-measuring methods generated the evidence for sky waves, they became tools for further probing the still mysterious ionosphere and its effects on radio waves. Appleton understood the scientific potential of using radio to study the ionosphere—sending radio waves to the upper sky to observe the patterns of their returns—and devoted himself to these efforts, setting aside his research on vacuum tubes and atmospherics.[32]

### *Appleton's Program of Ionospheric Sounding*

Two immediate topics concerned Appleton: the diurnal variations of the ionosphere's height and the mechanism of fading. With his recent experimental success, he no longer had to rely exclusively on instruments from Oxford. He became a core member of the new Radio Research Board (RRB) under Britain's Department of Scientific and Industrial Research (DSIR). He persuaded the board to turn a deserted wartime wireless direction-finding base in Dogsthorpe, Peterborough, into a radio research station with receiving apparatuses—a tuned antenna and amplifiers with flat frequency responses. For transmitting purposes, he arranged for the board to build him a special transmitter at the National Physical Laboratory (NPL) in the London suburb of Teddington (124 km from Peterborough) and the British Broadcasting Company (BBC) to let him use again its Bournemouth station (236 km from Peterborough). The transmitters generated wavelengths of 360–400 meters and could generate continuously changing frequencies to produce artificial fading. Their waves' shorter durations of continuous-frequency variations ensured that the change of receiving signals with time did not come from natural fading that had periods of several seconds. The receiver in Dogsthorpe recorded in oscillographs and then photographed the rapidly varying waveforms.[33]

Of course, Appleton did not work alone. Assistants helped him run his field-experimental network, which consisted of Bournemouth (BBC), Cav-

32. Ratcliffe mentioned that Appleton's list of projected research papers before 1925 fell into three categories: atmospherics, nonlinear tube electronics, and radio-wave propagation. After 1925, his list was exclusively about radio-wave propagation and ionospheric research. Ratcliffe, "Appleton" (1966), 10.

33. Appleton and Barnett, "On wireless interference" (1926–27), 450–53.

endish (Cambridge), Dogsthorpe (DSIR), Teddington (NPL), and his own lab at King's College, London, where he was still teaching. Helpers included the BBC's P. P. Eckersley, NPL's E. L. Hatcher and A. C. Haxton, and W. C. Brown in charge of the Dogsthorpe station. But the major experimenters were Appleton himself, Miles Barnett, and later another Cambridge student, John Ratcliffe. After forty years, Ratcliffe still vividly recalled how they performed experiments:

> He [Appleton], Barnett, and I used to travel over to Peterborough three times a week to conduct the experiments after midnight when there was no fear of our interfering with the broadcasting service. We were in continuous telephonic contact with Hatcher and Haxton at the N.P.L., when we were ready for an experiment we would ask for a change of wavelength. They would then swing the transmitter condenser back and forth while we photographed the "fringes" on a moving film. The experiment was repeated at intervals throughout the night. We could not always experiment for as long as we would have wished. Often we could not start until about 2 a.m. because a Spanish station was broadcasting dance music, and we frequently had to stop at 4 a.m. because the Germans started with physical jerks.

Such night work could be "very tiring." Yet Appleton cheered up these lonely lads by frequent visits, telling them stories, sharing his latest theories, and drinking strong tea with them.[34]

In this kind of milieu, Appleton and Barnett performed the height measurements in May and June 1926. They performed tests each night throughout the second half of the dark hours—the first half had too many broadcasting signals. Each test lasted fifteen seconds and consisted of three parts: a uniform change (increase or decrease) of wavelength for five seconds, a steady (highest or lowest) wavelength for five seconds, and a uniform change (decrease or increase) of wavelength for another five seconds. Altogether the men obtained about 400 records.[35]

The experimental results in 1926 confirmed Appleton and Barnett's earlier physical model. Figure 8.5 illustrates the data from the same night for signals from Bournemouth (farther from Peterborough) and Teddington (closer to

34. Ratcliffe, "Some memories" (1965), box "Radio Research Board Committees, 1920–40s," Historical Records, Rutherford Appleton Laboratory (Space Science Department), Didcot, England.

35. Appleton and Barnett, "On wireless interference" (1926–27), 450–58.

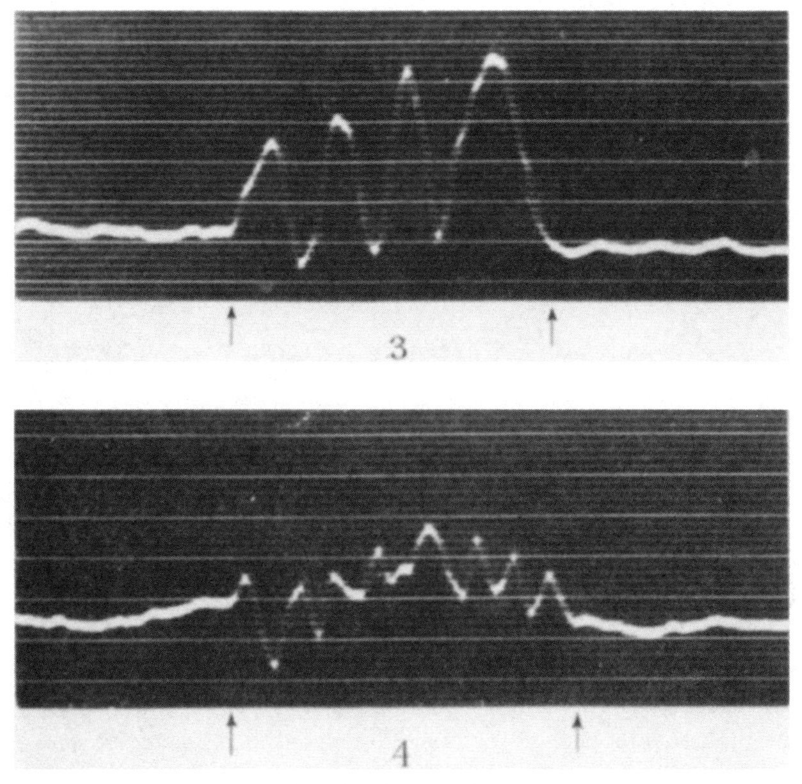


FIGURE 8.5. Frequency-changing data. Appleton and Barnett, "On wireless interference" (1926–7), plates 3-4. The horizontal axis denotes time, and the vertical axis signal strength.

Peterborough). With fixed height of the ionosphere, the path difference between sky and ground waves at a shorter distance was longer (cf. figure 7.4). That the fringes in the Bournemouth plate were fewer than those in the Teddington plate was thus consistent with the two men's model predicting more fringes for a longer path difference. Also figure 8.5 shows the signals from Bournemouth for a wavelength change of 5 meters and 10 meters. The number of fringes (four in plate 1 and eight in plate 2) was proportional to the wavelength change (5 meters in plate 1 and 10 meters in plate 2), consistent with the prediction of equation (7.1) that $N_m$ was proportional to $\delta\lambda$.

Appleton and Barnett also used the new experiments to study the variations of the ionosphere's height. From equation (7.1), the differential change of the fringe number over the differential change of wavelength $\delta N_m/\delta\lambda$ was proportional to the path difference $D$ between the sky wave and the ground

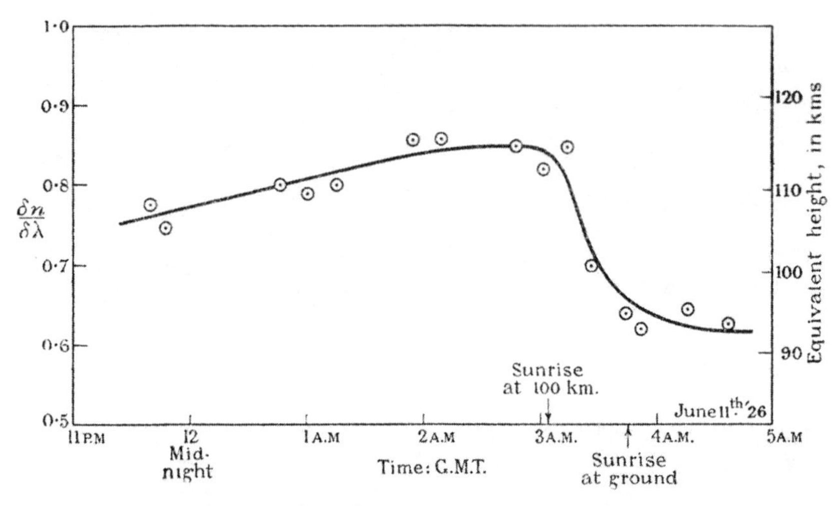


FIGURE 8.6. Measured $\delta \mathcal{N}m/\delta\lambda$ vs. time. Appleton and Barnett, "On wireless interference" (1926–7), 456.

wave. Since the path difference $D = (4h^2+d^2)^{1/2}-d$ increased with the ionosphere's height $h$ for a given transmitter-receiver distance $d$, the measure of the fringe number's sensitivity with wavelength change ($\delta \mathcal{N}_m/\delta\lambda$) at different times indicated the diurnal variation of the ionosphere's height. Figure 8.6 illustrates a sample of measured $\delta \mathcal{N}_m/\delta\lambda$ at different times of a night (signals from Teddington, 11 June 1926). The results showed that the ionosphere's height steadily increased from approximately 90 kilometers after midnight but fell considerably and quickly about sunrise, when it reached about 120 kilometers. This was consistent with the solar model of the ionosphere: the degree of ionization steadily diminished after midnight because of lack of sunlight but rose again as sunrise approached.[36]

In the meantime, throughout 1926 Appleton continued his and Ratcliffe's earlier measurements of polarization to investigate the interferences between ground and sky waves that caused natural fading. The team moved from Cambridge to the better-equipped Dogsthorpe station to receive signals from the BBC's London station and NPL's Teddington station. It improved the experimental method by *combining* the rod-loop set with continuous frequency change: waves from the vertical rod and the vertical loop now modulated with a continuous frequency change to create regular and periodic maxima and minima at the receiving signals. The artificial maxima and minima of the

36. Ibid., 456–58.

rod and loop signals allowed estimates of the sky wave's incident angle. The estimated heights of the ionosphere from the frequency-change method, the polarization measurements of natural fading, and the new method combining both were all between 90 and 120 kilometers. This consistency again confirmed the reality of the ionosphere.[37]

With the vertical rod, vertical loop, continuous frequency change, and adjustable antenna coupling, Appleton and Ratcliffe looked further into the interference that caused fading. There had been a consensus by 1926 that fading was a superposition of invariant ground waves and variant sky waves. But which parts of the sky-wave variations were more important: amplitude, angle of incidence, phase (the height of the ionized layer), or polarization? Appleton and Ratcliffe developed a series of arguments to infer the most likely cause(s) by identifying the covariance of intensity waveforms at different antennae. They concluded that sky waves' variations in intensity were the most frequent source of fading, a conclusion similar to Breit and Tuve's from the pulse-echo experiments. Yet phase variations also played a part.[38]

## *Discovery of the F Layer*

Appleton and his colleagues in the mid-1920s improved and developed experimental methods—building on the British approaches of frequency change and polarization finding—to better measure the height of the ionosphere. Like the Carnegie Institution of Washington, Britain's Radio Research Board focused on that dimension. Appleton et al.'s Teddington-Bournemouth-Dogsthorpe experiment in 1926 demonstrated its variation throughout the night, 90–130 kilometers (see figure 8.6), considerably narrower than Breit and Tuve's 88–211 kilometers. Why this difference? The Britons used much longer radio wavelengths (360–400 meters) than the Americans (20, 41, and 71 meters) and measured at night, unlike their counterparts. (This was a vicious circle for the Britons: they had collaborated with a broadcasting station that transmitted at lower frequencies and, working at broadcasting frequencies, had to operate late at night to avoid interference.) But why did these different experimental conditions lead to the different results of the British and American experimenters? Didn't their quite different results reveal something about the ionosphere?

A clue emerged from experimental results in mid-1926. In their filmed in-

37. Appleton and Ratcliffe, "On the nature I" (1927), 291–304.
38. Appleton and Ratcliffe, "On the nature II" (1927), 305–17.

terference patterns, Appleton and Barnett noted that sometimes the fringes resembled not simple harmonics, as the theory had predicted, but "secondary ripples" on top of the primary harmonics, especially during the few hours before sunset.[39] The deviation from a simple harmonic meant that the incoming signals had more than two—one ground and one sky—wave components. Quite likely, therefore, the secondary ripples corresponded to sky waves with two distinct ray paths.

This was hardly a surprise. Radio scientists had been well aware that waves reflecting from the ionosphere might have multiple paths. The split of "ordinary" and "extraordinary" rays in a geomagnetized ionosphere—which Breit and Tuve used to account for their "triple humps"—was an example. Even though ionic-magneto dispersion could not create sufficient phase differences in this case, another common mechanism—multiple reflections—could. A radio wave propagating from transmitter to receiver could bounce once from the ionosphere, but it could also bounce more times between sky and ground. For instance, the ray path might reach the sky first, reflect to the ground, reflect again to the sky, and finally arrive at the receiver on the ground. This "doubly reflective" sky wave would interfere with the singly reflective sky wave and the ground wave to create secondary ripples in the frequency-change experiment. (Higher-order multiply reflective waves also possibly existed. But the high loss owing to their longer paths much reduced their effect at the receiver.)

Were the secondary ripples results of double reflection from the ionosphere? There was a simple method to find out. From elementary geometry, the path length of a ray doubly reflecting from height $h$ equaled that of a ray singly reflecting from height $2h$. Therefore, when the transmitter-receiver separation was $d$, the path length of the ground wave was $d$, that of the singly reflective sky wave equaled $(4h^2+d^2)^{1/2}$, and that of the doubly reflective sky wave equaled $(16h^2+d^2)^{1/2}$. This phase relationship entailed that the number of subsidiary fringes (because the doubly reflective wave interfered with the ground wave) per primary fringe (resulting from the singly reflective wave's interfering with the ground wave) was a fixed ratio $[(16h^2+d^2)^{1/2}-d]/[(4h^2+d^2)^{1/2}-d]$.

At first, Appleton and Barnett did not doubt that the subsidiaries that they observed in mid-1926 could have other causes than multiple reflection. They only hinted at the phenomenon in the *Royal Society Proceedings* in December 1926. In early 1927, however, Appleton changed his mind and went back to the oscillograph records from 1926 to check their subsidiary-to-primary-fringe ra-

39. Appleton and Barnett, "On wireless interference" (1926–27), 457.

tio (this time with the help of Ratcliffe, for Barnett was returning to New Zealand). To his surprise, some records had the ratio considerably deviating from the theoretical prediction $[(16h^2+d^2)^{1/2}-d]/[(4h^2+d^2)^{1/2}-d]$. Moreover, the ripples looked sufficiently conspicuous and sometimes even overrode the primary fringes, even though doubly reflecting waves always suffered a heavier loss owing to their longer paths. In other words, the ripples in these records were probably not consequences of double reflections from the sky. They were, Appleton believed, rays singly reflecting from *another layer* higher than the ordinary ionosphere (90–130 km) that he and Barnett had identified.[40]

This bold conclusion suggested that the ionosphere actually consisted of more than one layer. The subsidiary ripples were not secondary by-products of the frequency-change radio sounding, but rather a means to detect this extra layer. Moreover, the method that he and Barnett used to measure the ordinary layer should work here: one could count the subsidiary ripples within a known range of frequency change and use equation (7.1) to determine the path difference between the extra sky wave and the ground wave.[41] (Figure 8.7 presents an example of the sounding data.)

With the same approach of active sensing and the same "direct" height measurement, Appleton identified a new scientific object and provided the same degree of certainty about its reality as the ordinary ionosphere. The radio sounding experiments showed that the ionosphere had a structure. It contained two layers. One layer, which first appeared in Appleton and Barnett's radio sounding in 1924, was 90–130 kilometers high. Most observers thought this layer to be the Kennelly-Heaviside layer proposed in 1902 to explain Marconi's transatlantic propagation experiment. Appleton also called it the "E region" or "E layer" of the upper atmosphere, where "E" stood for electricity. The other layer, which the 1926 sounding experiment uncovered, was higher. Appleton coined the name "F region" or "F layer" for this novel entity.[42]

Throughout 1927, Appleton spent time experimenting with the F layer. He and his assistants inspected the filmed sounding records flowing continually from the Teddington-Bournemouth-Dogsthorpe setup, marked the points where subsidiary ripples appeared strong, and determined the equivalent height of the F layer by counting the subsidiary ripples. According to

40. Appleton, "On some measurements" (1930), 549–55.

41. For this method to work, one had to assume that the secondary ripples peaked when the extra sky wave and the ground wave had constructive interference, regardless of the primary sky wave. This was approximately true.

42. Appleton, "The ionosphere" (1964), 81.

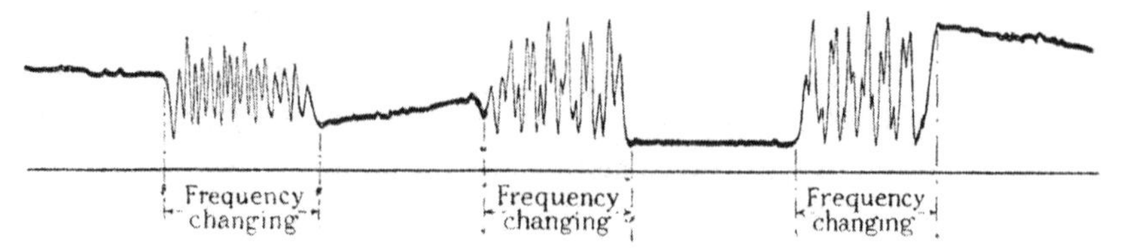


FIGURE 8.7. Sounding record with reflections from both layers. The low-frequency bumps correspond to the E layer, and the high-frequency bumps to the F layer. Appleton, "On some measurements" (1930), Figure 10.

their measurements, the F layer ranged from 250 kilometers to 350 kilometers high. The sounding records corresponding to this layer began to appear three to four hours before sunrise. Like the E layer, its measured height increased steadily in the second half of the night. But about 30–40 minutes before dawn, its height fell sharply, the subsidiary ripples quickly disappeared, and only the returns from the E layer remained on the oscillographs. In September 1927, Appleton reported his finding of the F layer in *Nature*.[43]

The discovery of the F layer solved the incongruence between Appleton and Barnett's height measurements in 1924 and Breit and Tuve's in 1925. The Britons obtained 90–130 kilometers, while the Americans had 88–211 kilometers. The new finding offered an explanation of the Americans' wider range of heights: they were actually measuring the E layer at the low side and the F layer at the high side. It was not that the ionosphere's height varied as much as Breit and Tuve had claimed. Rather, the DTM set received echoes sometimes from the E layer and sometimes from the F layer.

But this explication raised a further question. In what circumstances did radio sounding detect the E layer, and in what circumstances the F layer? In Breit and Tuve's experiment in 1925 (wavelengths 20, 41, and 71 meters, daytime measurements), more than half of the measured heights exceeded 190 kilometers (i.e., the F layer), while the level of less than 120 kilometers (i.e., the E layer) appeared only occasionally. In Appleton's experiment in 1926 (wavelength 400 meters, nighttime measurements), the F layer effect appeared only in the second half of the night. Such a comparison offered a clue: the detectability of the E and F layers had to do with radio frequency. In fact, one could deduce the exact relationship from the ionic-refraction theory: longer waves deflected easily in a medium with a low ionic concentration, while shorter waves required a greater level (cf. Larmor's formula in equation (6.1′)). Thus

43. Appleton, "The existence" (1927), 330.

Appleton's 400 meter waves reflected most of the time from the lower E layer, whose electron density was smaller than the F layer's but enough to refract back the long waves. Only when the E layer's electron density fell considerably after midnight were the long waves able to penetrate the E layer and reach the higher F layer. For the same reason, it was not a surprise that Breit and Tuve's waves penetrated the E layer during the daytime, for the wavelengths were much shorter.

The ionic-refraction argument was not a mere explanatory device. From it, Appleton learned an important lesson—that the frequency of a radio sounder determined how high its waves could reach: the higher the frequency, the greater the height. Hence Appleton suggested "the possibility that simultaneous observations between the same two stations on widely different wavelengths might enable us to *study the characteristics of both layers at the same time*."[44] In such a setup, the longer wavelengths (lower frequencies) could sound the E layer, while the shorter wavelengths (higher frequencies) could sound the F layer. A few years later, several Americans would turn this concept into the first sweeping-frequency pulse-echo radio sounder.

Appleton did not explore further the techniques of multifrequency radio sounders. But he managed to measure the F layer's height more systematically at different frequencies and locations. From 1928 to 1930, he, Ratcliffe, R. D. Gander, Alfred Leonard Green (a Cambridge student from Australia who became Appleton's substitute for Barnett), and W. F. B. Shaw conducted frequency-change radio-sounding experiments at three receiving stations in Cambridge, Dogsthorpe, and King's College, London, using the transmitter at NPL. The operating wavelengths included not only the original 400 meters but also 100 meters. With the shorter wavelength, Appleton et al. measured heights in the daytime. Their large amount of data corroborated Appleton's initial finding of 1926: the F layer's height ranged from 250 kilometers to 350 kilometers during the night, remained fairly constant from different locations in England, and dropped sharply about sunrise. The new data at 100 meters complemented the daytime observations: after dawn, the F layer maintained a height between 200 km and 250 km. This diurnal change of the layer's height was consistent with the assumption that solar radiation generated the most ions about noon and pushed the layer downward, and the fewest about midnight, when it pulled the layer upward.[45]

44. Ibid., 330 (my emphasis).

45. Appleton and Ratcliffe, "Some simultaneous observations" (1930), 133–58; Appleton and Green, "On some short-wave" (1930), 159–78.

The discovery of the F layer from the British researchers' frequency-change sounding revealed radio's utility as a probe of the upper atmosphere. Unlike the experiments of DSIR, DTM, and NPL in 1924–25, here the radio sounding was no longer an instrument just to prove the reality of the ionosphere. Rather, it was producing *new knowledge*—not only did the ionosphere exist, but it possessed at least two separate regions—and hence radio sounding might help uncover the structure of this still mysterious new scientific object. And the latter function returned to serve the former. The very fact that the ionosphere had a structure discernible to a sensing instrument increased the belief that this entity was real—a real object must have some features; the more we knew about it, the more we were sure about its reality.

Of course, people still did not know much about the ionosphere's properties. Appleton's radio sounding as a sensing probe was inadequate and crude around 1927: it could obtain only height information about the ionosphere. How did these ionized layers form, what were their electromagnetic properties, what was their chemical constitution, what were their thermo- and fluid dynamics, and how did they interact with the sun and the geomagnetic field? Radio-sounding data did not yet help answer these questions. Yet the sheer fact that one could "see" two distinct regions of the ionosphere on oscillographs pleased radio scientists who had attempted to bind wave-propagation behavior with the characteristics of the upper layers. It also excited geoscientists who had endeavored to determine the stratified structure of the atmosphere. In the following decades, radio sounding would improve considerably and retrieve a variety of the ionosphere's properties.

The "discovery" of the ionosphere from the three radio methods in 1924–27 constituted the last piece of a major transformation in the 1920s in radio science, geophysics, and wireless technology. In the 1900s and 1910s, scientists and engineers studying long-range radio-wave propagation—a subject linking all three areas—concentrated on the mathematical consistency of wave solutions and fitting theoretical predictions with the empirical Austin-Cohen formula. The Kennelly-Heaviside layer was a hypothesis for explaining observed propagation phenomena at wavelengths longer than 300 meters. The physical structure of this layer was irrelevant as long as it could reflect electromagnetic waves like a mirror. And the propagation "experiments" were no more than by-products of Marconi's or the U.S. Navy's tests of their transoceanic communications systems.

Things had changed since the early 1920s. At first, amateurs' discovery of long-range short-wave radio challenged the Austin-Cohen formula. Immediately afterwards, government and corporate researchers performed systematic empirical investigations on wave propagation at increasingly higher frequencies. Contrary to the old theory, they found that shorter waves reached a long distance along the earth's curvature, had a range minimum at the wavelength of 200 meters, and skipped over intermediate regions. A magneto-ionic refraction theory emerged to explain these experimental data. Instead of treating the upper atmospheric layer as a reflective board, the new theory stressed the layer's structure as causing various short-wave phenomena, including the skip zone and minimum range. In so doing, however, researchers were implicitly turning the upper layer from a hypothetical entity explaining observed phenomena of radio-wave propagation into an object whose properties they could infer from the results of propagation experiments such as skip-distance and range measurements. Theories were changing from explanatory devices or formal representations of quantities into conceptual tools for finding the natural environment's unknown features.

This implicit transformation became explicit with the development of some actual material tools to testify to the reality of the upper layer, the ionosphere. The results from experiments by Britain's Radio Research Board and the Carnegie Institution of Washington in the mid-1920s seemed to yield direct evidence for the ionosphere's existence. Underlying the three radio methods in these experiments—polarimetric, frequency-change, and pulse-echo—and the ontological claim they made was the concept of active probing. These methods, especially Appleton's frequency change and Breit and Tuve's pulse echo, consisted of sending specifically prepared radio waves to the sky, actively perturbing the upper atmosphere with these waves, and observing the resulting changes in the returning signals. Using such a radio-sounding approach, Appleton et al. not only directly saw the ionosphere on visual records, but also found an unexpected structural feature—the F layer.

With completion of this transformation, geophysics, radio science, and wireless technology now came together much more closely. And the corresponding ramifications and impacts were at experimental, practical, and theoretical levels. Practically, the further development of short-wave wireless communications systems relied on measurements, estimates, or predictions of the ionosphere's state and its influence on wave propagation. Waves shorter than 100 meters skipped over; and the extent of such a skip-zone effect depended on the state of the ionosphere (specifically, electron density), which varied with the strength of solar radiation. Consequently, only a range of frequencies was usable for a given geographical path of wireless communications, and such a frequency range changed diurnally, seasonally, and climatically. This phenomenon made the operations of wireless communications systems difficult, since the usable radio frequencies were shifting all the time. But propagation and ionospheric studies offered means to predict the usable frequencies for short-wave communications between two given locations at a given time. The prediction came primarily from measurements of the ionosphere's electron-density profile using the radio-sounding method that Appleton, Breit, and Tuve had begun to explore.

In addition, Appleton et al.'s early empirical investigations of the ionosphere in the mid-1920s led to the fully fledged development of the radio sounder that revolutionized experimentation vis-à-vis the upper atmosphere. Using the same idea of active sensing and probing, American and British scientists and engineers in the 1930s combined the frequency-change and pulse-echo features to make genuine multifrequency radio sounders that could easily measure the ionosphere's electron-density profile. This new experimental technology encouraged the boom of ionospheric stations and observatories in Australia, Britain, Canada, Germany, the Soviet Union, and the United States. Between airplanes/balloons of the 1910s and rockets/satellites of the 1950s, ground-based radio sounders were the single dominant experimental device for atmospheric research at altitudes of 20–400 kilometers.

Experiment needed theory, however. Signals arriving at a radio sounder required the magneto-ionic theory of wave propagation for interpretation and use. In response to the progress of radio-sounding technology, therefore, the theory as a conceptual tool to retrieve the unknown needed improvement. Starting with Eccles and Larmor and with supplements from Appleton, Nichols, and Schelleng, this theory gained common acceptance by the mid-1920s as the genuine description of radio-wave propagation in the ionosphere. But it required polishing and extension. To contemporary radio scientists, the theory ought to assign meanings to every feature of received signals, be it amplitude, frequency, phase, or polarization. It must withstand the trials of microphysics from which the very idea of ionic refraction derived. And it should be able to undergo testing not only in fields but somehow in laboratories. These were the tasks for the theorists in the following decade.

The rest of the book explores one of the three aspects of the new horizon of radio ionospheric studies that this transformation opened up. Specifically, it examines the elaboration of the magneto-ionic theory and attempts to bestow microphysical meanings on it in the 1920s and 1930s. I will leave the technological and experimental aspects to a subsequent book.

* 3 *

# *Theory Matters: 1926–35*

CHAPTER 9

# Consolidating a General Magneto-Ionic Theory

Opening *The Magneto-Ionic Theory and Its Applications to the Ionosphere*, we find a different treatise from previous classics on radio-wave propagation. Unlike P. O. Pedersen's *Propagation of Radio Waves along the Surface of the Earth and in the Atmosphere*, this book by John Ratcliffe did not attempt to tackle all the issues in radio ionospheric research from the surface diffraction formulae to the molecular constitution of the ionosphere. Unlike H. Bremmer in his *Terrestrial Radio Waves*, the author did not build on certain mathematical techniques for solving differential equations and their application to increasingly complicated boundary-value problems that simulated the earth and the atmosphere. Instead, the English scientist's goal was narrow—to deduce the magneto-ionic theory and to illustrate how to use it. Part 1 derived the theory's central equation, the so-called Appleton-Hartree formula. Part 2 taught the ways to interpret this equation. Part 3 showed how to apply the results in part 2 to study radio-wave propagation in some "model ionospheres." Part 4 addressed "miscellaneous considerations." In brief, *The Magneto-Ionic Theory* was a radio scientist's instruction manual for the Appleton-Hartree formula.[1]

Ratcliffe did not hide his instrumentalist attitude towards the magneto-ionic theory. On the first page, he claimed that this book dealt with the theory of electromagnetic waves passing through "a gas of neutral molecules

1. Ratcliffe, *The Magneto-Ionic Theory* (1959). Pedersen, *The Propagation of Radio Waves* (1927). Bremmer, *Terrestrial Radio Waves* (1949).

in which is embedded a statistically homogeneous mixture of free electrons and neutralizing heavy positive ions, in the presence of an imposed uniform magnetic field." This model ionosphere was contingent on several conditions that, as the author acknowledged, were not strictly real: the neutral gas, the homogeneous mixture, and the uniform geomagnetic field. Even so, he assured readers about the usefulness of such a medium in understanding wave propagation in the real ionosphere.[2]

But perhaps a more interesting question concerns *how* he used the magneto-ionic theory in his monograph. He did not apply the Appleton-Hartree formula to construct a mathematical or numerical relationship between the returning radio wave and the ionosphere's electron-density profile—an ideal function that a modern geoscientist would expect from the wave propagation theory. Instead, diagrams and graphs played a crucial part. From the formula, Ratcliffe plotted a propagating wave's refractive index (which determined its velocity) and attenuation (which determined its amplitude) as functions of electron density under various circumstances: ordinary or extraordinary polarization, wave-propagating direction as longitudinal, transverse, or any of their combinations vis-à-vis the geomagnetic field's direction. These diagrams offered guides to inferring the qualitative or semi-quantitative properties of the ionosphere's electron-density profile from returning radio waves that originally came from the frequency-change, pulse, or polarimetric instruments: how the wave's delay time varied with electron density, at what radio frequency the electron-density variation changed direction, how polarization depended on the geomagnetic field, and so on (figure 9.1). In other words, Ratcliffe's presentation of the magneto-ionic theory was tailored to the British and American researchers' sounding-echo experiments in the 1920s as visual tools for their data interpretation. Wave propagation theory served as machinery to assist systematic production of empirical information about the ionosphere via a particular experimental method.

Although *The Magneto-Ionic Theory* appeared in 1959, it synopsized the major developments of the radio-wave propagation theory in the second half of the 1920s, when a general theory came together. While Appleton, Nichols, and Schelleng in 1924–25 had formulated magneto-ionic refraction for special cases (in which the geomagnetic field was parallel or perpendicular to the wave-propagating direction), five individuals in 1926–27 (Appleton, Austrian maverick Wilhelm Altar, Cambridge student Sydney Goldstein, English mathematician Douglas Hartree, and German physicist Hans Lassen) extended the

2. Ratcliffe, *The Magneto-Ionic Theory* (1959), 1.

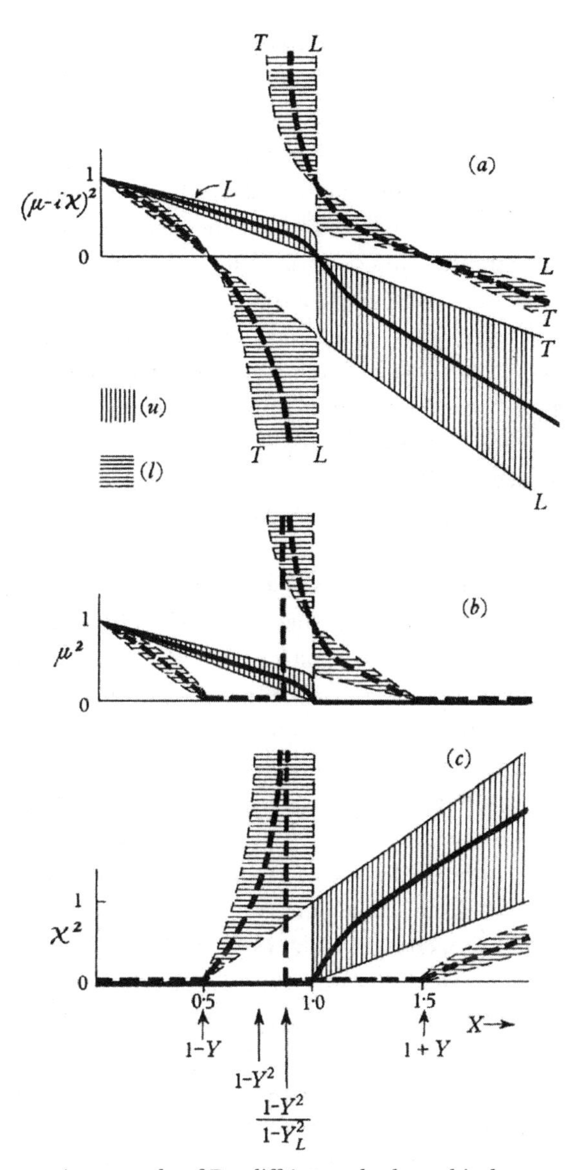


FIGURE 9.1. An example of Ratcliffe's standard graphical representation. The top diagram plots the square of the (complex) refractive index stipulated by the Appleton-Hartree formula, the second the refractive index, and the third the attenuation. $L$: longitudinal direction of wave propagation (parallel to the geomagnetic field); $T$: transverse (normal to the geomagnetic field). All curves that Appleton-Hartree predicted must lie within the shaded areas. Ratcliffe, *The Magneto-Ionic Theory* (1959), figure 6.8, reprinted with the permission of Cambridge University Press.

theory to general cases involving *all* wave-propagating directions, leading to the Appleton-Hartree formula.

Appleton's group in London and Ratcliffe's in Cambridge quickly identified a novel empirical implication of the general theory: radio waves' senses of polarization were mostly left handed in the northern hemisphere and mostly right handed in the southern. Field trials by the British ionospheric researchers and their Australian allies confirmed this prediction and crowned the Appleton-Hartree formula. However, Appleton and his British colleagues did not generalize the theory for logical completeness. They endeavored to turn the general theory into a paper tool to assist echo-sounding exploration of the upper atmosphere. For instance, Appleton commissioned a Cambridge mathematician, Mary Taylor, to devise, as we can see in Ratcliffe's 1959 monograph, a variety of graphical representations of the results calculated from the formula. These elaborations of the magneto-ionic theory are the subject of this chapter.

Historians and philosophers have urged us to go beyond the conventional view, whereby the role of scientific theory is to identify natural laws, and focus on its potentially diverse epistemic functions in scientific practice. Nancy Cartwright has pinpointed the essential discrepancy between the idealized principles and the actual theoretical constructions that scientists deployed to make sense of data. Her perspective has opened up philosophical investigations on the intermediate models between general laws and empirical data.[3] Recently, historians such as David Kaiser and Andrew Warwick have advocated viewing scientific theories not as sets of principles, laws, and explanations, but as *tool kits* for research. Theories could be calculating techniques, paper devices, and conceptual apparatuses that did not necessarily connote ontological commitments.[4] Similarly, Peter Galison has highlighted "instruments' theories"—the scientific theories crucial to the design and operation of experimental instruments.[5]

The history of the general magneto-ionic theory in the 1920s showed a clear transformation of the epistemic status of the theory of radio-wave propagation that resonates with the scholarly works above. In the 1900s and 1910s, the surface diffraction theory was the rigorous mathematical representation

3. Cartwright, *How the Laws of Physics Lie* (1983), For philosophical discussions on models, see Charkravartty, *A Metaphysics for Scientific Realism* (2007), chaps. 7–8; Chang, *Inventing Temperature*, (2004), chap. 4.

4. Kaiser, *Drawing Theories Apart* (2005); Warwick, *Masters of Theory* (2003).

5. Galison, *Image and Logic* (1997), 641–68.

of a highly simplified physical scenario, whereas the atmospheric reflection theory provided qualitative explanations for various wireless phenomena. In both cases, the upper layer either did not exist or served as a means to understand wave propagation. The ionic-refraction theory in the early 1920s began to generate information about the upper atmosphere, as it linked peculiar short-wave effects to the structure of the hypothetical ionic medium. But it was the introduction of the sounding-echo methods and the "direct" experimental confirmation of the ionosphere that altered the course of the theory of radio-wave propagation. Appleton, Barnett, Breit, and Tuve's trials in the mid-1920s eventually turned the magneto-ionic theory into a paper tool that helped the radio-sounding apparatuses to probe—produce empirical knowledge about the structural features of—the ionosphere.

The magneto-ionic theory in the 1920s was by no means the most accurate wave-propagation theory at hand. It ignored the statistical variation of the induced polarizing current in the ionic medium—unlike the concurrent statistical treatments of gaseous dynamics, magnetism, and metal conductivity. The theory itself derived from a physical-optics assumption that was only approximately true at high frequencies. But most scientists did not bother with its elaboration: its statistical version did not appear until 1960;[6] only a few pursued the "full-wave" theory of ionospheric propagation, to no avail.[7] To the majority, the magneto-ionic theory was an instrument's theory that was a good enough approximation to reality and, more important, the calculating device for retrieving ionospheric information from radio sounding data. In essence, its role in radio sounding was not different from acoustic propagation in sonar, Fourier optics in computed tomography, and nuclear resonance in magnetic resonance imaging (MRI).

## GENERALIZING THE MAGNETO-IONIC THEORY

When Appleton, Nichols, and Schelleng introduced the magneto-ionic theory in 1924–25, they followed Larmor, Lorentz, and other nineteenth-century magneto-optics researchers to focus on two special cases: the wave-propagating

6. Sen and Wyller, "On the generalization of the Appleton-Hartree magnetoionic formulas" (1960), 3931–50.

7. For works on the full-wave theory, see Epstein, "Geometrical optics" (1930), 37–45, and "Reflection of waves" (1930), 627–37; Eckersley, "Long-wave transmission" (1932), 158–73, "On the connection" (1931), 83–98, "Radio transmission problems" (1932), 499–527, and "Studies in radio transmission" (1932), 405–59.

directions normal to or parallel to the geomagnetic field. Within a few years, Appleton and Altar, Goldstein, Hartree, and Lassen independently developed a general theory that dealt with any arbitrary wave-propagating directions. Among the four versions, Appleton and Altar's—the "Appleton-Hartree formula," or the "Appleton-Lassen formula"—became the most popular among geoscientists.[8] In this section, we examine the work of Appleton, Altar, Goldstein, and Lassen, leaving Hartree's to the next chapter for its relevance to microphysics, the major theme of that chapter.

### *Appleton and Altar's Derivation*

Appleton did deserve much of the credit for the formula. He was the first to entertain the idea of generalizing the magneto-ionic theory. Parallel to his research in atmospherics, electronic tubes, and fading, he had been looking at the geomagnetic effect on radio-wave propagation since the early 1920s. In 1923, he corresponded with Lorentz about the rotation of a radio wave's plane of polarization by the earth's magnetic field. When Larmor gave a lecture at Cambridge on ionic refraction, Appleton pointed out that the theory neglected the geomagnetic effect, and he found how to incorporate it. Larmor suggested that he publish his finding.[9] So Appleton presented a conference paper on the special magneto-ionic theory in November 1925, about the time Nichols and Schelleng independently published the same results.

Appleton, Nichols, and Schelleng's special magneto-ionic theory yielded important outcomes such as the resonance at gyro frequency $\omega_g = eB_g/m$ that explained propagating range minima. But radio waves in general propagated in all directions, which restricted the special theory's applicability in most circumstances. In 1926, Appleton managed to derive the general magneto-ionic theory to cover all wave-propagating directions. He did so with the help of an Austrian physicist, Wilhelm Altar. Historian C. Stewart Gillmor has rediscovered this invisible figure, who, he has argued, may have done more than Appleton in the derivation and may have been the actual developer of the Appleton-Hartree formula.[10]

8. Mimno, "Physics" (1937), 21–24. Appleton's student Mary Taylor coined the term "Appleton-Hartree formula" in the 1930s. Working at Britain's Radio Research Board, she was familiar with the general magneto-ionic theory through the work of Appleton and Hartree, who was also serving as the board's consultant. See the end of this chapter for more about her work.

9. Gillmor, "Wilhelm Altar" (1982), 398.

10. Ibid., 404.

Born in Vienna, Wilhelm Altar studied theoretical physics at the University of Vienna and engineering at the Technische Hochschule of Vienna. The job market for academics in postwar Austria was poor. After obtaining his PhD and working briefly in a radio company, Altar went to London in 1925. Because wireless interested him, Professor A. O. Rankine at Imperial College, London introduced him to Appleton, who invited him to work on the generalization of the magneto-ionic theory. They met almost every day and made steady progress in six months. In 1925–26, the Austrian wrote "Wellenausbreitung in ionisierten Gasen unter dem Einfluß eines Magnetfeldes" ("Wave propagation in ionized gases under the influence of magnetic fields") which Appleton used to write journal articles. The "Appleton-Hartree formula" first appeared in a letter that Altar sent to Appleton in 1926.[11]

The Appleton-Altar approach was an exercise in Lorentzian magneto-optics. It began with an average ion's equation of motion: $md^2\vec{r}/dt^2 = e\vec{E} - gd\vec{r}/dt + e(d\vec{r}/dt) \times \vec{B}_g$ ($\vec{r}$ was the ion's spatial displacement, $g\,d\vec{r}/dt$ was friction due to its collisions with molecules in the ionic medium, and the ion was most likely an electron). In this model, each ion had an average spatial displacement $\vec{r}$ resulting from the radio wave's electric field and the geomagnetic field—from its "original" position (the position when these external fields did not disturb the ion). This displacement created an average dipole moment $e\vec{r}$. So the dipole-moment polarization (not the polarization associated with the direction of the wave's electric field), or the average sum of dipole moments per unit volume, was $\vec{P} = Ne\vec{r}$, where $N$ was the ion number density. From the ionic equation of motion and $\vec{P} = Ne\vec{r}$, a linear relation between the vector components of the electric field and the vector components of the polarization was derivable in a matrix form: $\varepsilon_0\vec{E} = \mathbf{c}\vec{P}$, where $\mathbf{c}$ was a $3 \times 3$ matrix with entries determined by $Ne^2/m\varepsilon_0$, $\omega_g = eB_g/m$, $g/m$, and the wave's angular frequency $\omega$.

The constitutive relation in electromagnetism connected electric field $\vec{E}$ with polarization $\vec{P}$ and electric displacement $\vec{D}$ via $\vec{D} = \varepsilon_0\vec{E} + \vec{P}$. The constitutive relation, together with Maxwell's equations, yielded another equation governing $\vec{E}$ and $\vec{P}$. Substituting this formula into the relation $\varepsilon_0\vec{E} = \mathbf{c}\cdot\vec{P}$ and canceling $\vec{E}$ led to three linear equations for $P_x$, $P_y$, and $P_z$, when the field had a form of plane wave $X_0\exp\{i[\omega t - k(a_x x + a_y y + a_z z)]\}$, ($k = 2\pi/\lambda$ was the wave number, and $a_x$, $a_y$, and $a_z$ were the $x$, $y$, and $z$ components of the wave-propagating direction). In tensor form, these equations were

11. Ibid., 405 and 423. Gillmor translated Altar's manuscript in ibid., Appendix B, 425–40.

$$(9.1)\quad \left[\delta_{ij}+c_{ij}+\frac{k^2}{\omega^2\mu\varepsilon}(a_i a_l-\delta_{il})c_{lj}\right]P_j=0$$

for $i = x, y, z$; $\delta_{ij}$ was Kronecker's delta ($\delta_{ij} = 1$ for $i = j$, $\delta_{ij} = 0$ for $i \neq j$). Solving equation (9.1) gave the wave number $k$ and hence the refractive index $n$ ($n^2 = k^2/\omega^2\mu\varepsilon_0$). Moreover, substituting the solution of equation (9.1) for $P_x$, $P_y$, and $P_z$ into the relation $\varepsilon_0\vec{E}=\mathbf{c}\cdot\vec{P}$ led to the solution for the wave's polarization (the change of the direction of $\vec{E}$ over time).[12]

Appleton and Altar did not immediately seek the general solutions to equation (9.1). They discussed the solutions under several special cases such as zero friction and waves propagating parallel to the geomagnetic field. They began to attack the problem in its general form after Altar failed to secure a job in England, returned to Austria in February 1926, and obtained a position at a radio receiver manufacturer in Vienna. Altar continued to exchange letters with Appleton; in correspondence they obtained two general solutions to equation (9.1) for the refractive index $n$:[13]

$$(9.2)\quad n^2=1+\frac{2}{2(\alpha+i\beta)-\dfrac{\gamma_T^2}{1+\alpha+i\beta}\pm\sqrt{\dfrac{\gamma_T^4}{(1+\alpha+i\beta)^2}+4\gamma_L^2}}.$$

where $\alpha = -\omega^2/\omega_0^{\,2}$, $\beta = (\omega/\tau\omega_0^{\,2})$, $\gamma_T = \omega_T\omega/\omega_0^{\,2}$, $\gamma_L = \omega_L\omega/\omega_0^{\,2}$, $\omega_0^{\,2} = (Ne^2/m\varepsilon_0)$, $\omega_T = eB_T/m$, $\omega_L = eB_L/m$, $B_L$ was the geomagnetic field's component along the wave-propagating direction, $B_T$ was its component normal to that direction, and $\tau = m/g$ was the time an ion traveled within the mean free path.[14] Later Appleton and Altar also deduced the corresponding ratios of polarization for electric field:

12. Note that their approach differed slightly from Nichols and Schelleng's. While Nichols and Schelleng reduced the wave equation to an equation for electric field, they reduced it to an equation for dipole-moment polarization, wherein it was not necessary to solve the inverse matrix of **c**. In addition, they incorporated friction in the ionic equation of motion, while Nichols and Schelleng did not. If the Americans had done so, then they would have had to solve the inverse of an even more complicated matrix **c**—a problem that Appleton and Altar avoided.

13. The general solution first appeared in Altar's letter to Appleton on 8 May 1924. Gillmor thought that perhaps Altar had already derived it when he was in England; ibid., 408.

14. Note that Lorentz modeled the collision effect in a dielectric material with $g = 2m/\tau$; Appleton revised Lorentz's coefficient to $g = m/\tau$ for a difference he conceived between an ionic and a dielectric medium.

$$(9.2') \quad \frac{E_x}{E_y} = \frac{-i\gamma_L}{\frac{\gamma_T^2}{2(1+\alpha-i\beta)} \mp \sqrt{\left[\frac{\gamma_T^2}{2(1+\alpha-i\beta)}\right]^2 + \gamma_L^2}},$$

assuming the waves propagating along direction $z$. Since the 1930s, the usual name for equations (9.2) and (9.2′) has been the "Appleton-Hartree formula" (see next chapter for Hartree's contribution). The formula indicated that a radio wave propagating with an arbitrary direction split into two components with different phase velocities and polarizations. When $\omega < \omega_0$, the component associated with the upper sign (+ for refractive index and − for electric polarization) was the "ordinary wave," and the one with the lower sign "extraordinary," since the former deviated less from the geomagnetic-free case. The notation was the opposite when $\omega > \omega_0$ for the same reason.

Altar perhaps actually derived the Appleton-Hartree formula, but Appleton incorporated the equation into radio ionospheric research. Formal generalization as such did not interest him; his aim in asking Altar to derive the formula was not the mathematical completeness of the magneto-ionic theory. Instead, he wanted to use the general formula as a tool to answer empirical questions about the conditions for the refractive index to have "strange" or critical values and about the physical meanings for these values. Appleton took the questions to his former mathematics professor at St John's College, Cambridge, Ebenezer Cunningham, an authority on relativity.[15] Cunningham answered them: neglecting the frictional loss, equation (9.2) implied two critical values for $n^2$: $n^2 = 0$ and $n^2 = \infty$. When $n^2 = \infty$, the radio wave had a resonance with the geomagnetic field; its phase velocity was infinitely small so that it took forever for the radio wave to reach any distance. In this case, its propagating range was infinitely short. When $n^2 = 0$, electromagnetic energy did not transmit in the medium, and the ionosphere totally reflected the radio wave.[16]

Appleton also wanted to know the frequencies corresponding to $n^2 = 0$ and $n^2 = \infty$. He, Nichols, and Schelleng had found a resonance at the gyro frequency $\omega_g = eB_g/m$ for the special cases and had given different numerical estimates—the Americans used the entire geomagnetic field of 0.5 gauss to

15. For Cunningham, see Warwick, *Masters of Theory* (2003), chap. 8.

16. Cunningham to Appleton (letters), 1927 (undated), MS 2370, C 217, Edward Appleton Papers (H37), Special Collections, University of Edinburgh Libraries, Edinburgh.

obtain a resonance wavelength of 214 meters, the Briton used only its horizontal component of 0.18 gauss to obtain a resonance at 580 meters. Thus Appleton wondered whether to use the entire geomagnetic field or only a part of it to calculate the resonance and other critical frequencies in the general case. Cunningham found that $n^2 = 0$ if and only if $1+\alpha = 0$ or $1+\alpha = \pm\omega_g\omega/\omega_0^2$. So the frequencies for total reflection depended on the total geomagnetic field $B_g$, not its components. The condition for total reflection was independent of the wave-propagating direction; even a vertical ray could be reflected.[17] But the resonance frequency for $n^2 = \infty$ depended on the direction. Appleton found from equation (9.2) that $n^2 = \infty$ if and only if $\alpha^2 = \gamma_L^2 + \alpha\gamma_T^2/(1+\alpha)$, which depended on both $\gamma_L$ and $\gamma_T$. Only when $|\alpha| >> 1$ (low degrees of ionization) did the total geomagnetic field determine the resonance frequency via the relation $\alpha^2 = \gamma^2$. Therefore, Appleton argued, the critical resonance effect, such as a significant reduction of propagating ranges, occurred at about 200 meters only at a low degree of ionization. Cunningham cautioned Appleton that even the low-ionization condition was insufficient—$|\alpha| >> 1$ and $\alpha^2 = \gamma^2$ implied that the geomagnetic field also had to be large.[18]

The original search for resonance flowed from the U.S. Naval Research Laboratory's (NRL's) empirical observations that the wave-propagating range was lowest at about the wavelength of 200 meters. Yet a range experiment by German electrical engineer Alexander Meißner at Telefunken in 1926 challenged NRL's data; Meißner found no significant reduction of the propagating distance at about the wavelength of 200 meters.[19] Did the resonance really exist? This was another question that Appleton addressed in his magneto-ionic theory. He noted that for a given frequency a resonance could occur only at one of the two components in equation (9.2). Although propagation of the resonating component ceased because of its infinitely small velocity, the other component still propagated along, preventing the apparent resonance at 200 meters. Its immediate corollary was that a wave arriving at about the resonance frequency would have identical polarization to that of the longer-reaching component. Crucial empirical evidence for the magneto-ionic theory therefore lay in polarization measurements.

In the General Assembly of the International Union of Radio Science (French acronym URSI) in Washington, DC, in October 1927, Appleton re-

17. Ibid., Cunningham to Appleton (letters), 13 and 31 May 1927, ibid.

18. Ibid.

19. Meißner, "Hat das Erdfeld einen Einfluß auf die Wellenausbreitungsvorgänge?" (1926), 321–24.

ported the theoretical exercise that he and Altar had done.[20] He presented equation (9.2) and his discussions with Cunningham on the refractive index's critical values. This was the first public appearance of the general magneto-ionic theory.

## *Lassen's Derivation*

Not everyone read the proceedings of the URSI conference, however. Unaware of Appleton and Altar's work, German physicist Hans Lassen in the late 1920s independently tried to generalize the magneto-ionic theory. Born in Alsen, Denmark, and growing up in Sonderburg, Germany, Lassen studied electrical engineering and physics at the University of Jena and obtained his PhD in 1924. He joined the University of Cologne in 1925 to work for physicist Karl Försterling (a Göttingen graduate working on electromagnetic waves at all frequencies and a pioneer of German radio ionospheric research)[21] and electrical engineer Hans Rukop (the author of an influential textbook on radio engineering). Lassen later became assistant professor and stayed in Cologne until the Nazis expelled him from the university in 1935. Then he worked at Siemens until after World War II. He was a founder of the Physical Institute of the Humbolt University and a founder of the Free University, both in Berlin.[22]

On arriving in Cologne in 1925, Lassen started to work on radio-wave propagation at the request of his supervisor, Försterling. He investigated first the daily variation of the ion density in the Kennelly-Heaviside layer. Lassen estimated different ion-density profiles at different times of a day from a theory of ion generation-recombination and concluded that ion density fluctuated more considerably at night than during the day.[23] This research directed him to the magneto-ionic theory. Nichols et al. had indicated that the geomagnetic field split a propagating wave into two, and the refractive-index difference between the two waves increased with ion density. If so, Lassen asked, then at what time of day and at what height of the ionosphere was the ion density high enough to engender a conspicuous geomagnetic effect on the refractive-index difference, and how did this critical height vary with the direction of the

20. Appleton, "The earth's magnetic field" (1927), 2–3.

21. Weinmeister, *J. C. Poggendorff's*, vol. 5 (1922), 771.

22. Historischen Kommission bei der Bayerischen Akademie der Wissenschaft (ed.), *Neue Deutsche Biographie* (1982), 674; Stobbe (ed.), *J. C. Poggendorff's*, vol. 6 (1931), 1470–71.

23. Lassen, "Die täglichen Schwankungen" (1927), 174–79.

wave? Not knowing of Appleton and Altar's formula, he developed his own general theory in 1927 to answer these questions.[24]

Lassen also began with the equation of the average ionic motion in a geomagnetic field: $md^2\vec{r}/dt^2 = e\vec{E} - gd\vec{r}/dt + e(d\vec{r}/dt)\times\vec{B}_g$. He followed Nichols and Schelleng's process to solve the equation of motion and to express $\vec{r}$ in terms of $\vec{E}$, contrasting Appleton and Altar who expressed $\vec{E}$ in terms of $\vec{r}$. This approach carried the burden of solving the equation of motion; its solution was cumbersome. To simplify matters, Lassen assumed that $(g/m\omega) << 1$, meaning that the radio wave was much shorter than one kilometer, as $2m/\tau$ replaced $g$ (in contrast to Appleton's supposition that $g = m/\tau$). Here Lassen plugged in $m$ with the electron's mass and collision frequency $1/\tau$ with the value of $1.5\times10^5$ (1/sec) from his early estimate for the ionic and molecular densities in the ionosphere. Under such a condition, a simpler form could approximate the expression for $\vec{r}$

Lassen substituted the approximate expression for $\vec{r}$ into the polarizing relation $\vec{P} = Ne\vec{r}$ and the constitutive relation $\vec{D} = \varepsilon_0\vec{E}+\vec{P}$ to deduce an expression for $\vec{E}$ in terms of $\vec{D}$ only. He then plugged this new expression for $\vec{E}$ into Maxwell's equations. After algebraic manipulations, such relations led to two linear equations for the two mutually orthogonal components of the wave's magnetic flux density $B_x$ and $B_y$ (both normal to the wave's direction). By solving the two equations, he obtained two solutions for the refractive index and polarization $B_x/B_y$.

In a strict sense, Lassen's magneto-ionic theory was not identical to Appleton and Altar's: it was approximate while theirs was exact; and the approximation held only when $(g/m\omega) << 1$ or, equivalently (according to Lassen's estimate), when the radio wave was much shorter than one kilometer. In addition, Lassen's expression was algebraically more complicated than Appleton and Altar's. But it is safe to say that both versions of the theory were approximately equivalent to each other in many, if not all, practical cases.

Lassen tried to explore further his magneto-ionic theory by examining the variations of the refractive index and polarization with wave direction and by calculating the refractive-index profiles over height for waves propagating along and normal to the geomagnetic field. He and Försterling planned to construct the paths of the splitting waves moving in a nonuniform geomagnetized ionosphere. They partially realized their plan. For instance, they analyzed the pole and zero structures of the refractive-index formula and probed

24. Lassen, "Über den Einfluß" (1927), 324–34.

their effects on the variations of the refractive index over some parameters.[25] Like Appleton, Lassen and Försterling also applied the general theory to empirical questions in radio ionospheric propagation. Unlike the English application, which focused on the qualitative or numerical behavior at more easily observable critical conditions, the German stressed more the closed-form mathematical solutions for the complete wave path or refractive-index profile. Both approaches could work in principle. Unfortunately, history did not give the Germans a chance. In 1935, the Nazis forced Lassen to leave the University of Cologne, interrupting his radio ionospheric research.[26]

### *Goldstein's Derivation*

After talking to Appleton in 1924, Larmor at Cambridge revised his lecture notes to include the geomagnetic effect on radio-wave propagation. Taking inspiration from his discussions with Appleton, Ebenezer Cunningham gave a question in the Mathematical Tripos of 1928 asking students to deduce the refractive index and polarization of a radio wave propagating along any direction vis-à-vis the geomagnetic field. Sydney Goldstein took the challenge.[27]

Like all the great wranglers from Kelvin to Larmor, Sydney Goldstein bore the trademark of Cambridge mathematical physics. His reputation rested on his work on the boundary-layer theory in fluid dynamics, which he pursued throughout a career at Cambridge, Manchester, NPL, Haifa, and Harvard. A native of Hull, England, Goldstein studied mathematics at St John's College, Cambridge, in the 1920s. He learned the special magneto-ionic theory from Larmor's course on electric waves and encountered the issue of its generalization in Cunningham's Tripos question, when both men were teaching at St John's. He took the challenge as a mathematical question, solved it, and published the solution in 1928.[28]

Goldstein also began by considering the ionic equation of motion. But he neglected the effect of collisions, so that friction $g = 0$. In so doing, he was able to reduce the equation of motion, the polarizing relation, the constitu-

25. Försterling and Lassen, "Kurzwellenausbreitung" (1933), 26–60.

26. *Neue Deutsche Biographie* (1982), 674.

27. Goldstein, "The influence of the earth's magnetic field" (1928), 261–63.

28. Ibid., 260–84. For Goldstein, see MacTutor History of Mathematics Archive: http://www-history.mcs.st-and.ac.uk/Mathematicians/Goldstein.html (last accessed on 3 December 2012).

tive relation, and Maxwell's equations to two linear equations for $E_x$ and $E_y$, the wave's electric-field components. Solving the two equations gave two refractive indices and polarizations. His results were identical to Appleton and Altar's equations (9.2) and (9.2′) in the special case when $\beta = 0$. Unlike Lassen's, Goldstein's solution was exact, not approximate.

In his article, Goldstein attempted to engage his general magneto-ionic theory through empirical research, as he discussed the Radio Research Board's experimental results. But radio ionospheric propagation was not his topic. After 1928, he focused on special functions in mathematical analysis and fluid dynamics and rarely published on propagation. To him, the work on the general theory was simply a neat analytical exercise to solve a mathematical problem.

Altar's fadeout, Lassen's forced interruption, and Goldstein's indifference to radio ionospheric research prompt us to rethink the origins of the general magneto-ionic theory. Indeed, Appleton and Hartree were neither the first nor the only independent derivers of the Appleton-Hartree formula. Gillmor was right. We should give due credit to Altar, Lassen, and Goldstein. Nevertheless, the theory's origin comprised not only the derivation of its central equations. Equations did not function alone. The theory's important and complex part was not deducing formulae but interpreting them and yielding from them predictions amenable to experimental verification. If the theory's "real life" started from its *use* as a paper apparatus for addressing experimental questions and retrieving empirical information about radio ionospheric research, then Appleton was its midwife.

In fact, even Appleton did not know exactly how to make the theory experimentally relevant when he and Altar derived the Appleton-Hartree formula. Radio ionospheric researchers did not touch it until Appleton and his fellow experimenters recognized its implications for radio-wave polarization in 1927. Their polarimetric measurements in England and Australia, as it later became clear, empirically confirmed the general theory.

## MEASURING POLARIZATION IN BOTH HEMISPHERES

Radio experiments preoccupied Appleton far more than the mathematics of the magneto-ionic theory when the generalized formula was emerging. In 1926, he performed trials on the diurnal variations of the ionosphere's height and fading at BBC transmitting stations around the country, the Cavendish Laboratory, King's College, London, the National Physical Laboratory, and the Radio Research Board's Peterborough and Slough stations. A group of protégés and assistants helped him, including the young John Ratcliffe, who entered Sidney Sussex College, Cambridge, in 1920 and started to work with him in 1924.[29] Appleton's fading experiments with Ratcliffe combined two apparatus schemes from the discovery of the ionosphere—the frequency-changing method of producing artificial phase variations and the multiple-antenna arrangement for measuring the directions of electric and magnetic fields (chapter 7). The combination discerned the direction, intensity, and ray-path differences between sky and ground waves.

Appleton and Ratcliffe's results opened up further investigations. A sky wave's electric-field components on and normal to the plane of propagation had similar intensity fluctuations during fading, implying a connection between fading and the sky waves' changes of polarization. Also, since the errors in the loop-type direction finders resulted from the electric-field component normal to the plane of propagation, one might expect a correlation between direction-finding errors and fading for certain polarized downcoming waves. Both implications called for direct measurements of a sky wave's polarization.[30]

## *A Polarimetric Experiment*

The method of directly measuring radio-wave polarization did not exist in 1926. All the previous polarization experiments with multiple-antenna setups, including those of Smith-Rose, Barfield, Appleton, and Barnett, measured only the time-average intensities of a wave's field on and normal to the plane of propagation and their ratios, not the trace of the overall field's direction over time (i.e., the temporal variation of the ratio of the two field components) that defined wave polarization. So Appleton and Ratcliffe combined multiple antennae and frequency alteration to measure polarization directly. Their setup consisted of three vertical loop antennae—one on the plane of propagation, one tilting forward at 45°, and one tilting backward at 45° (figure 9.2) Analysis showed that the time-average currents at the three loops were functions of four independent variables: $H_1/H_0$, $H_1'\cos\theta/H_0$, $\phi$, and $\phi'$. $H_0$ was the ground wave's magnetic field magnitude, $H_1/H_1'$ was the ratio of the magnitudes of the sky wave's magnetic-field components on and normal to the plane of propagation, $\phi$ and $\phi'$ were the phase angles of those components

29. Ratcliffe would become Cambridge's major radio scientist and a leading trainer for Britain's ionospheric researchers after World War II. See Budden, "Ratcliffe" (1988), 671–711.

30. Appleton and Ratcliffe, "On a method" (1928), 576–77.

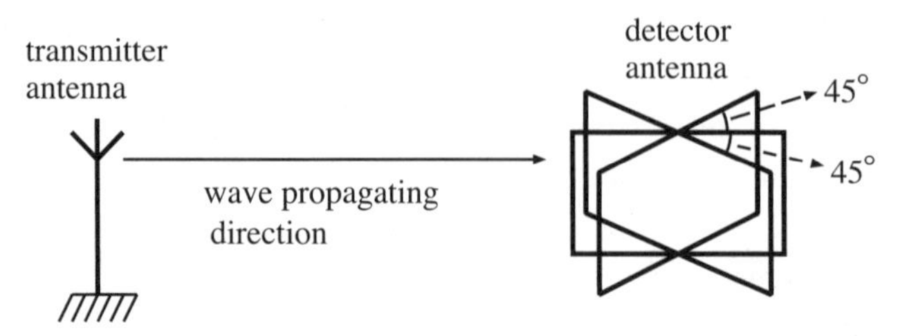


FIGURE 9.2. Appleton and Ratcliffe's antenna setup for direct polarization measurements.

relative to the ground wave, and $\theta$ was the sky wave's angle of elevation. The sky wave's polarization by definition found full specification in the magnitude ratio $H_1/H_1'$ and the phase difference $\phi-\phi'$ between its two components. If the currents at the three loops could determine the four quantities $H_1/H_0$, $H_1'\cos\theta/H_0$, $\phi$, and $\phi'$, then one could obtain the sky wave's polarization by calculating $(H_1/H_0)/(H_1'\cos\theta/H_0)$ and $\phi-\phi'$ ($\cos\theta$ was determined from another experiment).

Unfortunately, three measurables could not determine four unknown quantities. Enter the frequency-change method. Appleton and Ratcliffe argued that one could figure out the four quantities if one measured the waveforms of the three antenna signals with a continuous and uniform change of radio frequency, not the time-average signal intensities. Changing the frequency uniformly yielded a sinusoid-like waveform with a maximum level and a minimum level at each antenna loop. Appleton and Ratcliffe showed that the ratios of maximum to minimum at the three loops were expressible as functions of three variables: $H_1/H_0$, $H_1'\cos\theta/H_0$, and $\cos(\phi-\phi')$. Therefore, their measured results could determine $H_1/H_0$, $H_1'\cos\theta/H_0$, and $\cos(\phi-\phi')$. This specified the "sense" of polarization. The measured ratios could determine the value only of $\cos(\phi-\phi')$, which corresponded to two values of $\phi-\phi'$ with the same magnitude but opposite signs. The sign of $\phi-\phi'$ was critical information—it determined the wave's sense of polarization (right- or left-handed). Appleton and Ratcliffe argued that one could obtain this figure by carefully comparing the phase relations of the sinusoidal waveforms with the continuous frequency change at the three loops. They implemented the setup by switching alternately the receiver connection among the three antenna loops (figure 9.3).

Appleton and Ratcliffe performed this polarization experiment in 1927. They built a three-antenna receiver with oscillographic recorders at the board's station in Dogsthorpe, Peterborough. Their frequency-modulating

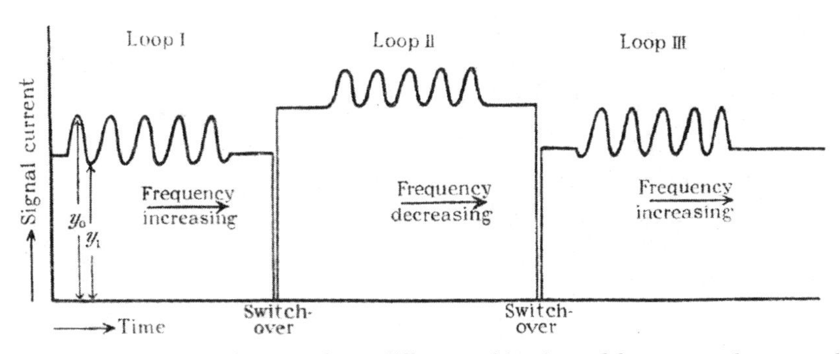


FIGURE 9.3. Appleton and Ratcliffe's combination of frequency change and multiple antennae. Appleton and Ratcliffe, "On a method" (1928), figure 3.

transmitters (wavelength 400 meters) were at the BBC's Birmingham station and NPL in Teddington. The transmission from Teddington to Dogsthorpe was roughly south to north, while that from Birmingham to Dogsthorpe was west to east. The experiment started in March and continued until June. Observers took all the measurements at night to ensure stronger sky waves.

The results were extremely regular. Most of the time, the ratio $H_1/H_1'$ was very close to 1, and the phase difference $\phi-\phi'$ was close to 90°, no matter whether the transmission was south-to-north or west-to-east.[31] This result implied that the incoming waves' polarizations were mostly left-handed circular. Sometimes $H_1/H_1'$ deviated from 1 and $\phi-\phi'$ deviated from 90°, indicating an elliptical polarization. But the phase difference $\phi-\phi'$ *never* went below 0° or above 180°, which implied that the sense of polarization was *always* left-handed.

Why was the sense of wave polarization always left-handed for both south-to-north and west-to-east propagation? Appleton and Ratcliffe based their account on the magneto-ionic theory—because only one of the two splitting waves arrived. The splitting waves in the geomagnetized ionosphere had different attenuations and polarizations. When the wave path was long enough, only the component with the smaller attenuation had adequate energy to reach the receiver. Thus the incoming signal's polarization became that of the dominant component. Appleton, Nichols, and Schelleng had shown that for a wave propagating along the geomagnetic field, the less absorbed component had a left-handed circular polarization (i.e., $E_x/E_y = -i$, or the magnitude of $E_x/E_y$ was 1 and its phase was 90°). Yet this special case could not explain

31. Ibid., 584–85.

why waves propagating in various directions had the same sense of polarization. The explanation, Appleton and Ratcliffe believed, needed the general magneto-ionic theory.

The task of developing such an explanation fell on the shoulders of two Australians: Sydney-trained electrical engineer William George Baker and Appleton's former student Alfred Leonard Green. Baker and Green explored and elaborated the general magneto-ionic theory's implications for polarization. From Appleton and Altar's formula in equations (9.2) and (9.2′), they showed that the polarization $E_x/E_y$ of the less attenuated wave component depended only on the angle between the geomagnetic field and the wave-propagating direction. When the angle was smaller than 90°, the phase of $E_x/E_y$ was between $-180°$ and $0°$, meaning the sense of polarization was left-handed. When the angle was larger than 90°, the phase of $E_x/E_y$ was between 0° and 180°, making the sense of polarization right-handed.[32]

The factor that determined whether a downward sky wave was left- or right-handed was therefore whether the angle between its propagating direction and the geomagnetic field was smaller than 90°. The geomagnetic field did not point from south to north on the horizon—its direction had a vertical component. At high latitudes in the northern hemisphere (such as England), it tilted downward with a large angle vis-à-vis the horizon. Also, a short transmitter-receiver distance (such as Birmingham or Teddington to Dogsthorpe) created a large angle between the sky wave's downward trajectory and the horizon. The vertical inclinations of both directions reduced the angle between them. Consequently, the angle between the geomagnetic field and the downcoming ray at relatively high latitudes in the northern hemisphere was smaller than 90° for both south-to-north and west-to-east transmissions (figure 9.4) In fact, the angle between the field and the wave was smaller than 90° for most transmitter-to-receiver directions in the northern hemisphere (especially at high latitudes). Therefore incoming sky waves in the northern hemisphere should have a left-handed polarization for most cases. For the same reason, most sky waves in the southern hemisphere should have right-handed polarization, since the geomagnetic field there inclined upward.

Appleton and Ratcliffe's discovery in 1927 signified the importance of polarization experiments. Baker and Green's theoretical explication further led to an unexpected prediction: the magneto-ionic theory, originally explaining the apparent resonance in range data, resulted in an assertion that the

32. Baker and Green, "The limiting polarization" (1933), 1103–31.

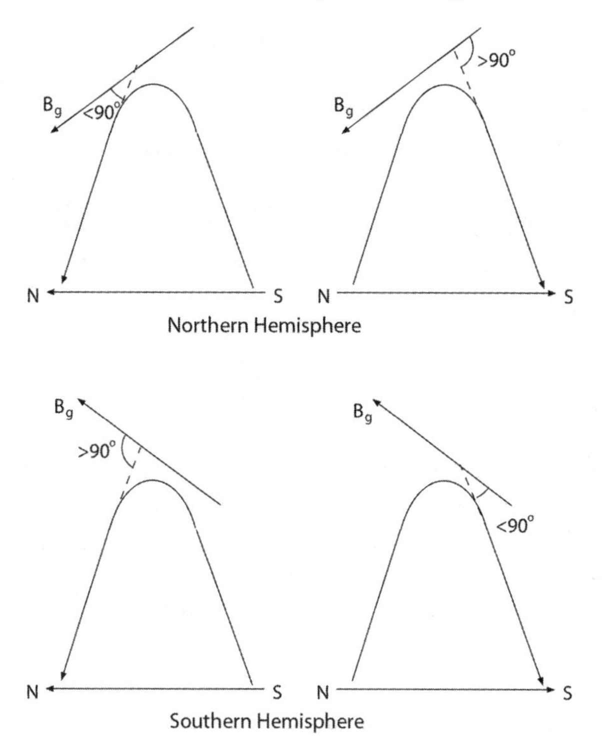


FIGURE 9.4. Geometry of the geomagnetic field and the downward sky waves in the Northern and Southern Hemispheres.

polarimetric behavior of radio waves was different in different hemispheres. The experimental verification of this prediction would therefore offer independent evidence for the general theory. Such verification required two sets of measurements—one in the Northern Hemisphere, as Appleton and Ratcliffe had performed, and the other in the Southern Hemisphere, which no one had yet done.

### *More Polarization Research in Both Hemispheres*

Australia's new Radio Research Board provided an opportunity to do the experiments. Historians Rodney Home and Aitor Anduaga have pointed out that Australia, as a British dominion, was under strong British influence in its interwar development of science and technology. In the 1920s, it emulated British scientific organizations, and its key research personnel studied or trained at English and Scottish universities, NPL, and government research stations.[33] Following the United Kingdom's DSIR, the Australian government established a Council for Scientific and Industrial Research (CSIR) in 1923. The CSIR concentrated on manufacturing of agricultural products, the country's major source of overseas income. But John Madsen, professor of electrical engineering at the University of Sydney, suggested that CSIR form a Radio Research Board, which it started in 1926 and that was fully functional in 1929.[34]

In the beginning, most research officers at the board were British immigrants who had studied in Britain. Among them was Alfred Leonard Green. Born in London, Green went to King's College, London, in the early 1920s, studied with Appleton, and participated in the radio experiments at King's, the Cavendish Laboratory, and NPL. After obtaining his BS in 1926, he continued to work for Appleton as a scientific assistant at Britain's Radio Research Board. As its Australian counterpart was open in 1929, he moved to Sydney and became a research officer there.[35]

Green was familiar with Appleton and Ratcliffe's polarization experiment and the need to reproduce it in the Southern Hemisphere. His first task in Sydney was thus to replicate their measurement in Australia. To make a faithful comparison, the experimental procedure, the measuring method, and the instrumental setup had to be identical. The radio frequencies and the transmitter-receiver distance also had to be the same. Yet the horizontal direction from transmitter to receiver should be symmetrically opposite to that in the Northern Hemisphere to make the angle between the downcoming ray and the geomagnetic field identical in both cases. Since the transmitter was to the south of the receiver in England, it had to be to the north in Australia.

Green secured the help of a radio station in Coogee, Sydney, for frequency-modulating transmission. With the government's help, he obtained a site at the Naval College in Jervis Bay, 143 kilometers south of Coogee, to install the receiver. Measurements took place from May to October 1930. The results were unambiguously positive—the sense of polarization was right-handed throughout, which confirmed the magneto-ionic theory's prediction.[36]

33. Home, "To Watherloo and back" (1994), 149–60; Anduaga, *Wireless and Empire* (2009), chap. 3.

34. Evans, *History of the Radio Research Board* (1973), 8–29.

35. "Green, Alfred Leonard," in Home, *Physics in Australia to 1945* (1990).

36. For Green's experiments, see Green, "The polarization of sky waves" (1934), 324–43; Evans, *History of the Radio Research Board* (1973), 66–75.

Appleton immediately heard the results from Green and thought them a major verification for the general theory. He wrote a survey of his work since 1924, including the magneto-ionic theory that he developed with Altar.[37] His article, which appeared in 1932, presented the generalized formulae for refractive index and polarization—equations (9.2) and (9.2′). This was the first systematic exposition of (Appleton's version of) the Appleton-Hartree formula in print.[38]

At the same time, scientists continued to improve the techniques of polarization measurement. In 1933, Ratcliffe and Appleton's students Eric L. C. White and F. W. G. White at the Cavendish designed a new apparatus to measure the polarization of sky waves,[39] using two mutually orthogonal vertical loops instead of three, but no frequency change. A tuning mechanism in the coupling circuit between the two antennae suppressed ground waves, and signals from the two antennae went to the horizontal and vertical axes, respectively, of an oscillographic recorder to trace their coevolution over time (the Lissajours figures). The trace's shape—circular or elliptical—determined the shape of polarization, and its evolving direction—clockwise or counterclockwise—indicated the sense of polarization. Without frequency change, the new visual method of measuring polarization was simpler.

Appleton went one step further: he began to use Breit and Tuve's pulse-echo device in the polarimetric and other ionospheric experiments. With the new instrument came another discovery. In 1933, when Appleton and his student Geoffrey Builder at King's College, London, measured the intensity and polarization of trains of pulses returning from the sky, they received three distinct groups of pulses during the night (figure 9.5). They argued that the three groups of pulses corresponded to the ground wave G and refraction from two distinct ionospheric layers—F1 and F2—which American engineers at NBS had identified early in the year. Both the F1 and F2 echoes had two pulses, the leading one with right-handed circular polarization and the lagging one with left-handed circular polarization.

Appleton and Builder interpreted this observation as another confirmation of the magneto-ionic theory: the F1 and F2 layers at night were high enough that the collision-induced absorption for both ordinary and extraordinary waves was low, and thus both waves appeared in the return. From the theory,

37. Gillmor, "Wilhelm Altar" (1982), 416–17.

38. Appleton, "Wireless studies of the ionosphere" (1932), 642–50.

39. Ratcliffe and White, "The effect of the earth's magnetic field" (1933), 125–44; Ratcliffe and White, "The state of polarization" (1933), 423–41.

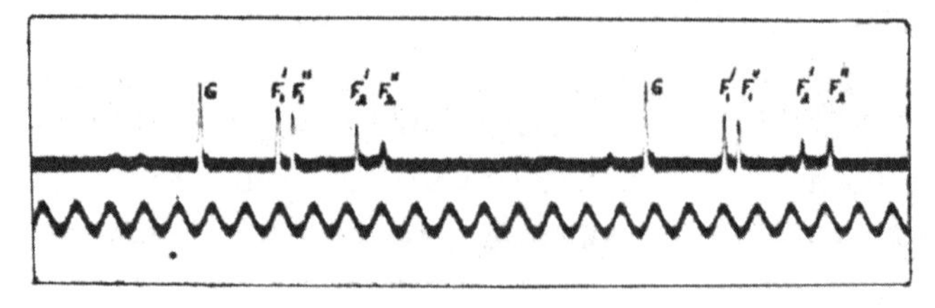

FIGURE 9.5. Multiple ionospheric layers from pulse-echo measurements. G: ground wave; F1′, F1″: F1 layer; F2′, F2″: F2 layer. Appleton and Builder, "The ionosphere" (1933), figure 2. The horizontal axis denotes time, and the vertical axis the strength of returning pulses.

these two waves had different velocities and different states of polarization (one right-handed and the other left-handed). So they corresponded to two pulse trains with different times of arrival and different polarizations. In addition, the polarizations were close to circular because the sky wave and the geomagnetic field had a small angle relative to each other, as both inclined vertically in this case.[40]

The British and Australian experiments with radio-wave polarization in the 1920s and 1930s revealed the use of the magneto-ionic theory to engage empirical data. The theory originally emerged to account for the variation of short-wave propagating range with frequency. Yet the inclusion of the geomagnetic field introduced a critical side effect: a propagating radio wave split into two—ordinary and extraordinary—components with different speeds and polarizations. Thus any prediction from the theory involved the comparison and synthesis of the effects that the two distinct components created. Generalizing the theory led to a prediction of this kind on the senses of wave polarizations in different hemispheres. The empirical confirmation of this prediction not only marked the theory's success but also vindicated a type of reasoning regarding what scientists could do with the polarimetric data. They could understand the shape and sense of polarization in terms of the geometric relationship between the signal's transmitting direction, geomagnetic field, and latitude and the height that the radio waves reached in the ionosphere.

This consideration addressed only the polarimetric aspect of the theory, however. The complicated Appleton-Hartree formula connoted information about the speed of propagating waves and ultimately the electron-density profile of the ionosphere as well. How to engage data on radio propagation not

40. Appleton and Builder, "The ionosphere" (1933), 208–20.

only with the polarization formula (9.2′) but also with the refractive-index formula (9.2)? How to infer the ionosphere's electron concentration from echo sounding? To turn the general magneto-ionic theory into machinery for exploring the ionosphere, scientists needed to examine more closely the mathematical properties of the Appleton-Hartree formula.

## A MATHEMATICAL INQUIRY INTO THE GENERAL MAGNETO-IONIC THEORY

The success of the polarization experiments in 1927–33 encouraged radio scientists to explore further the implications of the general magneto-ionic theory. As the developers and first users of such a theory, the Britons were especially eager to look into the formal structure of the Appleton-Hartree formula. Researchers such as Appleton recognized this project's significance to radio ionospheric studies. But their experiments were too absorbing for them to dedicate time to analysis that required not only mathematical skills but also tedious numerical computations. They needed a theorist to crank numbers, plot diagrams, analyze curves' behavior, and supply quantitative predictions for further experiments. The task went to Appleton's former student Mary Taylor.

It is claimed that Mary Taylor was the first woman to take up the study of radio as a profession. She entered Girton College, Cambridge University, in the late 1910s to study mathematics and natural sciences, obtained a B.S. in 1920, and stayed at Girton to teach mathematics and pursued an M.S., which she received in 1924. During this period, she was introduced to radio physics by Appleton and conducted researched on vacuum tubes with him at Cavendish Laboratory. After graduating from Cambridge, she studied at the Mathematics Institute at the University of Göttingen, was awarded a Ph.D. in 1926 with a dissertation on magneto-ionic theory, and stayed in Göttingen for three more years as a research fellow. In 1929, Taylor returned to Britain to take a position as a scientific officer at the Radio Research Board's Slough Research Station. Taylor's scientific career was forced to stop in 1934, when she got married and the U.K. Civil Service Rules required her to resign.[41]

Taylor's work at Slough comprised theoretical analysis and numerical computation. In 1930, Appleton and the station's superintendent, Robert Watson Watt, asked her to perform mathematical analysis for the general magneto-

41. Haines, *International Women in Science* (2001), 308.

ionic theory.[42] Her task was to calculate the variations, with respect to several parameters, of the complex refractive index and polarization in equations (9.2) and (9.2′), as well as to give physical interpretations to her analytical and numerical results. She did not aim to develop a complete theory of radio-wave propagation to predict radio-wave intensity at any given distance or to predict the wave propagating range for a given wavelength, which required knowledge of the still immeasurable ion-density profiles. Neither did anyone ask her to predict the full wave behavior along the entire propagating path. Rather, her mathematical analysis was to make sense of some qualitative and quantitative features of the data from the Slough station's vertical ionospheric sounding, an apparatus that sent propagating radio waves vertically upward and detected their downward returns. The questions her theoretical work addressed included whether the right- or left-handed wave arrived at the receiver first and what changes of returning signals meant for the ionosphere. In other words, her mathematical endeavor related closely to experimental practice under a particular setup—the Slough station's vertical sounding equipment. And theory helped her to infer the ionosphere's physical state from the radio data. To her, the magneto-ionic was an instrument's theory.

With this goal in mind, Taylor simplified her problems by fixing the geomagnetic field to its value in Slough and by considering vertically downward waves. In so doing, she fixed the values of $\omega_g$, $\omega_L$, and $\omega_T$ and determined the refractive index and polarization in equations (9.2) and (9.2′) (for which she coined the name Appleton-Hartree formula) by three independent variables: radio frequency $\omega$, electron density $N$, and mean collision frequency $1/\tau$. At first, she considered the absorption-free case in which $1/\tau = 0$. The numbers $n$ and $E_x/E_y$ were therefore functions of $\omega$ (or wavelength $\lambda$) and $N$ only. She chose $N$ as a variable to follow the continuous variations of the refractive index and polarization with electron density that were similar to their supposed variations in the ionosphere. She calculated and plotted the dispersion curves of $n^2$ versus $N$ for numerous wavelengths from 5 meters to 18,000 meters ($n^2$ was real for zero absorption). These curves were useful for retrieving ionospheric properties from radio sounding. For instance, if the wave returned later, then the wave had slowed and the refractive index $n^2$ had increased. The plotted curves could indicate that such an observed change corresponded to an increase or decrease in the ionosphere's electron density.

42. Meeting Minutes, Committee on the Propagation of Waves and Directional Wireless, Radio Research Board, 14 April 1930, box "Radio Research Board Committee P. 2., Propagation of Waves, 1933–39," Historical Records, Rutherford Appleton Laboratory.

The curves that Taylor constructed were complicated—$n^2$ went to 0, $\infty$, and $-\infty$ at several wavelengths, and its change over $N$ differed considerably at different wavelengths as well as between ordinary and extraordinary waves (e.g., figure 9.6). She identified four classes of dispersion curves. Their behavior was sometimes strange—$n^2$ approached infinities from time to time and jumped immediately from one infinity to another—and the difference between ordinary and extraordinary waves was conspicuous in these cases. In general, waves shorter than 300 meters behaved strangely only with electron density much lower than its typical value in the ionosphere. This and other properties implied that ordinary and extraordinary waves differed most for wavelengths between 80 meters and 250 meters.[43]

Criticism ensued as soon as Taylor announced this work in 1931 to a few British radio researchers. The main objection consisted of certain "unphysical" features in the curves that she constructed. In these curves, the extraordinary wave's $n^2$ did not reduce to 1 with zero ionization $N = 0$ at the gyrofrequency $\omega_g$. Also, an abrupt change of $n^2$ occurred as the propagating direction deviated slightly from the geomagnetic field, as Builder found.[44] More important, the absorption-free assumption yielded unrealistic results such as abrupt jumps of $n^2$ from one infinity to another with an infinitesimal change of $N$. In fact, Appleton had known the shortcoming of the zero-absorption assumption; it was a problem even without the geomagnetic field. Larmor's absorption-free theory showed that $n^2$ in an ionic medium decreased with $N$ monotonically from 1 to $-\infty$ (cf. equation (6.1′)). Mathematically, a negative value of $n^2$ signified attenuation that prohibited a wave's propagation in the medium; but the physical ground of such attenuation was inconceivable, since the medium was absorption free. Rather than trying to find a physical interpretation for the negative $n^2$, Appleton contended, it was better to include a small friction in calculating the refractive index. Thus $n^2$ would decrease asymptotically to 0 instead of $-\infty$, and the unphysical behavior would disappear.[45]

Taylor knew the shortcoming of the zero-absorption assumption. After 1931, she focused on calculations with nonzero absorption. In 1932 and 1933, she steadily increased the collision frequency $1/\tau$ in her computation of $n^2$.[46] As she expected, the earlier infinities and discontinuities disappeared with

43. Taylor, "The Appleton-Hartree formula Part 1" (1933), 245–65.

44. Appleton and Builder, "The ionosphere" (1933), 217–18.

45. Appleton, "Note on dispersion problem," MS 2370, C 205, Edward Appleton Papers.

46. Taylor to Appleton (letters), 16 Feb. 1932 and 1 April 1933, Edward Appleton Papers.

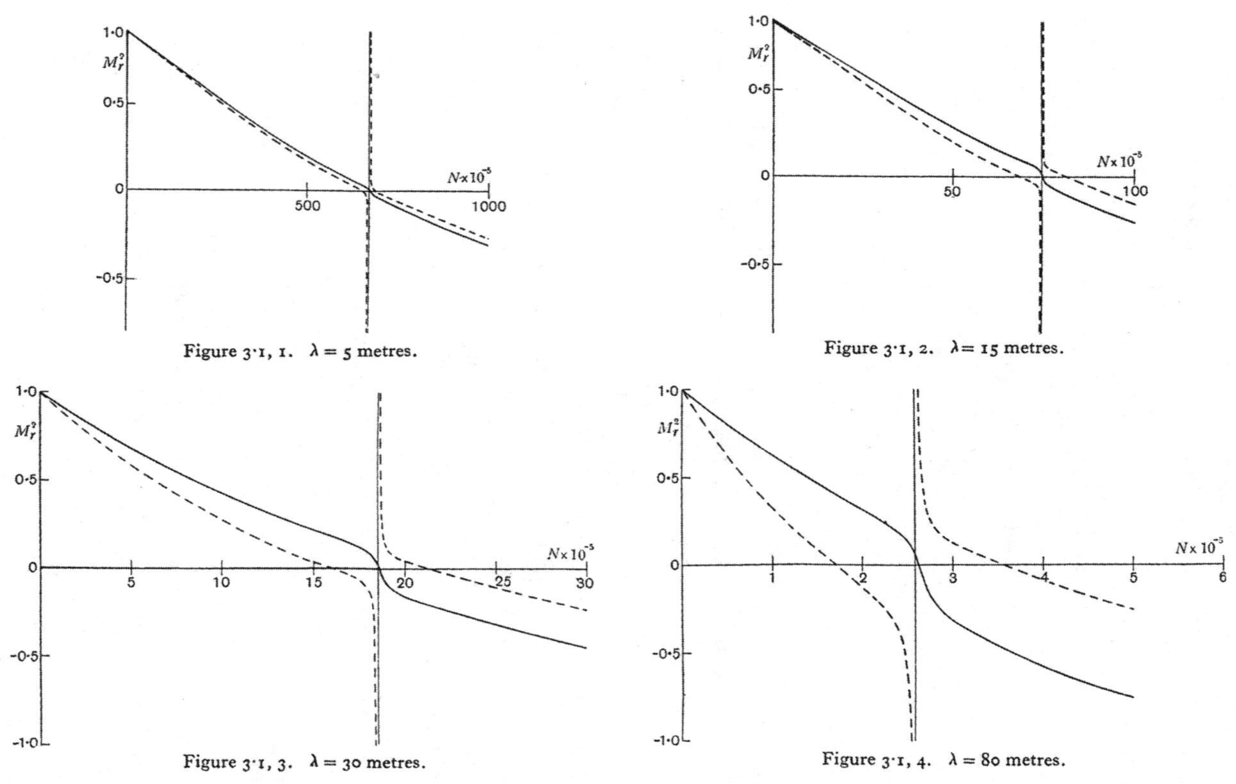


FIGURE 9.6. An example of Taylor's curves for refractive index vs. ion density. Taylor, "The Appleton-Hartree formula part 1" (1933), figures 3.1.1–4.

absorption. Also propagation in general was closer to that of waves normal to the geomagnetic field (transverse waves) when $1/\tau < \omega_T^2/\omega_L$ and to that of waves parallel to it (longitudinal waves) when $1/\tau > \omega_T^2/\omega_L$. Finally, she confirmed that in Slough a right-handed polarized wave attenuated more in general than a left-handed. [47]

Taylor's absorption-free curves might seem an unlucky excursion, but why did she choose to do it? Zero attenuation was clearly not a physical assumption for radio ionospheric propagation. Taylor could, for instance, have started constructing the curves by supposing a nonzero absorption coefficient $g$. But she did not do so until the problem with infinities came up. Appleton's advice then was equally intriguing: instead of plugging a more empirically adequate value to $g$ as a good experimenter should, he suggested values barely enough to remove the infinities. These treatments of the idealization $g = 0$ revealed how Slough's experimental program appropriated the general magneto-ionic theory through Taylor's work. Precise numerical predictions, testing the theory's efficacy, or iterative mutual correction between Appleton-Hartree and radio sounding data were not in her agenda. Instead, her task was to turn the theory into a paper tool that determined the *qualitative* trend of the ionosphere's electron density from specific features of radio sounding data, including delay times and polarizations of returning waves.

Mary Taylor's work epitomized expanding scholarship on the magneto-ionic theory. In the 1930s, radio scientists and ionospheric researchers, especially those in Cambridge, London, Slough, and Teddington, developed more and more comprehensive physical interpretations, analytical approximations, and numerical calculations for the Appleton-Hartree formula for refractive index and polarization. By the end of the decade, Ratcliffe proposed a nomenclature and a graphical representation that eventually became the standard language for the theory.[48] Based on Taylor's analysis, the scheme represented the variations of refractive index and attenuation with electron density in a Cartesian coordinate system, plotted separately the curves for ordinary waves and those for extraordinary waves, and drew the boundaries between longitudinal-like and transverse-like propagation. The diagrams and notations in Ratcliffe's 1959 monograph (e.g., figure 9.1) were exactly the culmination of the mathematical inquiries into the general magneto-ionic theory in the 1930s.

47. Taylor, "The Appleton-Hartree formula part 2" (1934), 408–19.

48. Ratcliffe, *The Magneto-Ionic Theory* (1959), 7–10; Mimno, "Physics" (1937), 21–27.

The decade after the "discovery" of the ionosphere witnessed substantial development of the magneto-ionic theory. Researchers generalized it to cases in which a radio wave could propagate along any direction with respect to the geomagnetic field. Such a generalization entailed predictions regarding the wave's polarization that field experiments in England and Australia vindicated. Moreover, efforts were under way to characterize the mathematical behavior of the general theory's central formula, the Appleton-Hartree formula. Much of this research concerned the theory's utility in bridging radio data and ionospheric inquiries. As a wave-propagation model, it differed both from the surface diffraction theory, which had focused on mathematical consistency, and from the atmospheric reflection theory, which aimed to explain some general wireless phenomena. Instead, the new theory evolved into a paper tool to assist radio sounding.

However, that does not mean that scientists in the 1920s and 1930s ignored the ontological status of the magneto-ionic theory. While some accepted the theory's use value and postponed questions about its factuality, others attempted to revise the theory to make it more real. The latter perhaps needed to add more terms to the Appleton-Hartree formula. Their basis was microphysics.

CHAPTER 10

# Handling Microphysics

The boom of short-wave radio, the discovery of the ionosphere, and sounding-echo experiments with the upper atmosphere helped consolidate the magneto-ionic theory. As a much more plausible model of radio-wave propagation to scientists and engineers of the 1920s, magneto-ionic refraction was distinct from surface diffraction and atmospheric reflection for its deeper reliance on wave-material interaction. In the two earlier theories, geophysical environments such as the earth and the Kennelly-Heaviside layer served primarily as boundary conditions to reflect or guide radio waves. By contrast, the magneto-ionic theory drew its explanatory power from the mutual influence between radio waves and the ionic medium in the upper sky. According to the latter theory, the propagation of radio waves above the earth depended on how the waves interacted with the ionosphere as they deflected *in* it. And the deflecting waves carried information about this wave-medium interaction that revealed the ionosphere's structure.

This implication endorsed the raison d'être of sounding-echo experimentation and made the magneto-ionic theory a powerful tool for radio probing of the ionosphere. But it opened a Pandora's box, too. The studies of electromagnetic fields in various media had been a central topic in physical optics and electrodynamics since the mid-nineteenth century, and it sat squarely at the center of fin-de-siècle physicists' preoccupation with the relationship between ether, field, and matter. All these prior explorations had called for a further understanding of the microscopic constitution of the materials through which electromagnetic fields traversed. Such "microphysics" turned out to be highly complicated and controversial.

At the end of the century, Dutch physicist Hendrik Antoon Lorentz synthesized a theory of electrons for electrodynamics, gas discharge, optics, and spectroscopy. In his model, a medium comprised many atoms, which contained negatively charged electrons. An electric or magnetic field in this medium drove the bounded atomic electrons into motion that in turn produced a new field. Adding the medium-excited secondary field thus modified the original field. To evaluate the medium's effect on the propagation of electromagnetic waves, Lorentz formulated the equations of motion for the electrons, solved the equations to obtain the electrons' positions, incorporated those positions into the displacement current in Maxwell's equations, and accordingly obtained a new wave equation for radiation.[1]

Although the development of atomic research and quantum mechanics after 1900 quickly overshadowed Lorentz's microscopic model, his electronic theory of electromagnetic-wave propagation survived, especially in studies of electromagnetic waves' "classical" behavior in dielectric materials. As Appleton, Eccles, Hartree, Larmor, Lassen, Nichols, Schelleng, and others were building the ionic-refraction theory for radio-wave propagation in the ionosphere, Lorentz's model became a convenient resource. These researchers borrowed the Dutch master's mathematical structure and physical interpretation but replaced his bounded atomic electrons in the dielectric medium with free electrons in the ionized plasma.

While such a microphysical approach enabled the magneto-ionic theory, it also exposed a gap of reasoning in its early versions. The scheme's core was a recursive interaction between field and matter. When one introduced an electromagnetic field into a material, it drove the electrons into mechanical motion. The electronic motion then generated an additional field. This secondary field in turn altered the electrons' motion, which produced another field, and so on. As a result, the electromagnetic field governing an individual electron's equation of motion should be the superposition of the original input field *and* the induced field *from all the other electrons* in the material. This was not in the early radio scientists' model, however. When Appleton, Eccles, Larmor, Lassen, Nichols, and Schelleng derived the refractive-index formulae—equations (6.1), (6.2), and (9.2)—for the ionic-refraction or magneto-ionic theory, they all assumed *free electrons* that were driven only by the input radio wave entering the ionized medium. They ignored the mutual interaction between ionic electrons in the form of a secondary field. The omission of

1. Lorentz, *The Theory of Electrons* (1952). The book appeared first in in 1909, and the author based it on his lecture notes.

the interaction between electrons might be computationally convenient, but if Lorentzian microphysics was serious, it had to include this interaction in order to form a more rigorous theoretical treatment.

The attempt to incorporate interelectron interaction into the magneto-ionic theory nonetheless triggered debates and controversies among radio ionospheric researchers. In the late 1920s and the 1930s, radio scientists and engineers posed two challenges to the microphysics of wave propagation: should they include a "Lorentz correction term" in the Appleton-Hartree formula for the ionosphere's effective refractive index, and was there a "quasi-elastic force" that influenced each electron's motion in the ionosphere? Both problems concerned how to incorporate the interaction between electrons. Both were difficult to resolve: the Lorentzian microphysical arguments were flexible enough to accommodate starkly different conclusions, while many researchers considered experiments either irrelevant or misleading. Nor did formal settlements arise during the period. The debate over the Lorentz correction term lost momentum in the late 1930s and did not come to life until after the war, whereas the quasi-elastic force faded away quietly. The handling of microphysics in the magneto-ionic theory of the 1920s and 1930s constituted one of the most perplexing and anti-climactic episodes in the history of radio-wave propagation.

The stories of the Lorentz correction term and the quasi-elastic force attest to different relationships between theory and experiment from that in radio ionospheric sounding. Here the microphysical modifications of the Appleton-Hartree formula and objections to them did not really translate into theoretical machinery that helped radio sounding experiments infer ionospheric properties. The debate over the Lorentz correction term was mainly theoretical. Whether to include that term in the magneto-ionic theory had little bearing on the reduction and interpretation of radio ionospheric data in the 1920s and 1930s. Although radio scientists proposed several ionospheric sounding experiments to test the theory, the data were ambiguous or imprecise and led them either to opposite conclusions or to agnostic reservation. The fact is, empirical testing of the Lorentz correction required more information than radio sounding could offer. All the frequency change, pulse-echo, and polarimetric instruments measured the qualitative change of the ionosphere's electron density, not its absolute value. Yet the inclusion of the Lorentz correction term would not affect such qualitative attributes—under the same condition, for instance, the electron density increased (or decreased) with the radio wave's delay time regardless of the correction term. Thus radio sounding in the period did not provide experimental evidence to settle the hypothesis.

Unlike the Lorentz correction term, the debate over the quasi-elastic force originated from experimentation; but the empirical testing was distinct from radio ionospheric sounding, which sent and detected probing signals in open space. Instead, it came from the tradition of tabletop "mimic experiments" inside laboratories. Such experiments aimed at reproducing a natural phenomenon in miniature using lab devices. It appeared attractive for its promise of creating a morphologically similar form of nature, as well as its potential to supplement observational science with active control and manipulation. Throughout the nineteenth and early twentieth centuries, researchers had used mimic experiments to study a variety of astronomical, geological, and meteorological effects, ranging from cloud formation and continental drift to the solar corona and comet tails.[2] As ionic refraction became a popular model for long-range radio-wave propagation, some scientists also began to simulate the behavior of radio waves in the ionosphere via electromagnetic or optical experiments with ionized gas tubes.

In the mid-1920s, a group of French physicists in Nancy observed a peculiar effect when they measured an ionized gas's permittivity, which they identified as the impact of a certain "quasi-elastic force" on the free electrons of any ionized medium, including the ionosphere. This unexpected ionic force could profoundly affect radio-wave propagation—for example, adding resonance frequencies. For a while, the hypothesis attracted many radio ionospheric researchers, but they remained wary. Like skeptics concerning mimic experiments in other areas, they worried that the results of tabletop trials reflected the instrumental setup, not the actual effect in nature.

The stories of the Lorentz correction term and the quasi-elastic force remind us of the openness of the magneto-ionic theory. Although the theory had become a paper tool for radio probing of the ionosphere and was shaping sounding-echo instruments by the mid-1920s, it was not yet totally packed in a "black box." An instrument's theory did not have to be unproblematic, stable (and perhaps boring) machinery to read experimental data and crank out predictions. It could turn problematic, stimulate new questions, invite revisions, and even open a new research frontier. In the case of the magneto-ionic theory of the 1920s and 1930s, microphysics provided the seed of constructive

2. For emulating cloud formation and the associated development of the cloud chamber, see Galison and Assmus, "Artificial clouds" (1989), and Galison, *Image and Logic* (1997), chap. 2. For emulating the mechanism of continental drift, see Oreskes, *The Rejection of Continental Drift* (1999), chap. 4. For mimic solar corona, see Schaffer, "Where experiments end" (1995), 257–99. For mimic comet tails, see Hedenus, *Der Komet in der Entladungsröhre* (2007).

instability. The further investigation into the microscopic basis of radio-wave propagation in ionized media inspired scientists to pay closer attention to interelectron interaction and hence to contemplate the possibility of modifying the original theory. Here wave propagation was not just a backstage crew supporting ionospheric study; it stood in the spotlight.

## THE LORENTZ CORRECTION

In Lorentz's theory of electrons, the actual force on a point with unit charge in a material comprised not only the external electric field but also a secondary field from the surrounding particles as the external field excited them. For a dielectric material, the secondary field was the sum of the neighboring atoms' induced dipole moments. The additional term modifying the actual electric force in a material—the Lorentz correction—raised controversy as the theory expanded from dielectric to ionic media. In the 1920s and 1930s, physicists debated whether to incorporate the Lorentz correction in the ionic refraction theory of radio-wave propagation. Consensus was difficult to reach among both theorists and experimentalists. The story of this controversy reveals the complexity of incorporating microphysics into radio ionospheric research.

### *Origin of the Concept*

The Lorentz correction term stemmed from a revised definition of electric field in media. In the original definition that Faraday gave in the 1850s, electric field at a point A in space was the electric force that a point mass with unit electric charge would receive if one placed it at A.[3] In principle, this stipulation should hold in vacuum as well as in materials. But with the introduction of microphysics into electrodynamics, Faraday's definition yielded an electric field that might change spatially with the atomic scale, which was too fine for most problems in classical physics. When Lorentz developed his theory of electrons in 1892, he redefined the electric field at a point A in a material (comprising atoms) as the *average* of the electric force over all points within a volume surrounding A. He offered this redefinition to smooth the electric force's fine-scale variations resulting from the local atomic constellation surrounding A. Thus the volume should be large enough to contain many atoms, yet not too large to eliminate the innate spatial variation of the external force (when the electric force was a wave, this volume should be much smaller than

3. Baigrie, *Electricity and Magnetism* (2007), 88–89.

the cube of the wavelength). The electric field at A according to this definition equaled the actual electric force at A if the material was a vacuum or continuum.[4]

Yet there was a systematic difference between the actual electric force and the electric field in a material comprising discrete atoms. In a dielectric material of this type, an external electric field drove the electrons bound to atoms into constrained motions, changed the relative positions of electrons within atoms, and induced atomic dipole moments. Each induced dipole moment created a tiny electric force. So the actual electric force at a point in a dielectric material was the external electric field plus the induced atomic dipole force.

The computation of the actual electric force was particularly challenging for contributions from the region extremely close to A. To calculate the actual electric force at A in a dielectric material, Lorentz borrowed William Thompson's technique to scoop out a spherical volume S surrounding A: the actual electric force at A was the sum of the external electric field, the induced force from discrete dipoles within S, and the induced force from discrete dipoles exterior to S. Since S was comparable to the volume used in defining the electric field, the atoms exterior to S were all far away from A (in atomic scale), implying that the force of their discrete dipole moments on A approximated the dipole force from a continuum. This continuous dipole force equaled the continuous dipole force from the whole space *minus* that from S. Therefore the actual electric force $F$ at A was the external electric field $E^{(i)}$, plus the continuous dipole force $E^{(0)}$ from the entire space, minus the continuous dipole force $E'$ from S, plus the discrete dipole force $E''$ from S: $F = E^{(i)}+E^{(0)}-E'+E''$. Moreover, the sum $E^{(i)}+E^{(0)}$ was the actual electric force at A when the material was a continuum, and this equaled the defined electric field $E$ at A. Thus $F = E-E'+E''$ (figure 10.1). The actual electric force differed from the electric field by $-E'+E''$.[5]

The values of $E'$ and $E''$ were expressible in simple forms. Finding the dipole force $E'$ from a continuum within sphere S was a classical problem in electrostatics; its well-known solution was $E' = -P/3$, where $P$ was the dipole-moment polarization at $P$ (as a field, $P$ was defined in terms of averaging over finer scales, too). The dipole force $E''$ from discrete atoms within S was more difficult to calculate, since it depended on atoms' locations in S. Lorentz found that it was zero when the atomic constellation was cubical grids and thereby argued that it was small ($E'' = sP$, where $s$ was a small

4. Lorentz, *The Theory of Electrons* (1952), 132–34.

5. Ibid., 137–38.

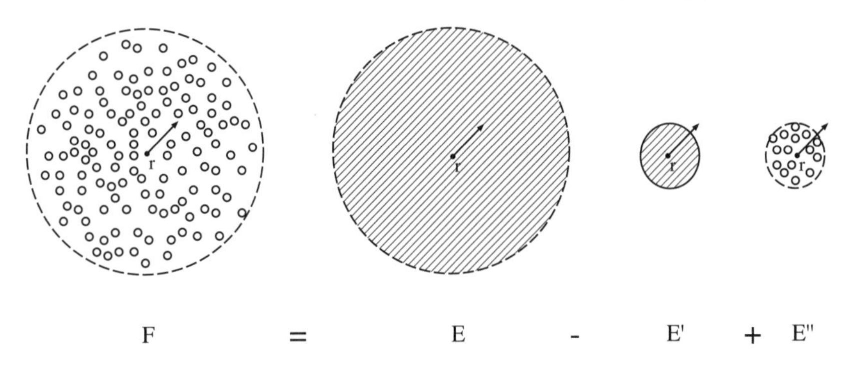


FIGURE 10.1. Geometry in Lorentz's technique of electric-force calculation.

number) when the material was isotropic in general, such as a fluid, gas, or glass. In other words, $F = E+(1/3+s)P \cong E+P/3$.[6] The term $P/3$ corresponded to the Lorentz correction term.

### *Douglas Hartree and the Lorentz Correction in Ionic Refraction*

Lorentz's correction modified the actual electric force in a dielectric material. The equation of motion for an electron, he argued, therefore needed revision: instead of $md^2x/dt^2 = eE - \kappa x$ ($\kappa x$ was the elastic constrained force that the atom employed on the electron), it should be $md^2x/dt^2 = e(E+P/3) - \kappa x$. The expression for the dielectric constant in terms of electronic force also had to change, as one substituted the solution to the new equation of motion into Maxwell's equations to obtain a different relationship between $P$ and $E$. In other words, the Lorentz correction revised the microscopic theoretical prediction for the dielectric constant.

Should we expect the same correction for ionic media? Eccles, Larmor, Nichols, and Schelleng did not ask this question when they developed the ionic-refraction theory. To them, the Lorentz correction did not exist; they simply took $F = E$. Not until the late 1920s did the English mathematician Douglas Hartree raise the issue. A native of Cambridge, Douglas Rayner Hartree entered St John's College in 1915. The British army drafted him during World War I. He returned to Cambridge thereafter, completed his PhD in mathematics in 1926, and took a chair at the University of Manchester. Although Hartree was also a product of Cambridge pedagogy in mathematical physics, he looked at the world differently from wranglers of the previ-

6. Ibid., 138–39, 305–8 (notes 54, 55).

ous generation. Growing up in a period when Maxwellian electrodynamics gave way to the new quantum mechanics and relativity, Hartree did not fear embracing microphysics. Niels Bohr's 1921 lectures on quantum physics at Cambridge influenced him: he tackled various problems in wave-matter interactions, including crystal X-ray diffraction and atomic wave mechanics, and thus explored techniques of solving differential-integral equations, including automatic computing.[7]

Hartree began to work on radio in the mid-1920s when he served as a consultant to the Radio Research Board. A friend of Appleton's and Taylor's, he was familiar with the recent theories of radio-wave propagation in the ionosphere.[8] In general, such theories were similar to theories of light in crystalline materials, a topic he had encountered earlier in his work on X-ray diffraction. Thinking that his optical approach might offer a different perspective from radio scientists', he examined the problems of ionic refraction in 1928.

Hartree's method closely followed C. G. Darwin's theory that dealt with light in material. Grandson of the great evolutionist and son of a prominent astronomer, Charles Galton Darwin was himself a renowned physicist. He made contributions to quantum mechanics (suggesting spin's existence), statistical mechanics (revising the concept of partition function), and optics (crystalline X-ray diffraction).[9] When Darwin worked on X-ray crystallography at the University of Manchester in the 1910s, he found from the diffraction patterns that the rays were a collective effect of scattering waves from all atoms in a crystal. From this observation, he developed a microscopic theory of wave-matter interaction in 1924.[10] Without directly appealing to Maxwell's equations, Darwin formulated a propagating wave as the sum of the incident wave and the wavelets from all the atoms in the material. An atom irradiated by the propagating wave yielded a minuscule oscillating dipole moment that in turn radiated a wavelet. Assuming a linear relationship between the incident electric field and the induced atomic dipole moment, he obtained each wavelet's Hertzian potential and summed all of them to get the macroscopic field.

Darwin's method inspired Hartree's treatment of radio ionospheric propagation. In a paper that he read to the Cambridge Philosophical Society in October 1928, Hartree argued that the Darwinian approach led to the Lorentz

7. For the life of Hartree, see Darwin, "Hartree" (1958), 103–16.

8. Hartree to Appleton (letters), MS 2370, C177, C218, Edward Appleton Papers.

9. For C. G. Darwin's life, see Thompson, "Darwin" (1963), 69–85.

10. Darwin, "The optical constants of matter" (1923–28), 137–67.

correction term in ionic refraction.[11] His goal was to find the equations for electromagnetic waves propagating in a stratified medium that modeled the ionosphere. To do so with the microscopic method, he first had to obtain the expression for the wavelet radiating from the induced dipole moment of an individual corpuscule.

The dipole moment in Hartree's ionized gas nonetheless had a different meaning from that in Lorentz's dielectric material. While bound electrons in a dielectric material formed static dipoles with the atoms' nuclei, free electrons in an ionized gas did not form static dipoles with positive ions when an electric field was applied. Yet free electrons from a time-variant electric field exhibited oscillatory motions that one could understand as "oscillatory dipole moments," which were sources of tiny amounts of electromagnetic radiation. The process whereby an incident wave induced an oscillatory motion on a free electron, which in turn radiated a wavelet, was equivalent to scattering. So the induced oscillatory dipole moment was a "scattering moment."

Hartree assumed that the scattering moment was proportional to the actual electric force. When the actual electric force on a point with unit charge at location $r_0$ was $F(r_0)$, the induced scattering moment $dP$ in a tiny volume $dV_0$ around $r_0$ was $dP = \sigma_s \cdot F(r_0)dV_0$ ($\sigma_s$ was the scattering coefficient). The Hertzian potential of the wavelet that $dP$ excited at $r$ was $d\Pi = [\sigma_s \cdot F(r_0)] \cdot \exp(-ik|r-r_0|)/(4\pi|r-r_0|) \cdot dV_0$. In a continuum, the overall Hertzian potential of the propagating wave was the integral of $d\Pi$ over the entire volume.

Hartree's focus was rather on an ionized gas with discrete free electrons than on a continuum, however. To calculate the induced wave in the discretized material, he used Lorentz's technique in evaluating the induced static field in a dielectric medium. He scooped out a sphere S around the point of observation $r$ and considered separately the field contributions from inside and outside the sphere. The actual electric force $F$ at $r$ equaled the incident wave's electric field $E^{(i)}$, plus the induced electric field $E^{(0)}$ from the entire volume of the continuum approximating the ionized gas, minus the induced electric field $E'$ from the continuum within S, and plus the induced electric field $E''$ from the discrete free electrons within S; i.e., $F = E^{(i)} + E^{(0)} - E' + E''$. Hartree took $E^{(i)}$ as given and obtained $E^{(0)}$ from its associated Hertzian potential $\Pi^{(0)}$ by integrating $d\Pi$ over the whole volume. To evaluate $E'$, he argued that a much shorter radius of S than the wavelength permitted a quasi-static approximation in which the induced scattering moment in S was in a

11. Hartree, "Stratified medium" (1929), 97–120.

form similar to the induced static dipole moment in S. Following Lorentz, he obtained $E' = -(1/3)\sigma_s \cdot F$. The quantity $E''$ was difficult to solve analytically. Hartree did not bother discussing it, except borrowing Lorentz's assumption that $E'' \cong 0$.

Combining the expressions for $E^{(i)}$, $E^{(0)}$, $E'$, and $E''$, Hartree obtained an expression and hence a wave equation for $F$. This equation was not the wave equation for the electric field in a stratified medium deduced from Maxwell's equations via the macroscopic approach. Another field $L = F-(1/3)\sigma_s \cdot F$ had a wave equation identical to the macroscopic equation. Therefore, he contended, $L$ should be the canonically defined electric field $E$; that is, $E = F-(1/3)\sigma_s \cdot F$. From the definition of $\sigma_s$, $\sigma_s \cdot F$ was the (oscillating) dipole-moment polarization $P$. This led to $F = E+P/3$, the same as Lorentz's conclusion. This conclusion, along with the formula $P = \sigma_s \cdot F$ and the constitutive relation $D = \varepsilon_0 E+P = n^2 \varepsilon E$ ($n$ was refractive index), rendered the refractive index in terms of the scattering coefficient: $n^2 = 1+\sigma_s/(1-\sigma_s/3)$.

Hartree's work in 1928 implied that the Lorentz correction term existed in an ionic medium. Moreover, this term predicted a macroscopic physical effect. The medium's refractive index was $n^2 = 1+\sigma_s/(1-\sigma_s/3)$ with the Lorentz correction and $n^2 = 1+\sigma_s$ without it. The current ionic refraction theory needed revision via incorporation of the Lorentz correction term into the refractive-index formula.

Hartree carried through the idea further. In 1930, he extended his previous theory by looking at wave propagation in a magnetized ionic medium.[12] With an external magnetic field, the induced scattering moment in an ionic medium was different along different directions, and the scattering coefficient $\sigma_s$ became a tensor. Hartree evaluated the scattering tensor from an electron's induced oscillating dipole moment by solving the electron's equation of motion in a magnetic field. The refractive index that he obtained from the solution of the scattering tensor was identical to Appleton and Altar's formula in equation (9.1)—he was thus the fifth person to develop the generalized magneto-ionic theory.

As for the Lorentz correction, he now considered more carefully the value of $E''$, the electric field induced by free electrons within the sphere S. He still believed $E'' \cong 0$ in an ionized gas, for a medium of this kind was isotropic. But he reserved the possibility of $E'' \neq 0$ and expressed $-E'+E''$ with $(1/3+s)P$, not $P/3$, although $s$ was still small.

12. Hartree, "Magnetic field" (1931), 143–62.

## *A Microphysical Controversy without Microphysics*

Hartree's argument provoked controversy. In July 1933, physical chemist Lewi Tonks at General Electric's Schenectady Research Laboratory in New York State wrote to *Nature*. Tonks contended that the Lorentz correction as "a 'polarisation' force could only exist when there was some detailed arrangement of the negative with respect to the positive charges, so that a volume element of the polarized medium was distinguishable from a volume element of the unpolarised medium."[13] Only well-defined dipole moments yielded the Lorentz correction; such moments did not exist in an ionic medium.

Tonks's critique was off the mark, since the "dipole moments" in Hartree's theory were oscillatory motions of free electrons that produced radiation, not separations between positive and negative charges that generated a static electric field. But a new critique followed quickly. In October, Kenneth A. Norton from the U.S. National Bureau of Standards wrote to *Nature* suggesting another possible shortcoming.[14] Norton criticized Hartree's estimate of $E''(r)$, the electric field at $r$ induced by the free electrons in sphere S, by offering a calculation of his own. Because the actual value of $E''(r)$ was difficult to obtain, he calculated the spatial average of $E''(r)$—i.e., the integral of $E''(r)$ over the space within S divided by the volume of S. The average value of $E''(r)$ turned out to be $-P(r)/3$, which cancelled $E'(r) = P(r)/3$, the electric field induced in a continuum within S. Consequently, the spatially averaged electric force at $r$ was $F = E^{(i)} + E^{(0)}$, equaling the canonically defined electric field at $r$. The Lorentz correction did not fit.

Norton's critique had a problem, too. The role of the Lorentz correction was precisely to distinguish between the actual electric force and the spatially averaged electric force. That Norton averaged all the field quantities indicated his confusion about this point. His equation $F = E^{(i)} + E^{(0)}$ was a tautology—that the spatially averaged electric force equaled itself. It did not imply identity between the *actual* electric force and the spatially averaged electric force, a necessary condition for leaving out the Lorentz correction.

Yet a third critique came immediately. Tonks wrote to *Nature* again in November.[15] He no longer held that the lack of genuine dipole moments in an ionized gas was a problem. Rather, he pointed out that Hartree's calculations did

13. Tonks, "Ionisation density" (July 1933), 101.
14. Norton, "Ionisation" (1933), 676.
15. Tonks, "Ionisation density" (Nov. 1933), 710–11.

not include the effect of positive ions. A neutral ionized gas consisted of free electrons and positive ions with the same amount of total charges. Hartree calculated only the electric force that electrons induced. His results were partial.

Tonks proposed a theory to include positive ions' effect. Positive ions in an ionized gas were protons or larger nuclei, much heavier and less mobile than electrons. The fixed positive ions created a potential field that affected the free electrons. When an electron moved away from its equilibrium position, which the potential determined, the positive ions' total Coulomb force tended to pull it back to its equilibrium position and hence amounted to a net "force of restoration." Consequently, the actual force on a free electron should be $eF = e(E^{(i)}+E^{(0)}-E'+E''+E_+)$, where $E_+$ was positive ions' net Coulomb force, a force of restoration.

The force $E_+$ was a localized phenomenon—only positive ions near the electron contributed significantly to its value; the Coulomb force of other ions fell quickly with the square of distance. So only the contributions from those positive ions in the sphere S needed consideration in the calculation of $E$. Tonks simplified the problem by replacing the discrete ions (with number density $N$) with a continuum of uniformly distributed positive charge (with charge density $Ne$) in S. From the quasi-static approximation, the force on an electron at distance $r$ from the center of the sphere with uniform charge density $Ne$ was $f_+ = -Ne^2r/3$, and the displaced electron created a polarization $P = Ner$. So the force of restoration on a unit charge at $r$ was $E_+ = f_+/e = -P/3$. This value of $E_+$ canceled $E'$, implying that $F = E^{(i)}+E^{(0)}$ (since $E'' = 0$). Therefore positive ions' effect balanced the Lorentz correction; the term $P/3$ did not belong there.

Hartree understood the significance of Tonks's second critique. In December, he wrote to *Nature* to address the issue of positive ions.[16] He agreed that his previous calculations were incorrect because they did not consider the Coulomb force of positive ions. But he disagreed with Tonks's evaluation of positive ions' effect. The problem, he held, was Tonks's assumption equating a constellation of discrete positive ions with a continuum with uniform charge distribution. The two conditions yielded different forces of restoration. In the continuum, it was indeed $E_+ = -P/3$. Yet in a body of discrete positive ions, Hartree suggested without comprehensive reasoning, $E_+$ was still zero because the electron did not overlap in space with the positive ions and the pulling forces from randomly locating positive ions canceled each other. Hence the Lorentz correction should stay in.

16. Hartree, "The dispersion formula" (1933), 929–30.

Up to this point, the controversy between Hartree, Norton, and Tonks exhibited an interestingly contradictory feature. Although their exchange concerned a microphysical modification of a wave-propagation theory, they tried their best to *avoid* detailed microphysical computations and considerations. Instead, they endeavored to find proper approximations to replace problems involving discrete atoms with classical scenarios that were straightforward to handle: they approximated the particle-particle interactions far away (outside sphere S) with the effect of a continuum. Within S, such interactions either resembled another continuous effect or were zero for the seemingly isotropic, random microscopic constellation. The scientists did not discuss the implications of the atomic structure for scattering wavelets, justify the isotropic assumption, or treat the interatomic constellation with more rigorous techniques from statistical mechanics (all were the hallmarks of microphysics at the time). They undertook a microphysical controversy without really engaging the analytical tools of microphysics.

But an exception came up soon. As the author of Hartree's microscopic approach to wave propagation, Darwin had found the Lorentz correction problem fascinating and had been working on a consistent theory since early 1933. At the end of the year, he publicly responded to Hartree's remark.[17] Darwin agreed with Hartree that Tonks's replacement of discrete positive ions with a continuum of positive charge was problematic. Yet he proposed a different method to calculate the scattering moment. Rather than following Lorentz's approach of creating a fictitious cavity around every electron in a material, he found it easier to take a small isolated sphere consisting of the material under consideration and to calculate the waves that the sphere scattered. He showed that this approach would give the same results as Lorentz's. Using the new approach to calculate scattering moment, he found that the Lorentz correction was zero in an ionized gas comprising free electrons and fixed positive ions if every electron experienced many collisions during the wave's oscillating period. Tonks was right in his conclusion, but wrong in reaching it.

Darwin's reasoning proceeded as follows.[18] Aiming to resolve the debate over the Lorentz correction, he pointed out that the issue was to select between two formulae describing the relationship between the scattering coefficient and the refractive index of a given medium—the so-called Sellmeyer formula $\sigma_s = n^2-1$ (after German physicist W. Sellmeyer, for his optical work in 1870) when $F = E$ and the Lorentz formula $\sigma_s = 3(n^2-1)/(n^2+2)$ when

17. Darwin, "Refraction" (1934), 62.

18. Darwin, "The refractive index" (1934), 17–46.

$F = E+P/3$; the latter was equivalent to $n^2 = 1+\sigma_s/(1-\sigma_s/3)$, a result that Hartree had obtained. The selection was difficult, for "it is too easy to find arguments, quite convincing as many of those always accepted in theoretical physics, which lead to either of the two contradictory formulae."[19] To prevent this shortcoming, his approach was to start with fundamentals—viz., the equations of motions for *all* electrons in the system, not the equation of motion for a representative individual electron.

Doing this created a problem. It was impossible to consider a finite piece in a material as a self-contained dynamic system, since each electron's equation of motion contained the radiating force from other electrons retarded by the finite speed of wave propagation; a retarding force of this kind prevented a Hamiltonian formulation and hence a dynamical solution. Thus Darwin considered a sphere S of an ionized gas with its radius much shorter than wavelength (so that the retarding effect was negligible) but much larger than the atomic dimension (so that electrons and ions interacted enough). Both the refractive index and the scattering coefficient of the ionized gas in S were expressible with the scattering moment of S. Thus calculating the scattering moment of S determined the relationship between the refractive index and the scattering coefficient.

Tonks et al. had shown the difficulty in dealing with a sphere of ionized gas that contained an astronomical number of electrons and positive ions. Like them, Darwin simplified the issue by first considering a model that embedded free electrons in a continuum of positive charge. From electrostatics, Darwin obtained the Coulomb force that the positive continuum exerted on each electron in S. Then he formulated the electron's equation of motion, which comprised the force from the continuum, the Coulomb forces from all other electrons, and the force from the applied electric field. Summarizing for all electrons in S, Darwin could cancel the electron-electron forces and obtain the overall equation of motion for the sphere's scattering moment $P = e\Sigma_f x_f$ ($x_f$ was the distance between an electron's location and the center of S). He then substituted the solution into the scattering coefficient $\sigma_s$ and refractive index $n$. Darwin's first result confirmed Tonks's previous conclusion: the relationship between the two quantities of an electron-continuum gas followed the Sellmeyer formula $\sigma_s = n^2-1$; the Lorentz correction did not belong in this kind of medium.

The real question, however, involved the relationship between the scattering coefficient and the refractive index of an ionized gas with free electrons

19. Ibid., 18.

and *discrete* positive ions, not an electron-continuum gas. The total equation of motion for such a medium turned out to be

$$(10.1) \quad m\sum_{f}\ddot{x}_{f}=\sum_{f}\sum_{p}\frac{\partial}{\partial x_{f}}\frac{e^{2}}{\left|x_{f}-X_{p}\right|}+n_{e}eE\sin(\omega t),$$

where $X_p$ was the location of positive ion $p$, the first term corresponded to the forces of positive ions, and the second term corresponded to the incident field. Darwin found that equation (10.1) was much more difficult to solve. It was almost impossible to obtain its analytical solution. The only hope, he contended, was to obtain the average solution of $P = e\Sigma_f x_f$. To reach that, he considered two averages of $P$. The first he took over time, and the second over all possible initial conditions for electrons' positions and momenta. A leading expert on statistical mechanics, Darwin recognized that this step was identical to the average over the entire Gibbsian canonical ensemble with a fixed temperature. He then transformed this ensemble average over initial conditions into an integral over electrons' positions $x_f$ and momenta $p_f$ at the current time $t$.

Darwin found that the difference between the scattering moment $P$ for the electron-continuum gas and the averaged scattering moment $<P>$ from equation (10.1) lay in the integral of $<P>$ over specific regions. The regions in the phase space $\{x_f, p_f\}$ that contributed to this difference were only the small neighborhoods around the positive ions—i.e., the regions in which $x_f \cong X_p$. In these small "regions of collisions," positive ions drastically changed the momenta of free electrons. Darwin discovered that the contributions from all the regions of collisions to the phase integral were negligible when (i) the regions did not overlap and occupied only a small fraction of the gas's entire volume, (ii) the time an electron took to traverse such a region (the duration of a collision) was much shorter than the external field's period of oscillation, and (iii) an electron entering such a region had almost the same speed as it had infinitely far away from the region (i.e., the region was sufficiently large). Without the contributions from the regions of collisions, therefore, the results from the electron-continuum model also applied to ionized gases. The scattering coefficient and refractive index followed the Sellmeyer formula; the Lorentz correction was not necessary.

The three conditions underlying Darwin's argument against the Lorentz correction led to the requirement that an electron experience many collisions during the external field's period of oscillation. This requirement, implying that electrons in such a medium were always pretty close to dense positive ions, was true for metals, but questionable for the ionosphere. After receiving Darwin's earlier, unpublished version of the argument against the Lorentz correction, Hartree wrote to Appleton: "I think he [Darwin] is right, as far as I can follow him, but am not satisfied that his argument applied to the Heaviside layer conditions."[20] He believed that the low ion density in the ionosphere might invalidate Darwin's theory. Developing a Lorentz-correction theory proper for thin ionized gases, however, needed "extremely careful thinking." Hartree lamented that if he "could go and think about nothing else, then [he] might get something toward it." Unfortunately, he could not. His teaching at the University of Manchester and other research problems consumed him. He did not publish further on the Lorentz correction in an ionic medium.

### *Where Was the Empirical Evidence?*

Hartree's doubts about the applicability of Darwin's theory to the ionosphere made sense. But given Darwin's poignant observation that it was too easy to find reasonable arguments to support both Sellmeyer and Lorentz's formulae, a theoretical settlement appeared unlikely. The direct route, it seemed, lay through radio ionospheric experimentation. Where was the empirical evidence underlying the controversy? How much observable and measurable difference did inclusion of the Lorentz correction term incur in radio ionospheric propagation? Was there a "crucial experiment" to select between the two theories? How did the experimental radio scientists react to the dispute?

The radio ionospheric researchers kept an inconspicuous but notable interest in the issue. Appleton was one of those who accepted Hartree's work on the Lorentz correction. In his correspondence with Altar in 1925 and 1926, Appleton (like others) equated the actual electric force in the electronic equation of motion to the radio wave's electric field. He was still ambiguous about including or not including the term in his URSI paper in 1927.[21] Yet his 1932 paper formally incorporated the Lorentz correction.[22] He replaced the electric field $\overline{E}$ in the equation of motion for an electron with $\overline{E}+\overline{P}/3$ ($\overline{P}$ was polarization) and obtained the same form as the Appleton-Hartree formula in (9.1), except that $\alpha = -\omega^2/\omega_0{}^2-1/3$, not $-\omega^2/\omega_0{}^2$. Throughout the 1930s, Appleton remained convinced of Hartree's theory.

20. Hartree to Appleton (letter), 4 April 1933, MS 2370, C 218, Edward Appleton Papers.
21. Appleton, "The earth's magnetic field" (1927), 2.
22. Appleton, "Wireless studies" (1932), 642–50.

Did this modification of the magneto-ionic theory yield any testable differences? In 1933–34, Ratcliffe and Zenneck's Munich student Georg Goubau independently proposed a radio sounding experiment to determine between Sellmeyer and Lorentz. With the inclusion of the Lorentz correction term, they observed, the Appleton-Hartree formula predicted a very large refractive index—meaning a long delay for returning echoes—for the extraordinary wave at some frequency (the so-called Lorentz frequency) lower than the geomagnetic gyrofrequency. So an observation of significantly long delays in radio sounding records at the sub-gyrofrequency region would validate Lorentz's theory. With the invention of scan-frequency ionospheric sounders, radio scientists could eventually try this test.

In 1938, Lloyd Berkner from the Carnegie Institution of Washington, Henry Booker from Cavendish Laboratory at Cambridge University, and David Martyn and G. H. Munro from the University of Sydney identified unusually long delays in sounding records. The Americans, with the endorsement of Appleton's team, took this as clear corroboration of the Lorentz theory; the Australians, in contrast, remained skeptical. According to Martyn and Munro, Berkner and Booker's estimate of the Lorentz frequency was contingent upon an ad hoc hypothesis about heavy ions in the ionosphere; their account did not fully consider the effect of absorption at various frequencies; and the waves with long delays did not exhibit extraordinary polarization in measurement. Instead of resonance at the Lorentz frequency, the Australians claimed, the long delays actually corresponded to returning signals from unusually high boundaries in the F layer. Berkner and Martyn engaged in a stimulating correspondence when the former visited Western Australia. But none could persuade the others.[23]

Not long afterwards, Ratcliffe considered another empirical test for the Lorentz correction. Since the magneto-ionic interpretation was too ambivalent to settle the debate, he now focused on ionic refraction without geomagnetism. In 1939, he began to investigate the possible effect of the Lorentz correction on vertical and oblique sounding. A quick calculation indicated that the correction did not create the notable difference in the curves of virtual height vs. frequency that vertical sounding indicated. By contrast, oblique sounding raised some hope. According to Ratcliffe's theoretical estimate, the Sellmeyer and the Lorentz formulae led to curves of propagating range vs. limiting frequency (the highest frequency that could reach the specified

23. Booker and Berkner, "Constitution" (1938), 562–63; Martyn and Munro, "The Lorentz 'polarization' correction" (1938), 1159–60; Home, "To Watherloo and back" (1994), 157.

range) that differed from each other by as much as one percent. In principle, this difference could be detected by measuring the limiting frequency.[24]

In principle, but not in practice. The radio sounding experiments with the new equipment of Britain's Radio Research Board, Ratcliffe reported, could achieve an accuracy of 0.5 percent for frequency. This resolution was far from sufficient for the empirical differentiation between the two theories (which had a maximum difference of only one percent). Worse, geomagnetism would easily produce effects as large as this one-percent difference due to the Lorentz correction. In the end, Ratcliffe concluded that it was "impossible to settle the problem of the Lorentz term by comparing vertical-incidence and oblique-incidence propagation through region F."[25]

The debate over the Lorentz correction term complicated the epistemic status of the magneto-ionic theory. The highly theoretical origin of this debate indicated that the canonical doctrine of radio-wave propagation—albeit a paper tool for ionospheric sounding now—was not docile, inert, and closed. Scientists in the 1920s and 1930s were investigating various microphysical assumptions and implications underlying the Appleton-Hartree formula that did not relate directly to its utility in sounding. Although they restricted themselves to the classical Lorentzian scenarios and their analysis rarely engaged quantum and statistical physics (the "serious" microphysics at the time), their curiosity about the theory's ontological basis was unquestionable.

But the maintenance of such a research interest still had to rely on radio ionospheric sounding. The theoretical deliberations on the Lorentz correction were too ambivalent to settle the dispute. To scientists of the 1930s, the only hope lay in radio experimentation. If successful, experiments of this kind could not only help select between Sellmeyer and Lorentz's theories but also provide an opportunity to incorporate all the microphysical discussions into the paper tools for radio sounding. Unfortunately, the differences between these two theories' empirical predictions were either too difficult to separate from other factors or too small to detect with current apparatuses. Throughout the 1930s, none of the experimental attempts to test the theory of the Lorentz correction with radio sounding reported success. This inevitably diminished interest in the topic. In the 1940s and the early 1950s, research publications and open discussions on the subject were scarce. Ratcliffe quietly dropped out the Lorentz term in his expression of the Appleton-Hartree formula in his monograph on the magneto-ionic

24. Ratcliffe, "Lorentz polarization" (1939), 747–56.

25. Ibid., 756.

theory. Not until the mid-1950s did geoscientists' new attention to a previously under-noticed phenomenon—the propagation of very low-frequency electromagnetic waves ("whistlers") in the ionosphere—revive interest in the Lorentz correction.[26]

## THE QUASI-ELASTIC FORCE

While the controversy over the Lorentz correction originated from theoretical extrapolation of a microphysical model, another debate involving microphysics came from the interpretation of a puzzling experimental observation. In the 1910s and 1920s, physicists and engineers tried to mimic radio ionospheric propagation with electromagnetic processes in laboratory-produced ionized gases. In particular, they expected the measured effective dielectric constant of such a "tabletop ionosphere" to decrease with frequency as a consequence of Eccles and Larmor's ionic-refraction theory $\varepsilon_{\mathrm{eff}} = \varepsilon_0(1-Ne^2/m\varepsilon_0\omega^2)$ (cf., equation (6.1′)). The results nonetheless disobeyed that prediction. Instead of decreasing to zero, the measured dielectric constant reached a minimum at a certain frequency and then bounced back.

In the late 1920s, French physicist Camille Gutton and his disciples in Nancy suggested that a "quasi-elastic force" from positive ions on electrons resulted in this anomaly of dielectric-constant measurements. Gutton's idea excited a serious dispute concerning whether the experimental anomaly implied a modification of the fundamental equation of motion for electrons in an ionic medium or turned out to be an artifact of laboratory setups. Microphysics built a relation with the issues of laboratory experiments' applicability to the studies of radio-wave propagation in macroscopic nature.

26. Since the 1910s, radio engineers had observed atmospheric noise at very low frequencies. The audible noise with frequencies as low as 400 Hz appeared as music-like whistlers at the receiver output. Many people believed that whistlers were the extremely low-frequency components of wide-band electromagnetic radiation from lightning flashes or ion-carrying solar winds in the upper atmosphere. A prerequisite for this hypothesis was that the very long waves were able to propagate in the ionosphere. In 1934, Thomas Eckersley discovered that while Sellmeyer's theory supported propagation of electromagnetic waves with very low frequencies in the ionosphere, Lorentz's theory did not. (See Eckersley, "Musical atmospherics" (1935), 104–5.) Hence whistlers were empirical evidence against Lorentz's theory. The experiments and theories on them remained underdeveloped before World War II. Only in the 1950s did systematic measurements, comprehensive predictions of their quantitative behavior, and consequently formal denials of the Lorentz correction theory in ionic refraction become available. See Budden, *The Propagation of Radio Waves* (1985), 40–42, 376–80.

### *Measuring the Dielectric Constant of Ionized Air*

The first attempt to verify the ionic refraction theory with tabletop experiments dated back to the 1910s. After Eccles reported his theory at the 1912 meeting of the British Association for the Advancement of Science, John Ambrose Fleming recognized its potential in making predictions not only on the upper atmosphere but also on artificial ionized gases. If laboratory work confirmed these predictions, then the theory could gain empirical support independent of radio measurements in open space. Fleming asked Edwin Barton, professor of physics at University College, Nottingham, to perform the experiment. In 1913, Barton and his assistant Walter Kirby designed a lab setup, ionized air in a glass tube with X-rays, and measured the effective conductivity (not dielectric constant) of the gas through its resonance with a resistive inductor. Their results were satisfying—their measurement of the resonating frequency seemed to follow nicely the prediction of Eccles's theory.[27]

Things did not always proceed as smoothly as that. At the end of the decade, Balthasar van der Pol made another attempt to test the ionic refraction theory in the laboratory. A close observer of radio-wave propagation studies, midwife of Watson's theory, and friend of Appleton's, van der Pol believed in the validity of the ionic refraction theory and sought evidence from lab work. In 1917–20, the Dutch visitor performed an experiment at the Cavendish Laboratory to measure the change of ionized air's dielectric constant with ionization.[28] His setup consisted of an air-filled glass tube and a measuring circuit. High discharge voltage ionized the air, and a pump thinned it. Like Barton and Kilby's design, van der Pol measured the air's capacitance (an index of its dielectric constant) at radio frequencies through a resonance circuit known as "Lecher wire"—a rectangular loop with a sliding bridge whose movement adjusted the device's inductance.

Van der Pol's experimental data contradicted Eccles's theory. The measured air capacitance (and hence dielectric constant) indeed decreased with ionization (the discharge voltage) at low and high voltages. In an intermediate region, however, it *increased* with the discharge voltage, indicating that the air's dielectric constant *increased* with ionization. Van der Pol did not see this apparent anomaly as evidence against ionic refraction. Instead, he explained (away?) the effect by invoking the nonuniformity of the air between the con-

27. Barton and Kilby, "The effect of ionization of air" (1913), 567–68.

28. These experiments later became the major part of his doctoral dissertation at Utrecht: "De Invloed van een Geioniseerd Gas" (1920).

denser plates. His original interpretation of data assumed that the air had uniform dielectric constant and conductivity. But in reality, he now supposed, the air between the plates consisted of two layers with different thicknesses, dielectric constants, and conductivities. So the air's actual impedance, including its capacitance and conductance, needed correction.[29]

Van der Pol's experiment caught the attention of Appleton and his collaborator A. G. D. West, Rutherford's assistant and the BBC's first research engineer. In 1922, the two Britons performed a new experiment to measure the dielectric constant of an ionized gas. Their setup was similar to van der Pol's, and so were their results. Capacitance decreased with ionization when the latter was low; as ionization increased further, capacitance rose; only at high discharge voltages did the gas's capacitance decrease again. This suggested that the strange effect that van der Pol had observed was not the result of erroneous practice; it was reproducible. Yet Appleton and West did not consider their finding as evidence against Eccles's theory, either. Like van der Pol, they believed that the anomaly was the fault of instruments.[30]

### *A Propagation Theory from Tabletop Experiments*

To other researchers, however, the anomaly was more than an instrumental artifact. In the 1920s, a group of French scientists proposed a theory of the quasi-elastic force to explain the effect, calling for a fundamental change of electrons' equation of motion in an ionized gas. The leader of this group was Camille Gutton.

Gutton was a major radio scientist in France. Born in Nancy in 1872, Camille Antoine Marie Gutton obtained his bachelor's degree in physics at the École Normale Supérieure and entered the laboratory of physicist René Blondlot (the "discoverer" of the infamous N ray that was later shown to be nonexistent) at the University of Nancy. There Gutton studied the new Hertzian waves—he accurately measured their velocity between two conductors and found a slight difference between the phase and the group velocity. This work became his doctoral dissertation in 1899.[31]

Gutton stayed in Nancy after graduation. As wireless technology advanced,

29. For van der Pol's experiments, see ibid., 32–64; Bergmann and Düring, "Dielektrizitätskonstanten" (1929), 1044–46; Appleton and Childs, "On some radio-frequency properties" (1930), 970.

30. Appleton and Childs, "On some radio-frequency properties" (1930), 970–79.

31. Ponte, "Camille Gutton," (1963), 2584–85.

he gradually turned to radio research. In 1915, the French army drafted him; he worked at the Military Telegraph Services, where he joined Gustav Ferrié's team on tactical radio and became his primary assistant. The group attracted young radio technologists from all over the country. They developed transatlantic wireless telephony between Paris and Arlington, Virginia, direction finders, radio espionage, and antennae.[32] Gutton did the first successful test of ground-to-air radio communication in France.[33]

Ferrié and his protégés played a crucial part in promoting radio in postwar France. Many left the military and applied their wireless expertise to civilian sectors. They initiated electronic-tube manufacturing, formed amateur societies, and advocated for broadcasting. Ferrié became inspector general of Military Telegraph Services and founded a Laboratoire National de Radioélectricité (LNR) in 1919.[34] Gutton returned to Nancy to assume a professorship of physics, maintained a lifelong friendship with Ferrié, and served as secretary of the Société des Amis de la T.S.F. (Telegraphie sans fils).

Like other radio scientists and wireless amateurs in the 1920s, Gutton focused on short waves. The atmosphere's effects on short-wave propagation interested him. He began to pay attention to the electrical properties of ionized media in the early 1920s as he studied gases' electric discharge at high frequencies, which glowed with light. Gutton and his students measured the gas-glowing voltage at various pressures[35] and found that such voltage did not change monotonically with pressure; it had a minimum, which could imply resonance in the gas. This resonance, Gutton thought, hinted at an explanation for the puzzling increase of a gas's dielectric constant with ionization.

Gutton asked his son Henri and his doctoral student Jean Clément at Nancy to examine ionized gases' dielectric properties. In 1927, the young men designed a setup to measure an ionized gas's dielectric constant. Their ap-

32. Telegraphie Militaire, "L'essais de téléphonie sans fil entre la Tour Eiffel et Arlington (États-Unis)," 4 June 1915, WWI Materials (6N 21), Service Historique de l'Armée de Terre, Paris, France.

33. Duchet-Suchaux, "Gutton" (1989), 371.

34. At the beginning, LNR was in an old barracks in les Invalides in Paris with four young engineers and several technicians. And it was not formally functional until 1926, when it affiliated with the Post, Telegraphy, and Telephony Services (PTT). Yet in the late 1920s and early 1930s, LNR became an organization comparable with the radio section of the U.S. National Bureau of Standards and Britain's National Physical Laboratory. Amoudry, *Ferrié* (1993), 207–12.

35. Gutton, "Sur la décharge électrique" (1924), 467–70; Gutton and Gutton, "Sur la décharge électrique" (1928), 303–5; Camille Gutton to Le Duc (letter), 7 March 1922, Special Collection, Libraries of Radio France.

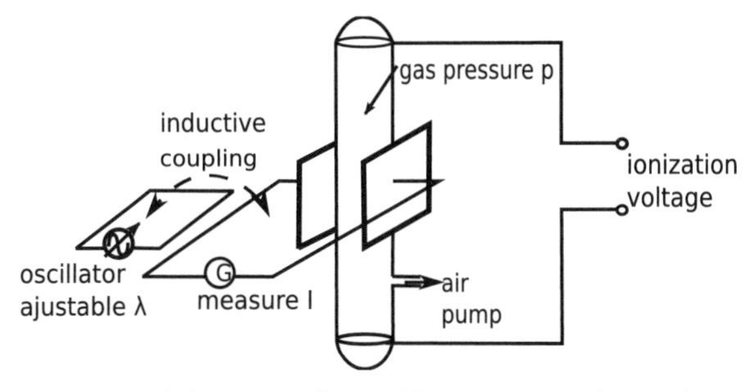


FIGURE 10.2. Henri Gutton and Jean Clément's experimental setup.

paratus had the same layout as van der Pol's: a gas-filled glass tube connected with a high ionizing voltage source and an air pump, a pair of metal plates clamping the tube, and a circuit to measure the capacitance between the plates (figure 10.2). The measuring circuit comprised a variable-frequency oscillator and a loop inductor. The loop and the tube formed a resonator that significantly amplified the oscillating current at the resonating frequency. By tuning the oscillator, one could measure the resonating frequency, from which one could calculate the tube's capacitance. A galvanometer measured the current at the resonator, and a Lecher wire the oscillating signal's wavelength. Gutton and Clément determined the resonating frequency for various discharge voltage levels from zero to maximum, and they repeated this procedure for several distinct levels of gas pressure. They plotted the data in a system of Cartesian coordinates in which the abscissa was the resonating wavelength and the ordinate the square of the maximum galvanometer reading. Data sets at different gas pressures formed different curves on the Cartesian plane.[36]

Gutton and Clément performed the measurements in February 1927. They selected the dimensions of the loop inductor and air condenser so that their resonating wavelength was 408.5 centimeters for nonionized air. At about that wavelength, they obtained three curves corresponding to gas pressures 0.015, 0.035, and 0.100 mmHg (figure 10.3).[37] The results at low pressure replicated the pattern that van der Pol had identified. At 0.015 mmHg, the resonating wavelength first decreased, then increased, and finally decreased with ionization, while the maximum current first decreased, then remained constant, and finally increased with ionization (following the curve from A to B along the arrow direction). The resonating wavelength was a measure of the

36. Gutton and Clément, "Sur les propriétés diélectriques des gaz ionisés et la propagation" (1927), 137–43.

37. Gutton and Clément, "Sur les propriétés diélectriques des gaz ionisés" (1927), 441–43.

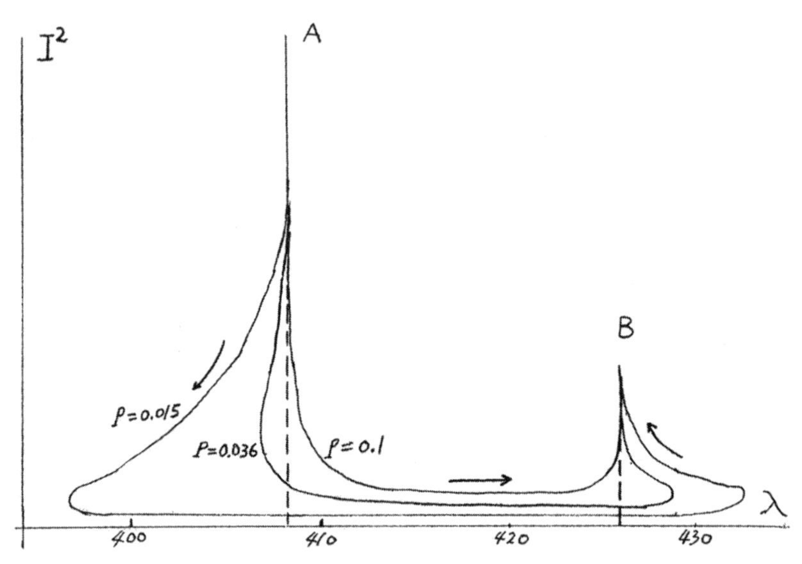


FIGURE 10.3. Henri Gutton and Jean Clément's experimental results in February 1927. Gutton and Clément, "Sur les propriétés diélectriques" (1927), 442.

condenser's dielectric constant, since it was proportional to the square root of the condenser's capacitance. At a low gas pressure, therefore, the ionized gas's dielectric constant first decreased, then increased, and finally decreased again with ionization, a pattern identical to van der Pol's. The gas's behavior was even more anomalous at higher pressure. At 0.100 mmHg, the resonating wavelength increased with ionization throughout all discharge voltage levels, and thus the gas's dielectric constant never decreased with ionization. Gutton and Clément immediately carried out another experiment in which the resonating wavelength at null ionization was 217.3 centimeters. The outcome was the same for this different range of frequency: a dip at low pressure and less obvious decrease at high pressure. The patterns that Gutton and Clément had noted now appeared around two frequencies; the effect was robust.

Gutton and Clément's finding was not new. Van der Pol, Appleton, and West had also observed an increase of the dielectric constant with ionization. Unlike their predecessors, however, the French researchers held that this phenomenological contradiction to the ionic refraction theory was not the fault of instrumental conditions. Instead, it revealed a more fundamental physical mechanism. As Camille Gutton had conjectured in the gas-discharge experiments, they believed that the minimum of the measured gas capacitance at a certain degree of ionization implied the existence of a natural resonance in the ionized gas.

Using this concept of natural resonance, Gutton and Clément proposed to modify the microphysical condition of ionic refraction by including a "quasi-elastic" force. In the Eccles-Larmor theory, the equation of motion for an electron in an ionized gas was $m(d^2x/dt^2)+g(dx/dt) = eE$. (Recall that $g$ was the frictional coefficient). Such an electron was free except for the influence of the external field $E$ and the frictional/colliding force $g(dx/dt)$. Wrong, Gutton and Clément contended. It should not be free. In an ionized gas, an electron's equation of motion should be

$$(10.2) \quad m\frac{d^2x}{dt^2}+g\frac{dx}{dt}+\phi x=eE\cdot$$

The additional force $-\phi x$ here was proportional to the opposite of the electron's displacement $x$ and hence was an elastic force. Equation (10.2) was similar to Lorentz's equation of motion for a bound electron in a dielectric material, and the "quasi-elastic force" $-\phi x$ resembled the binding force on an electron inside an atom.

Gutton and Clément argued that the physical foundation of the quasi-elastic force was the action of an electron's neighboring positive ions. An electron in an ionized gas was in a sea of relatively immobile positive ions. Each positive ion exerted a Coulomb force on the electron. At equilibrium, the positive ions' overall Coulomb force on the electron was zero. An electric field displaced the electron from its equilibrium, and its net Coulomb force—especially that from its neighboring positive ions—was no longer zero. This force tended to drive the electron back to its equilibrium. So it was a restoring force with its direction opposite to and its strength proportional to the electron's displacement: the electron further from equilibrium experienced a stronger restoration. Hence the force was $-\phi x$.

Gutton and Clément solved this new equation of motion (10.2) and plugged the solution into Maxwell's equations. In so doing, they obtained the revised formulae for an ionized gas's effective dielectric constant and conductivity (see chapter 6 for notations):

$$\varepsilon_{eff}=\varepsilon_0\left[1-\frac{Ne^2/\varepsilon_0}{(\phi-m\omega^2)^2+g^2\omega^2}(m\omega^2-\phi)\right]$$

$$(10.3)\quad \sigma_{eff} = \frac{Ne^2\omega^2 f_r}{(\phi - m\omega^2)^2 + f_r^2\omega^2}.$$

According to them, the dielectric-constant formula in equation (10.3) could explain their experimental results. In the Eccles-Larmor theory, $\phi = 0$ and the dielectric constant decreased linearly with the ion density $N$. It would be the same if the elastic coefficient $\phi$ was a constant. Yet $\phi$ should have increased with $N$, since a higher ion density led to a larger net Coulomb force and thus a stronger restoration. Consequently, the variation of $\varepsilon_{eff}$ with $N$ in equation (10.3) was not monotonic. For small $N$, $\phi$ was small, $m\omega^2 - \phi > 0$, and hence $\varepsilon_{eff} < \varepsilon$, which was consistent with Eccles's prediction. Yet when $N$ exceeded a critical value $N_c$ that satisfied $m\omega^2 = \phi(N_c)$, $m\omega^2 - \phi < 0$, and hence $\varepsilon_{eff} > \varepsilon$, contrary to Eccles's prediction. Thus the dielectric constant was minimum at $N = N_c$. This explained why the measured dielectric constant did not decrease indefinitely with ionization but kicked back at a minimum.[38]

At the same time, the theory of the quasi-elastic force entailed a natural resonance in an ionized gas. The equation of motion (10.2) implied a resonance at $m\omega^2 = \phi$ for the frictionless case; the value of $\phi$ that satisfied $m\omega^2 = \phi$ for a given frequency $\omega$ was exactly the value of $\phi$ at $N = N_c$. At the resonance, the effective dielectric constant was infinite, meaning that the refractive index was also infinite and electromagnetic-wave propagation infinitely slow. Gutton and Clément believed that this finding would change studies of short radio-wave propagation over long distances: the quasi-elastic resonance stopped radio-wave propagation in the ionosphere and created a minimum range at a certain frequency (or frequencies). Perhaps, they suggested, the true reason for the observed range minimum at about wavelength 200 meters was quasi-elastic resonance, not geomagnetism, as Appleton et al. had argued.[39]

In addition, Gutton and Clément performed a similar experiment in which they replaced the ionized gas with an electrolytic solution. The results were similar to the ionized-gas experiments.[40] This test on the liquid material further demonstrated the robustness of the observed anomaly: it was consistently reproducible at different frequencies and in both gas and liquid phases. Such

38. Gutton and Clément, "Sur les propriétés diélectriques des gaz ionisés et la propagation" (1927), 146–48.

39. Ibid., 150–51; Gutton and Clément, "Propagation des ondes électromagnétiques" (1927), 677–78.

40. Gutton and Clément, "Sur les propriétés diélectriques des gaz ionisés et la propagation" (1927), 148.

robustness strongly suggested that the effect was inherent to ionic media in general. To explain it, a theory addressing the fundamental principles of ionic electricity was preferable to one involving apparatus errors.[41]

## *Anatomy of a Debate*

Similar to the Lorentz correction, the hypothesis of the quasi-elastic force was also an attempt at a microphysical modification of the (magneto-)ionic-refraction theory. Yet unlike the Lorentz correction, which resulted from conceptual and mathematical considerations, the quasi-elastic force was an inference from the results of experiments that aimed to reproduce radio ionospheric propagation via tabletop experiments. Moreover, unlike the Lorentz correction, which made little empirical difference in radio experiments, the hypothesis of the quasi-elastic force implied a natural resonating frequency that could transform the whole theoretical apparatus in radio sounding and wireless technology, from the determination of the ionosphere's gyromagnetic frequency to the prediction of the skip zones. This, along with the fact that the quasi-elastic force subverted longstanding belief in the freedom of electrons in ionic media, discomforted many physicists and radio engineers. They did not hesitate to attack the theory.

The first to denounce the quasi-elastic force was Peder Oluf Pedersen—an authority on electrical engineering, a disciple of the electric arc oscillator's inventor, Valdemar Poulsen, a professor at the Royal Technical College of Copenhagen, and the author of a major textbook on radio ionospheric propagation.[42] In this 1927 book, Pedersen devoted a chapter to criticizing Gutton and Clément's work.

Pedersen maintained that the quasi-elastic force was not necessary to explain Gutton and Clément's experimental results. Rather, the apparent anomaly was the outcome of the additional capacitance between the condenser plates and the wall of the gas tube. The French experimenters' setup included a gas tube, two metallic plates clasping the tube, and a wire loop coupling

41. In fact, the theory of quasi-elastic force had rich mathematical implications, which Gutton and Clément explored only partially. Using equation (10.3) and a simple circuit model, one could not only demonstrate the nonmonotonic variation of the gas's dielectric constant with ionization but also produce curves very similar to those in Figure 10.3, Gutton and Clément's experimental results. I confirmed this point with my own calculations.

42. "Peder Oluf Pedersen, Medal of Honor Recipient, 1930," *Proceedings of the Institute of Radio Engineers*, 18:11 (1930), 1984. The textbook that Pedersen published was *The Propagation of Radio Waves* (1927).

electromotive force from the oscillator. The wire loop could be modeled as a resistive inductor connecting in parallel to a capacitor. The ionized gas within the area of the metallic plates could be modeled as a leaky capacitor (as Gutton and Clément had shown). The element that made the most difference was the space between the tube wall and the metallic plates. Strictly, Gutton and Clément's gas condenser consisted of a body of ionized gas, a part of the glass tube wall, ordinary air between the tube wall and the metallic plates outside the wall, and the plates proper (figure 10.4). So the model of the gas condenser should include not only a leaky capacitor to represent the ionized gas, but also two additional capacitors to represent the capacitive effect of the space between the metallic plates and the tube wall that confined the ionized gas. And the two additional capacitors needed connecting in series to the leaky capacitor at its ends to emulate the plate-wall geometry. Accordingly, Pedersen's model for Gutton and Clément's setup was an RLC circuit

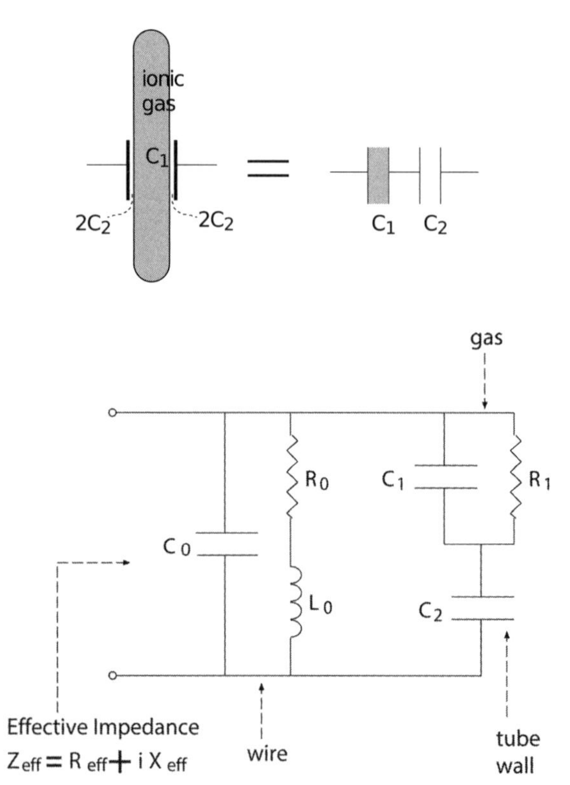


FIGURE 10.4. Pedersen's model of Henri Gutton and Jean Clément's setup (the upper panel) and its equivalent circuit (the lower panel).

in figure 10.4. $L_0$, $R_0$, and $C_0$ were the wire loop's inductance, resistance, and capacitance, respectively; $2C_2$ was the capacitance of the space between a metallic plate and the tube wall, and $C_1$ and $R_1$ were the ionized gas's capacitance and resistance, respectively. Also, $C_1$ was proportional to the gas's dielectric constant, and $R_1$ was inversely proportional to its conductivity.

Pedersen argued that the model in figure 10.4 and Eccles's ionic refraction theory predicted curves similar to Gutton and Clément's. In fact, if one included the effect of $C_2$, then the quasi-elastic force was not necessary. Pedersen interpreted Gutton and Clément's results in figure 10.3 as plots of maximum $1/R_{eff}^2$ versus $\lambda_{res}$ for the circuit in Figure 10.4, where $R_{eff}$ was the real part of the circuit's impedance that represented the galvanometer reading and $\lambda_{res}$ was the resonance wavelength at which $1/R_{eff}^2$ was maximal. The $1/R_{eff}^2 - \lambda_{res}$ relation, according to Pedersen, was parameterized by $R_1$ and $C_1$ (the ionic gas's resistance and capacitance, respectively), which in turn were functions of electron density $N$, following $C_1 \propto (1-K)$ and $R_1 \propto (1/K)$ from the ionic refraction theory ($K = [Ne^2/(m\varepsilon_0)]/(\omega^2 + g^2/m^2)$).

Pedersen examined several cases to observe the qualitative behaviors of the $\lambda_{res} - 1/R_{eff}^2$ curves with increasing $N$. (i) When $N = 0$, $K = 0$, which implied that $C_1 = C_1^{(0)}$, $R_1 = R_1^{(0)}$, and $1/R_{eff}^2$ was infinitely large. (ii) When $N$ reached the level so that $K = 1$, $C_1 = 0$. In this case, $1/R_{eff}^2$ was much smaller than that in case (i), and $\lambda_{res}$ was shorter than that in case (i), when $\omega^2 C_1^{(0)} C_2 R_1^{(0)} > 1$. If so, then the point corresponding to case (ii) on the $\lambda_{res} - 1/R_{eff}^2$ plane was on the lower left of the point corresponding to case (i), like Gutton and Clément's curves at low gas pressure. (iii) When $N$ increased further so that $K > 1$, $1/R_{eff}^2$ was smaller than that in case (ii), and $\lambda_{res}$ was longer than that in case (ii). So the point corresponding to case (iii) was on the lower right of the point corresponding to case (ii), consistent with Gutton and Clément's curves. Pedersen reproduced their curves without the theory of the quasi-elastic force.

The essence of Pedersen's argument was that the theory was not necessary for explaining the apparent anomaly in the dielectric-constant measurements. Rather, the anomaly was a consequence of the capacitance between the condenser plates and the gas inside the tube wall. The apparent contradiction to the ionic refraction theory was nothing fundamental—it was an artifact of the experimental apparatus's confining the ionized gas between the metallic plates. The quasi-elastic force did not exist.[43]

Henri Gutton disagreed, and he fought back. At the beginning of 1928, he

43. Pedersen, *The Propagation of Radio Waves* (1927), 90–94.

published a critique of Pedersen's argument. The junior Gutton's critique was twofold. First, Pedersen required that $\omega^2 C_1^{(0)} C_2 R_1^{(0)} > 1$ to make the resonance wavelength decrease from the case associated with zero ion concentration to that associated with the degree of ionization corresponding to $K = 1$. That is, the ionized gas's effective resistance $R_1^{(0)}$ should be large enough to yield a curve consistent with Eccles's theory at low ionization. Yet this condition was not empirically sound. Even at low gas pressure (e.g., 0.015 mmHg) at which the experimental data had no problem fitting Eccles's theory at low ionization, the ionized gas's energy loss still made the effective resistance $R_1^{(0)}$ considerably smaller than the condition $\omega^2 C_1^{(0)} C_2 R_1^{(0)} > 1$ required. Hence the necessary condition for Pedersen's theory did not hold in the experiments. Second, Gutton performed another experiment in which he moved the metallic plates of the gas condenser *inside* the wall of the glass tube. In the new arrangement, the capacitance between the ionized gas and the metallic plates should drop considerably, since the latter were now in direct contact with the former without the mediation of air and glass. Yet the experimental results showed little difference. Pedersen's hypothesis that the tube-plate capacitance was the major factor was thus problematic.[44]

Gutton's critique stimulated a further response. In July, Joergen Rybner—Pedersen's colleague at the Royal Technical College and his major assistant in preparing his 1927 book—wrote to *l'Onde électrique* regarding Gutton's points. Rybner remarked that Gutton's first critique was invalid, since Pedersen's model actually did *not* require $\omega^2 C_1^{(0)} C_2 R_1^{(0)} > 1$ to produce curves consistent with Gutton and Clément's. Pedersen had been wrong that the resonance wavelength at low ionization decreased with ionization only when $\omega^2 C_1^{(0)} C_2 R_1^{(0)} > 1$. In fact, his theory applied to an ionized gas with *any* value of conductivity, for what really mattered was the ordering of the capacitors. In the case when the gas's conductivity was zero, the overall capacitance of the circuit in figure 10.4 was $C = C_0 + C_1 C_2/(C_1 + C_2)$. This relationship between $C$ and $C_1$ implied that decreasing $C_1$ as a result of increasing ionization first reduced $C$ monotonically to $-\infty$ at $C_1 = -C_2$, then lifted $C$ instantaneously to $\infty$, and then reduced $C$ monotonically to $C_0 + C_2$ at $C_1 = -\infty$ (figure 10.5). As gas conductivity became nonzero, it smoothed the infinities in the $C$-$C_1$ relationship: with decreasing $C_1$, $C$ first decreased, *then increased*, and finally decreased (see figure 10.5). This behavior was independent of the gas's conductivity, which affected only the width of the region in which $C$ increased with $C_1$. So Pedersen's condition $\omega^2 C_1^{(0)} C_2 R_1^{(0)} > 1$ was not necessary. Violat-

44. Gutton, "Sur l'interprétation" (1928), 1–4.

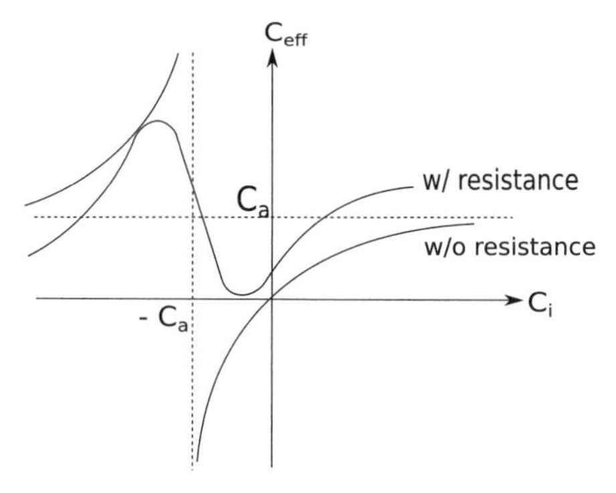


FIGURE 10.5. Rybner's *C*–*C*i curves with zero and nonzero gas conductivity.

ing it simply pushed the curve to its leftmost extreme, where it began to turn rightward before $K = 1$.[45]

Even so, the Frenchmen did not believe that the anomaly was an artifact instead of the reflection of a fundamental fact. Camille Gutton—Henri Gutton and Jean Clément's advisor in Nancy—joined the debate. He performed several experiments using the reasoning that the existence of the quasi-elastic force was identical with the existence of resonance. The Frenchmen had demonstrated resonance in ionized gases from the experiments that the Dane criticized as producing an artifact. The best way to respond was therefore to demonstrate the existence of resonance from experiments using other instrumental setups and theoretical principles.

Camille Gutton performed two experiments in April 1930. First, he put the gas condenser's metallic plates outside the tube wall but varied the separations between the plates and the wall. Pedersen and Rybner contended that the dominant factor creating the anomaly was the capacitance between the plates and the tube wall. If they were right, then a slight change of the wall-plate separation would alter significantly the tube-plate capacitance, which would affect the measured data noticeably. Nevertheless, Gutton did not find any recognizable change by varying the wall-plate separations. Pedersen and Rybner could not be right.[46]

Second, Gutton designed a new experiment to measure the conductivity of an ionized gas. From the theory of the quasi-elastic force, the gas's conductiv-

45. Rybner, "Note sur les expériences" (1928), 428–36.

46. Gutton, "Sur les propriétés" (1930), 844–47, and "Sur les propriétés des gaz ionizes" (1930), 7–8.

ity reached a maximum at the resonance $\omega = (\phi/m)^{1/2}$—equation (10.3). So a resonance-like phenomenon in conductivity measurements would be another empirical confirmation for the quasi-elastic theory. In the new experiment, he inserted an ammeter in the discharge circuit to measure the current from the discharge-voltage generator to the gas tube and took this reading as a measure of the gas's conductivity. He performed two measurements for a given discharge voltage (a given degree of ionization): the current ($C_0$) to the tube without additional devices and this current (C) with a nearby high-frequency (108 MHz, or wavelength 2.76 meters) oscillator coupled to the main circuit. The measured data at different degrees of ionization formed a curve on the $C_0$-C plane.

The measured data showed a salient resonance-like behavior (figure 10.6 here): for every curve, C jumped abruptly with almost an infinite slope at a certain $C_0$ (or equivalently, a certain degree of ionization). Gutton argued that a jump occurred when the gas's resonance frequency $(\phi/m)^{1/2}/2\pi$, a function of the quasi-elastic coefficient $\phi$ (and hence the degree of ionization), equaled the external signal's frequency 108 MHz. The data provided evidence for the existence of resonance.

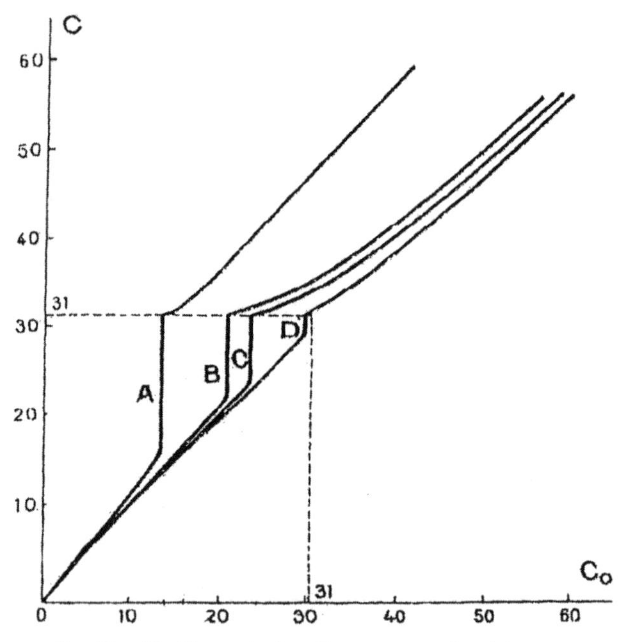


FIGURE 10.6. Camille Gutton's data from gas-conductivity measurements. Camille Gutton, "Sur les propriétés" (1930), figure 2.

Gutton also noted that other researchers' findings confirmed the existence of resonance in an ionic medium. Lewi Tonks and Irving Lagmuir at General Electric's Schenectady Laboratory found an electronic oscillation in an ionized gas in 1929.[47] They biased the air inside a glass tube with low-frequency discharge voltage and detected large current at certain high frequencies. Recognizing that Tonks and Lagmuir's discovery was similar to the conclusion from his second experiment, Gutton believed that their study also provided empirical confirmation for the existence of an ionized gas's innate resonance. Therefore Gutton offered three additional pieces of evidence for the quasi-elastic force (and resonance): measurements by varying plate-wall separations, measurements of gas conductivity, and Tonks and Lagmuir's finding.

The other Gutton provided additional evidence, too. As Camille left Nancy to take the directorship of the Laboratoire National de Radioélectricité in 1930,[48] Henri finished his PhD dissertation on the quasi-elastic force.[49] Henri Gutton's dissertation presented the results of his new experiments on ionized gases. At first, he fixed the oscillator's frequency $\omega$ ( $= 2\pi c/\lambda$), measured the discharge current i, and tuned the length L of the Lecher wire in the condenser-inductor circuit until the air condenser's current reached maximum D (figure 10.7). His measured data showed that the resonating air-condenser current D depended on the discharge current i and had a minimum at $i = i_{min}$. Observing i at different oscillating frequency $\omega$ (hence $\lambda$), he found that $\lambda^2 i_{min}^{3/4} =$ constant. Assuming that the electron density $\mathcal{N}$ was proportional to the discharge current i, the empirical regularity entailed $\lambda^2 \mathcal{N}_{min}^{3/4} =$ constant.

Gutton interpreted $\mathcal{N}_{min}$ as follows. The condenser-loop circuit consisted of a voltage source (the oscillator), a variable inductor (the Lecher wire), and a leaky capacitor (the ionized-gas condenser). Tuning the Lecher wire to obtain maximum current was matching the Lecher wire's impedance with the air condenser's so that the imaginary parts of the two canceled each other. At the matching condition, the overall circuit's impedance was real. And from equation (10.3), this matching impedance was minimum when $\phi = m\omega^2$. The electron density $\mathcal{N}_{min}$ that minimized D was therefore the one at which the "resonance" $\phi = m\omega^2$ occurred, viz., $\phi(\mathcal{N}_{min}) = m\omega^2$; $\mathcal{N}_{min}$ was the electron density $\mathcal{N}_{res}$ at the quasi-elastic resonance. So one can write the empirical relation $\lambda^2 \mathcal{N}_{min}^{3/4} =$ constant as $\lambda^2 \mathcal{N}_{res}^{3/4} =$ constant. From a different angle, one

47. Tonks and Lagmuir, "Oscillations in ionized gases" (1929), 195–210.
48. Duchet-Suchaux, "Gutton" (1989), 371.
49. Gutton, "Recherches" (1930), 62–129.

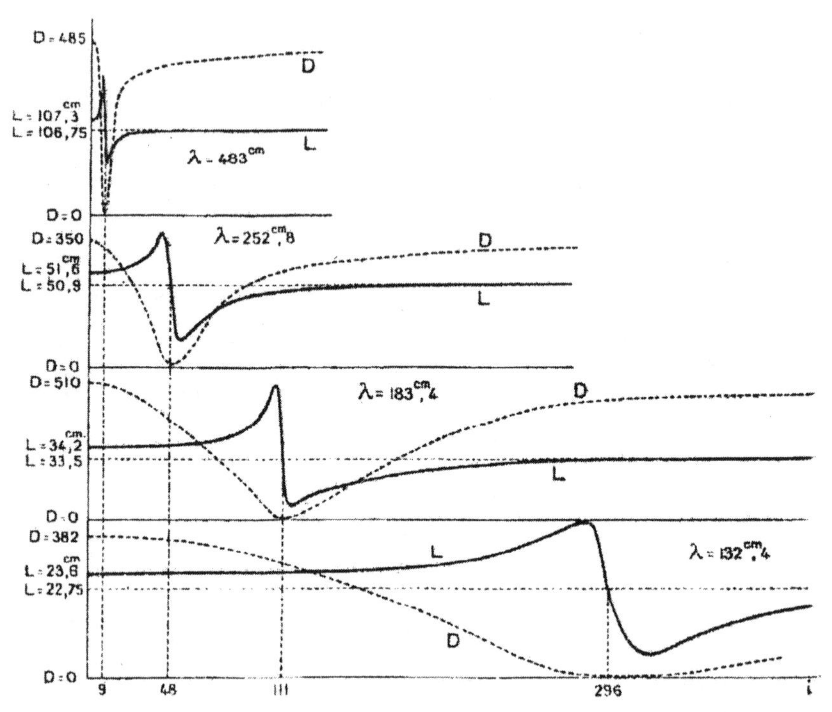


FIGURE 10.7. Henri Gutton's experimental results. Henri Gutton, "Recherches" (1930), figure 9. Horizontal axis: tube discharge current; vertical axis: maximum condenser current.

can also interpret it as $\lambda_{min}{}^2 N^{3/4}$ = constant, if the electron density remained fixed while the oscillating wavelength was tunable.

Furthermore, Henri Gutton observed the change of the measured curves in a magnetic field. He applied a constant magnetic field $B_0$ along the tube to the ionized gas. With the magnetic field, every i-D curve had two minima rather than one minimum. This strongly suggested that the magnetic field "split" the resonance in two, similar to the inverse Zeeman effect that Lorentz had discussed. Gutton rewrote the equation of motion for an electron in an ionic medium with the external magnetic field and found that the new equation entailed two resonance conditions $m\omega^2 = \phi \pm eB_0\omega$, contrasting $m\omega^2 = \phi$ without the magnetic field. So two resonance electron densities $N_1$ and $N_2$ were observable in a magnetized ionized gas. Hence there were two minima for D. Moreover, from the empirical regularity $\lambda^2 i_{min}{}^{3/4}$ = constant, the "Zeeman relation" $m\omega^2 = \phi \pm eB_0\omega$, and the measured data, Gutton deduced the electron's charge-to-mass ratio $e/m = 1.68\times10^7$, which was close to the ratio

$1.77 \times 10^7$ from Walter Kaufman's revision of J. J. Thompson's experiment. Hereby the theory of the quasi-elastic force gained additional support.

### *Death of a Theory*

The theory of the quasi-elastic force, if true, would have affected studies of radio-wave propagation in the ionosphere: it would have implied natural resonance without the geomagnetic field, resonance doubling with the geomagnetic field, and hence more complex frequency-dependent variations of propagation characteristics. The Guttons' advocacy of the force therefore encouraged more research into ionized gases' dielectric properties, but the radio scientists who admired their effort did not embrace their concept. Although the Guttons provided evidence from at least five distinct angles, none seemed absolutely convincing.

Experimenters eventually obtained a set of data wholly consistent with Eccles's ionic-refraction theory. In 1927–28, Ludwig Bergmann and Walter Düring at the University of Breslau set up a new experiment to measure the dielectric constant of an ionized gas. Their instrumental arrangement was similar to van der Pol's: they filled a glass tube with thin air ionized by high discharge voltage. They inserted metal plates inside the tube wall to form a condenser and connected it with a tunable Lecher wire. An oscillator generated high-frequency signals coupled to the Lecher wire. At a given frequency and discharge voltage, the experimenters adjusted the wire's length to maximize the air condenser's current; this length was inversely proportional to the ionized air's capacitance. Yet the arrangement differed from van der Pol's in two respects: the Germans used a better pump to reduce the air pressure much more—to $10^{-5}$ mmHg, one ten thousandth of van der Pol's. Also, their oscillating frequencies were much higher: their wavelengths were 1.10–2.40 meters, and his hundreds of meters.

The extremely low air density and high frequency made a difference. In Bergmann and Düring's measurements, the Lecher wire's length at resonance steadily increased with discharge voltage, meaning that the ionized air's effective dielectric constant decreased with ionization in all testing conditions. They did not see the "bouncing back" of the dielectric constant after a certain degree of ionization, as van der Pol, the Guttons, and Clément observed. Their new measurements under low gas densities and high frequencies completely agreed with Eccles's ionic refraction theory.[50]

50. Bergmann and Düring, "Dielektrizitätskonstanten" (1929), 1041–68.

Their conclusion did not necessarily directly contradict the Guttons and Clément's quasi-elastic force, since they used different conditions. That the ionized gas's dielectric constant did not have minima at extremely low air pressures and high frequencies did not disprove the existence of such minima at higher air pressures and lower frequencies. The theory could still be valid in specific, if not all, operating conditions. Nevertheless, Bergmann and Düring did show that the dielectric constant did not always bounce back. Behavior was consistent with the simple ionic refraction theory under some conditions. Although the quasi-elastic force still had a chance, its correctness would not be as general as it first appeared to be. New evidence was against it under the low-pressure high-frequency conditions.

In contrast with the Germans' indirect challenge, Edward Victor Appleton launched a direct attack. Studying radio-wave propagation in the laboratory, Appleton was among the first scientists to measure the dielectric constant of an ionized gas. His experiment with A. G. D. West in 1922 confirmed van der Pol's bouncing back. But he never accepted the quasi-elastic force. He believed from early on that the apparent anomaly was the result of an unexpected physical phenomenon relating to the instrumental setup, rather than a fundamental correction to the ionic-refraction theory. Specifically, he held, the cause was the formation of thin layers ("sheaths") on the surfaces of the condenser plates, a concept that he borrowed from Irving Langmuir and Harold Mott-Smith at General Electric.

Langmuir found an ionic sheath when experimenting on low-pressure gaseous electric discharge in 1923. He placed low-density air in a glass tube and ionized the air with a mercury electric arc. Putting a negative electrode on the path of the mercury arc, he found that the out-flowing current almost remained constant no matter how much negative voltage he applied. Why was the current independent of the applied voltage? Because, he argued, a sheath formed on the electrode's surface. When an electrode with a negative potential was in an ionized gas, it attracted positive ions and expelled electrons from its neighborhood. Around the electrode formed a layer containing only positive ions. This positive-ion sheath neutralized and shielded the negative electrode from electric forces. So the sheath consumed the potential drop between the electrode and the ionized gas. Also, from conservation of electric charges, the current from the sheath to the electrode equaled that from the ionized gas to the sheath, which was independent of the voltage applied to the electrode, since the voltage did not affect the potential difference between the sheath and the gas. So the current out of the electrode was independent of the applied voltage.[51] One could calculate the constant current of a biased electrode in an ionized gas from the space-charge equation that physicist Clement D. Child of Cornell University derived in 1911. From this equation, Langmuir obtained the relationship $I \propto V^{3/2}/x^2$ for the current $I$ out of the positive-ion sheath, the voltage $V$ across the sheath, and the sheath's thickness $x$.[52]

After 1923, Langmuir studied further with his colleague Harold Mott-Smith the sheaths that formed on electrodes in ionic media.[53] They found that sheaths existed not only on negative electrodes, but also on unbiased electrodes insulated from the discharge circuit because of the mobility difference between positive ions and electrons. In an ionized gas, an unbiased electrode captured both electrons and positive ions as the charged particles randomly hit its surface. Because an electron was much more mobile than a positive ion, the electrode captured more electrons than ions and hence charged the electrode negatively. This negative electrode had a sheath of positive ions on its surface. Thus thin layers of positive ions also existed on the condenser plates in the experiments of van der Pol et al.

Using Langmuir and Mott-Smith's study of ionic sheaths, Appleton and his student E. C. Childs at King's College, London, developed an alternative explanation for the anomaly in the dielectric-constant measurements. They argued that the medium between the two plates of an ionized-gas condenser included three parts: the two positive-ion sheaths attached to the plates' surfaces and the ordinary ionized gas between the sheaths (figure 10.8). The two sheaths consumed all the voltage across the plates. So the middle part containing the ionized gas had equipotential everywhere and functioned as a short circuit connecting the two sheaths. One could model the ionized-gas condenser as two leaky capacitors connected in series that represented the sheaths' effect (see figure 10.8). A sheath's capacitance was $C_s = \varepsilon_s A/x_s$, where $A$ was the plate's area, $\varepsilon_s$ the sheath's dielectric constant, and $x_s$ the sheath's thickness. The overall capacitance of the condenser comprising two sheaths in series was thus $C = C_s/2 = \varepsilon_s A/2x_s$. From the space-charge equation, $x_s^2$ was proportional to $V_s^{3/2}$ ($V_s$ was the voltage across the sheath and also half of the overall voltage $V$ across the condenser). Therefore, $C \propto 1/x_s \propto V^{3/4}$.

51. Langmuir, "Positive ion currents" (1923), 290–91.

52. Child, "Discharge from hot CaO" (1911), 492–511.

53. Langmuir and Mott-Smith, "Studies of electric discharges," part I (1924), 449–55; part II, 538–48; part III, 616–23.

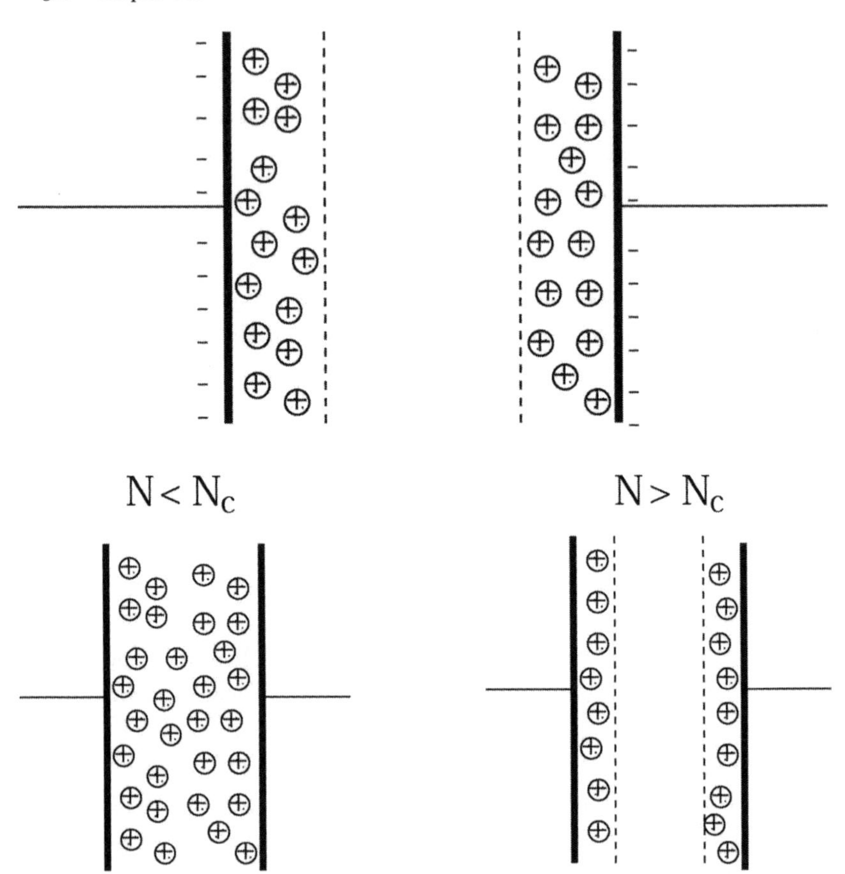


FIGURE 10.8. Appleton and Childs's model of positive-ion sheaths (upper panel) and sheath conditions when $N > Nc$ and $N < Nc$.

Appleton and Childs measured the gas condenser's capacitance and voltage bias and empirically confirmed the relation.[54]

The ionic sheaths explained the apparent anomaly in the dielectric-constant measurements: the current $I$ through the sheaths depended only on the rate of charge recombination at the sheath-gas interface, which was proportional to electron density $N$. When the bias voltage across the condenser was a constant, $N \propto I \propto 1/x_s^2$ from the space-charge equation. Suppose that a sheath's dielectric constant did not differ from an ordinary ionized gas's—$\varepsilon_s = \varepsilon_0[1-(Ne^2m/\varepsilon_0)/(m\omega^2+g^2)]$. The condenser's capacitance was thus $C \propto \varepsilon_s/x_s \propto [1-(Ne^2m/\varepsilon_0)/(m\omega^2+g^2)]N^{1/2}$. This implied that $C = 0$ when

54. Appleton and Childs, "On some radio-frequency properties" (1930), 980–84.

$N = 0$, which was incorrect, since a nonionized gas still had the capacitive effect. The problem was that $N \propto 1/x_s^2$ entailed an infinite $x_s$ at $N = 0$. When the gas was spatially confined, however, the sheath's thickness $x_s$ should not exceed the separation between the condenser plates. Suppose $N'$ was the electron density at which the sheath's thickness determined by $N \propto 1/x_s^2$ was half of the separation between the condenser plates. Between $N = 0$ and $N = N'$, the two sheaths filled the space between the condenser plates, and the condenser's capacitance was proportional to $[1-(Ne^2m/\varepsilon_0)/(m\omega^2+g^2)]$ only, equivalent to Eccles's ionic refraction theory. So the condenser's capacitance decreased with $N$. But when $N > N'$, the sheaths shrank, and the condenser's capacity became proportional to $[1-(Ne^2m/\varepsilon_0)/(m\omega^2+g^2)]N^{1/2}$, implying that $C$ increased with $N$ at first and then decreased with $N$ after reaching a maximum. In brief, the Appleton-Childs sheath theory predicted that the ionized-gas condenser's capacitance first decreased, then increased, and finally decreased with ionization. The apparent anomaly was explicable in terms of a hypothesis of an artifact from the measuring device—the sheaths.[55]

Nevertheless, this hypothesis had problems. Appleton and Childs assumed that the dielectric constant of a positive-ion sheath was not different from that of an ionized gas with equal numbers of electrons and positive ions. This assumption was baseless. The ionic refraction theory was valid when immobile positive ions and agitated electrons coexisted in a medium. Eccles's dielectric-constant formula did not apply to a sheath with only positive ions. Another problem was the assumption that the entire medium between the two condenser plates was a positive-ion sheath when ionization was low. This suggested that the entire space between the plates contained only positive ions, and the plates captured all the electrons. That did not seem possible.

Appleton must have soon found the hypothesis of positive-ion sheaths troubling. After 1930, he no longer stuck to it. Rather, in private communications with his British colleagues Childs, F. W. Chapman, and J. Goodier, he talked more about a revised theory of sheaths.[56] This theory returned to Pedersen and Rybner's model: the thin layers forming on the surfaces of the condenser plates were still responsible for the apparent anomaly of the dielectric-constant measurements. Yet the layers were no longer sheaths of positive ions. They were "depletion regions" where *both* positive ions *and*

55. Ibid., 969–94.

56. Appleton to Childs (letter), 15 Dec. 1931, C 169; Childs to Appleton (letters), 21 Feb. 1932, C 173; J. Goodier to Appleton (letters), 12 Dec. 1931, 1 Jan. 1932, C 174; F. W. Chapman to Appleton (letter), 13 Dec. 1931, C 176, MS 2300, Edward Appleton Papers.

electrons disappeared via recombination. So the material between the condenser plates consisted of three parts—two thin layers of nonionized gas and the ionized gas in between (figure 10.9). The two thin layers on the plates' surfaces resembled Pedersen and Rybner's residual capacitors, but they were not air gaps between the plates and the tube wall but rather depletion regions resulting from recombination of ions and electrons. Therefore Pedersen and Rybner's formal explanation of the anomaly of the dielectric-constant measurements still applied to Appleton's new sheath theory. Yet Camille Gutton's critique of Pedersen and Rybner was no longer valid: since the sheaths formed inside the glass tube, moving the condenser plates from outside to inside or changing the air-gap widths did not make a difference.

Appleton developed the neutral-sheath theory with J. Goodier of King's College and formulated it in a report to the Radio Research Board in 1931.[57] The two men not only used Pedersen and Rybner's analysis to account for the dielectric-constant anomaly but also proposed a quantitative empirical test between their theory and the quasi-elastic force. Henri Gutton had maintained that at "true resonance," in which the electron density minimized the maximum condenser current, electron density $N$ and the oscillator's wavelength $\lambda$ had a relationship $N^{3/4}\lambda^2 =$ constant. But Appleton and Goodier deduced a different rule that governed $N$ and $\lambda$ at "true resonance": each depletion sheath's capacitance was a constant $C_2$, and the middle ionized gas's capacitance was $[1-(Ne^2/\varepsilon_0\omega^2)]C_1$ from the ionic refraction theory. The condenser's capacitance (the three capacitances in series) was therefore $C = [1-(Ne^2/\varepsilon_0\omega^2)]C_1C_2/\{C_2+2[1-(Ne^2/\varepsilon_0\omega^2)]C_1\}$. The condenser circuit was at "true resonance" as its capacitance $C$ went to infinity or, equivalently, the denominator of $C$ became zero. The condition for "true resonance" was thus $1-(Ne^2/\varepsilon_0\omega^2) = -C_2/2C_1$. Since both $C_1$ and $C_2$ were constant, this relation entailed that $N\lambda^2 =$ constant at "true resonance."[58]

The different predictions—$N^{3/4}\lambda^2 =$ constant and $N\lambda^2 =$ constant—from the elastic-force theory and the neutral-sheath theory, respectively, offered the basis for a crucial experiment to select between the two. Appleton and Goodier performed such an experiment by measuring the degrees of ionization at "true resonance" for various wavelengths shorter than 100 meters. For each frequency, they fixed the discharge current (a measure of $N$) and tuned the Lecher wire until the condenser current reached a maximum, and

57. Appleton and Goodier, "The dielectric constant of ionized air," Report to Radio Research Board (1931), 1–18, MS 2300, C 174, Edward Appleton Papers.

58. Ibid., 7–9.

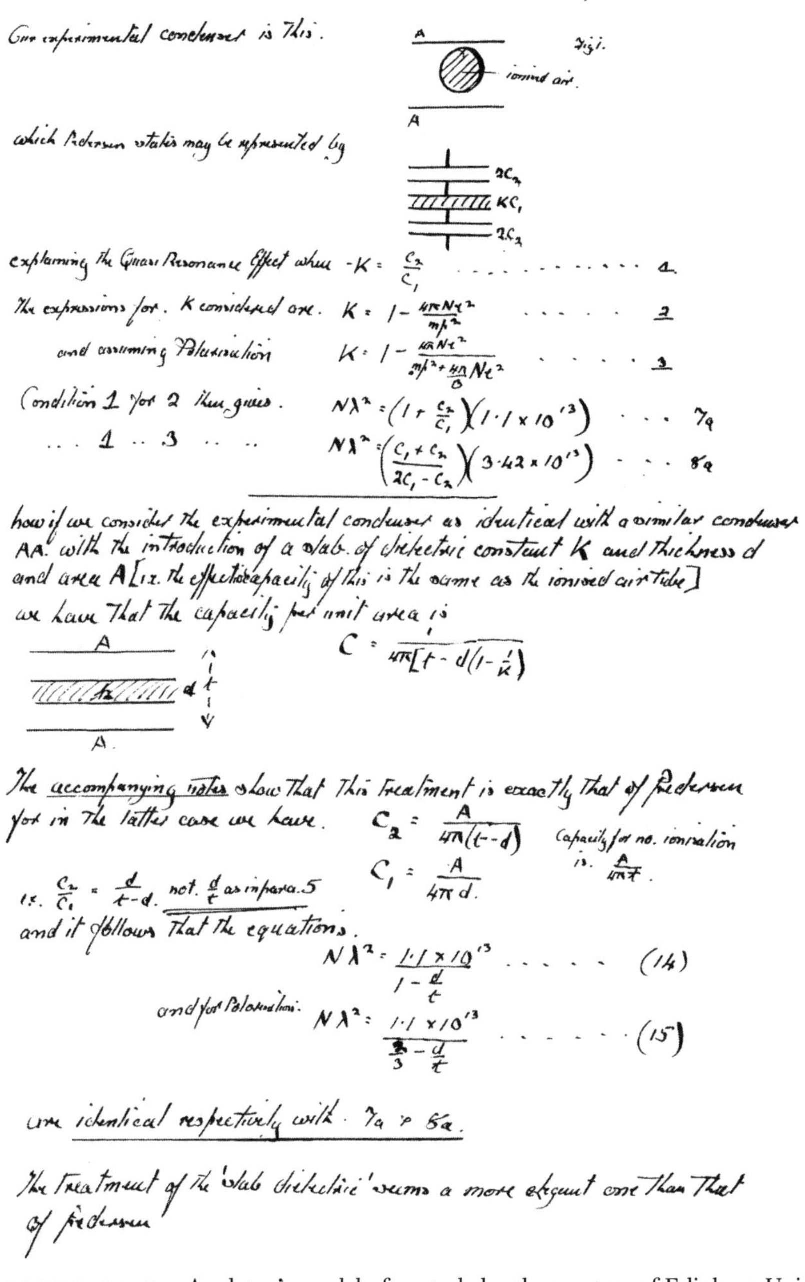

F.W. Chapman. 1.

Our experimental condenser is this.



which Pedersen states may be represented by



explaining the Quasi Resonance Effect when $-K = \frac{C_2}{C_1}$ . . . . . . . . . . . . 1.

The expressions for $K$ considered are. $K = 1 - \frac{4\pi Ne^2}{mp^2}$ . . . . . . 2

and assuming Polarisation $K = 1 - \frac{4\pi Ne^2}{mp^2 + \frac{4\pi}{3}Ne^2}$ . . . . . . 3

Condition 1 for 2 then gives. $N\lambda^2 = \left(1 + \frac{C_2}{C_1}\right)\left(1{\cdot}1 \times 10^{13}\right)$ . . . 7a

. . . 1 . . 3 . . . . $N\lambda^2 = \left(\frac{C_1 + C_2}{2C_1 - C_2}\right)\left(3{\cdot}42 \times 10^{13}\right)$ . . . 8a

Now if we consider the experimental condenser as identical with a similar condenser AA. with the introduction of a slab of dielectric constant $K$ and thickness $d$ and area $A$ [i.e. the effective capacity of this is the same as the ionised air tube] we have that the capacity per unit area is

$$C = \frac{1}{4\pi\left[t - d\left(1 - \frac{1}{K}\right)\right]}$$



The accompanying notes show that this treatment is exactly that of Pedersen for in the latter case we have.

$$C_2 = \frac{A}{4\pi(t-d)} \qquad C_1 = \frac{A}{4\pi d}.$$

Capacity for no ionisation is $\frac{A}{4\pi t}$.

i.e. $\frac{C_2}{C_1} = \frac{d}{t-d}$. not $\frac{d}{t}$ as in para. 5

and it follows that the equations.

$$N\lambda^2 = \frac{1{\cdot}1 \times 10^{13}}{1 - \frac{d}{t}} \quad \ldots\ldots \quad (14)$$

and for Polarisation: $N\lambda^2 = \frac{1{\cdot}1 \times 10^{13}}{\frac{2}{3} - \frac{d}{t}}$ . . . . . . (15)

are identical respectively with 7a & 8a.

The treatment of the 'slab dielectric' seems a more elegant one than that of Pedersen

FIGURE 10.9. Appleton's model of neutral sheath, courtesy of Edinburg University Library, Special Collections Department, Undated Manuscript C171, p. 2, MS 2300, Edward Appleton Papers.

they repeated the procedure with different levels of discharge current until the maximum galvanometer reading was minimal. All their data agreed better with $N\lambda^2 =$ constant than with $N^{3/4}\lambda^2 =$ constant. Therefore the neutral-sheath theory was preferable.

Strictly, the basis for Appleton and Goodier's crucial experiment was problematic. The empirical rule $N^{3/4}\lambda^2 =$ constant that Henri Gutton had found was not an integral part of the theory of the quasi-elastic force. Most results from the theory did not need this empirical rule. The only prediction that it affected was Gutton's estimate of an electron's charge-to-mass ratio from the measurements in magnetized cases. Otherwise, this rule was independent of his theoretical structure. What Appleton and Goodier compared was a formula that they deduced from their theory and an empirical rule independent of their enemy's theory. Also, their experimental conditions were different, as they operated at much higher frequencies. Whether Gutton and Clément's theory was totally invalid or valid under specific conditions was still a question.

But the French experimenters did not fight back. Unlike Appleton, Jean Clément, Camille Gutton, and Henri Gutton did not perform more experiments on ionized gases' dielectric constants. Neither did they defend their theory in public. Worse, Lewi Tonks, whose finding inspired Camille Gutton to argue for the existence of real resonance in an ionized gas, was not sympathetic. In a letter to Appleton in 1931, Tonks explicitly expressed his doubt:[59]

> It seems rather doubtful to me that any of the experimentally observed so-called ionic oscillations, except possibly those discussed by Dr. Langmuir and myself, are of the type for which we have developed the theory.

Moreover, the researcher at General Electric agreed with Appleton that such "resonance" required confinement of the ionized gas in a finite space:

> In answer to your question as to whether ionic oscillations require a boundary for their generation and are caused by the sheaths, I should say, if I understand you correctly, that a boundary is necessary . . . in order to establish a definite frequency.

If Tonks and Appleton were right, then the "resonance" that caused the apparent anomalous variation of an ionized gas's dielectric constant with ionization should not exist in the unbounded atmosphere. The anomaly was an

59. Tonks to Appleton (letter), 16 Dec. 1931, MS 2300, C 177, Edward Appleton Papers.

artifact, a product of the glass-tube boundary confining the ionized gas in the tabletop experiment. In the same year, Tonks remarked in a published article: "H. Gutton introduces an elastic restoring force of unknown origin which seems to be not only unnecessary but somewhat in contradiction with the actual experimental results, as has been pointed out by J. Rybner."[60] Although Gutton evoked Tonks's resonance in an ionized gas to support his quasi-elastic force, Tonks's resonance differed from Gutton's. The former appeared in an electrostatic equation, and the latter in an electromagnetic wave equation. Gutton was unaware of the subtle differences.

The theory of the quasi-elastic force faded away. In 1933, Camille Gutton began to drop the term in considering the equation of motion for a free electron in an ionic medium.[61] In 1934, he explicitly admitted that his previous experiments showed resonance for ionized gas in a finite-volume vessel.[62] At this point, his view was not different from Appleton's: the apparent anomaly in the dielectric-constant measurements was an artifact of the boundary between the material for testing and the instrument, not a natural fact that implied a quasi-elastic force in the ionosphere.

The controversy over the quasi-elastic force originated with an experimental anomaly when researchers were attempting to reproduce an open-space physical environment—the ionosphere—in laboratories. Some scientists explained the anomaly by adding a force term that changed the microscopic model of electromagnetic-wave interactions with ionic media. Others opposed this fundamental revision and insisted that the anomaly was an artifact of the experimental instruments. Unlike the story of the Lorentz correction, the debate over the quasi-elastic force was not simply theoretical. Active laboratory experiments took place to support the existence of the force. Yet "operational realism" did not prevail here: most radio scientists did not accept the force's reality even though some experimenters could show how to manipulate it. The hesitation reflected a problem perhaps common to all "mimicking experiments": whether an observed phenomenon was a real fact in nature or an artificial effect of the instruments was always a question scientists had to face. In the case of the quasi-elastic force, the instrumental factor lay not in any special design or arrangement, but in the simple fact that a bounded laboratory setup did not simulate well borderless nature—the boundaries yielded artifacts.

60. Tonks, "The high frequency behavior of a plasma" (1931), 1459.

61. Gutton, "Propriétés des gaz ionisés" (1933), 63.

62. Gutton, Galle, and Joigny, "Sur la réflexion des ondes radiotélégraphiques" (1934), 471.

# * 4 *

# *Conclusion*

CHAPTER 11

# A New Way of Seeing the World

## FROM WAVE-PROPAGATION STUDIES TO IONOSPHERE PROBING

Among the numerous reports that the Carnegie Institution of Washington (CIW) published on the ionosphere during the 1930s and 1940s was *Ionospheric Research at Huancayo Observatory, Peru, January 1938–June 1946* (1947). This document consisted of nothing but four hundred pages of big, long tables. It presented the measured data for the "critical frequency of ordinary wave component [of] F2-region; minimum virtual height of F-layer; critical frequency [of] F1-region; minimum virtual height of F1-region; critical frequency of normal E-layer; lowest frequency at which reflections are recorded"; and their median values. The tables recorded the radio ionospheric variables at CIW's Huancayo Magnetic Observatory at every hour of the day, on every day of the month, and in every month of the year from the beginning of 1938 to mid-1946.[1] Such railway timetable–like documentation witnessed the major transformation of radio ionospheric studies from a science of wave propagation to a science of the ionosphere. It also testified to the maturity of hardware and infrastructure for the use of radio echo-sounding to generate systematic, extensive data on the upper atmosphere.

Theoretical and experimental work on radio ionospheric propagation between 1900 and 1930 set the material, technical, and conceptual stage for such a transformation and maturity. But the actual establishment of radio iono-

1. Wells and Berkner, *Ionospheric Research at Huancayo Observatory* (1947), 33–35.

spheric sounding was contingent upon substantial instrumental and technological development in the 1930s and 1940s that deserves another book-length study, which I am undertaking. During this period, radio ionospheric sounding became a major geophysical measuring platform and a principal means of acquiring essential environmental data for an "ambience sensitive" engineering system of short-wave radio communications. Here is a brief overview of this historical development.

The first phase of this development came in the first half of the 1930s, when rudimentary echo-sounding apparatuses gave way to much speedier, more automated, and more "efficient" instruments. Here the central focus was the improvement of data-recording mechanisms, or what historians have called "inscription machines." From 1930 to 1935, some American, British, and German engineers and scientists replaced the conventional optical-mechanical oscillographic visual display with the cathode-ray oscilloscope to facilitate instantaneous data taking. They invented gadgets that automatically retrieved and recorded the virtual height of the ionosphere from the returning waveforms and thereby transformed data representation from faithful recording of echoes to abstraction of particular features of the echoes. Moreover, they introduced a measuring procedure that derived from the magneto-ionic theory—increasing the radio frequency until the virtual height rose abruptly—to determine the ionosphere's electron density, a property that conventional echo sounding could not demonstrate. And they encoded this theory-informed procedure into the design of automatic sounding equipment. The end product—the sweep-frequency sounder—became the dominant experimental instrument for radio ionospheric studies until the 1950s.

Automatic sounders disseminated swiftly in the 1930s. Soon after their invention, researchers at CIW and the U.S. National Bureau of Standards (NBS) incorporated them into the transcontinental network of ionospheric measurement that they were building. Similarly, British researchers at Slough helped to take them to observing posts in the British Isles and the Empire/Commonwealth. By the end of the decade, they were standard in the ionospheric-sounding stations that had gone up in Australia, Canada, France, Germany, Japan, New Zealand, and the Soviet Union. Their rapid spread went hand in hand with reification of their role in ionospheric research. To many engineers, geoscientists, and physicists, they seemed the only form of access to the upper atmosphere. Radio sounding shaped these experts' debates on the formation, constitution, and structure of the ionosphere. Not only did it provide the only empirical evidence for or against their arguments; their basic perception of the ionosphere as a region with a layered configuration (e.g., the D, E, and

F layers) rested on their interpretation of returning radio signals. The ionosphere manifested itself to them only in traces on radio sounding recorders.

Radio ionospheric sounding mattered not only to the scientific understanding of the upper atmosphere. As short-wave radio became prevalent and its interruption of communications because of the ionosphere-enabled skip effect increased, engineers found ionospheric data essential for improving radio technology. In the late 1930s, engineers at NBS and Slough came up with a method to predict the optimum radio frequency for a given communication path from relevant ionospheric sounding data. During World War II, this method developed into fully fledged "ionospheric forecasting" programs for both the Allies and the Axis powers. Engineers attempted to design infrastructure to facilitate acquisition, processing, and exchange of data, under great pressure from the military's culture of standardization. To facilitate speedy conversion of measurements into predictions, for example, American engineers at the Central Radio Propagation Laboratory in Washington developed routine procedures, graphic computing aids, and numerical tables. To ensure uniform exchange of data domestically and internationally, they issued a handbook on radio ionospheric propagation, offered training courses, and organized technical conferences with their Australian, British, and Canadian counterparts. These actions helped extend radio ionospheric sounding from a geophysical pursuit into a part of operating the technological system of short-wave radio. This close relationship between planetary science and communications technology remained crucial as space programs started up in the 1950s.

## A CASE HISTORY FOR ACTIVE SENSING

The story of radio ionospheric propagation formed part of a large technoscientific process in the twentieth century that continues to affect us today: It is one of the earliest cases for the emergence of active sensing as a way of seeing the world. About the time research on radio-wave propagation led to ionospheric sounding, several other schemes of active sensing were emerging. In the early 1910s, engineers and physicists conceived underwater sound echo ranging, resulting in construction of the first sonar systems during World War I. Within a decade, geologists and inventors employed similar principles to devise reflective seismological sensing for oil surveys. About the time of Nazi Germany's invasion of Poland in September 1939, British radio ionospheric researchers in Slough converted their sounding apparatuses into an invention that detected and tracked enemy aircraft, marking the début of radar's fully fledged operation. Like radio ionospheric sounding, radar, reflective seismology, and sonar profoundly affected warfare. And like the discovery of the ionosphere, these inventions either facilitated or built on the finding of otherwise invisible geoscientific, layer-like entities, such as the undersea deep sound channel and the Mohorovičić discontinuity between the earth's crust and its mantle.

The second half of the twentieth century saw the further spread and development of Michelson and Morley's interferometer in various geoscientific applications, the invention of laser-based light detection and ranging (lidar) for atmospheric experimentation, and the emergence of many different types of radar (e.g., Doppler radar, ground penetrating radar, synthetic aperture radar) in geology, meteorology, and physical geography. In laboratories and hospitals, it witnessed the adoption of principles of active sensing in making a variety of engineering, medical, and scientific instruments for observation. Powerful examples abound: accelerators and colliders in particle physics exposed the subatomic world; computer-aided tomography revealed the interior characteristics of objects by scanning them with X-rays from different angles; electronic microscopes exposed things at the atomic scale with high-energy electron beams; magnetic resonance imaging discerned particular molecular activities by coupling them with a weak radio signal under a strong magnetic field; ultrasound charted the inside of the human body via the reflection of high-frequency acoustic waves; and X-ray crystallography reconstituted molecular structures from the interference patterns of scattering X-rays. All these instruments visualized originally inaccessible entities—be they too far away, too small, or inside the body—by bombarding them with wave or particle beams and imaging them from the returning signals.

Each of these active sensing instruments has its own distinctive and complex story of origin. Thus the history of radio ionospheric propagation that concerns this book by no means "represents" all of them. Yet the thematic lessons from that history may offer some insight for understanding the common characteristics of active sensing. Above all, the discovery of the ionosphere and the development of radio ionospheric sounding show the primacy of propagation studies in the emergence of an active sensing scheme. The agent that illuminated the invisible—no matter whether it was acoustic wave, electronic beam, laser pulse, radio wave, or X-ray—was far from "transparent." It had its own peculiar characteristics that determined what people could see and in what form. While the agent's basic mechanism of propagation might have been familiar long before its use in active sensing, its propagation behavior in the physical environments relevant to such active sensing was still a mystery.

Unveiling propagation properties of this kind constituted the necessary condition for the development of the sensing system. In this book, for instance, we have seen clearly that ionospheric sounding originated from studies of long-distance radio-wave propagation on the earth's surface.

Moreover, active sensing brought new meanings to experimentation. Many active sensing systems, especially those for the open field, were operating in settings that rendered any laboratory-like control of objects impossible. But this barrier did not prevent practitioners from calling their work "experiment." The malleability of the probing agent was the key. Instead of manipulating the object of investigation, they changed the frequency, power, waveform, or other characteristics of the wave or particle beam that the active sensor emitted and observed how the object responded to the change in the returning echo. As the discovery of the ionosphere indicates, an immediate, clear-cut response in the echo seemed to be "direct" evidence for the invisible object. Doing experiments of this type became easier with the increasing trend to treat both the probing beam and the returning echo of an active sensor as signals and thus to turn their manipulation and analysis into signal processing with readily available toolkits.

In addition, although a propagation theory might originate from the need to explain puzzling phenomena or the formal exercise of representing certain observations with a mathematical model, it would evolve eventually into a "paper tool" as an active sensing system was emerging. Such a paper tool was a key to retrieve from the active sensor's data the physical state of the object under observation. Theory of this kind typically displayed four major elements: a clear structure—usually in the form of equations or formulae—that took the sensor data as input and quantitative information about the object as output; tight integration with the design and operation of the sensing apparatus; an impressive arsenal of computing aids; and a relatively "stable" status that did not need too much effort for revision. These instrumentalistic trademarks did not block scientists and engineers' quests for the theory's "foundation" or "reality," however. As the microphysical controversies over the magneto-ionic model show, the propagation theory that served as a paper tool for active sensing might still face strong challenges.

Finally, the process from research on propagation characteristics to development of an active sensor was interdisciplinary in nature. It involved many groups of people at various stages: field scientists, government bureaucrats, inventors and engineers, laboratory experimenters, mathematicians and theoreticians, military personnel, and technology enthusiasts and amateur researchers. This interdisciplinary landscape was not one comprising island empires; these people shared overlapping realms of actions as well as identities. Nor was it a fully connected network, for a group usually interacted with only a handful of other groups. As a matter of fact, a few key teams or individuals served as the main pushing force for propagation research and sensor development. They set their feet in different areas activity, translated their technical languages, and assembled their results into an integrated whole. They drove the narrative and glued the story together.

Eighty-six years after Appleton and Barnett's first radio sounding experiment on the ionosphere and about the time when NASA's Jet Propulsion Laboratory announced its second extension of the *Cassini* mission, I attended North America's largest annual engineering conference on signal processing. At a huge hotel convocation center in mild, springtime Dallas, thousands of computer scientists, electrical engineers, and mathematicians from all over the world presented papers about bio-imaging, electroacoustics, sensory arrays, and telecommunications.

What impressed me most after listening to session after session was how speakers treated the "signals" separately from the actual physical systems that carried them. No matter whether the subject was analysis of radar data, an electroencephalogram, performance enhancement for wireless antennae, or speech recognition, they often devoted no more than one slide to the relevant acoustics, auditory physiology, electromagnetism, or neurology, glossed it over quickly as several equations, conditions, or terms in the statistical signal model, and quickly moved on to deal with that model for the rest of the presentation. It was as if signals had a life of their own, and we could grasp their secrets by analyzing their innate patterns, without having to understand their physical generation and transmission. Let the hardware designers worry about the latter problems. Perhaps this sharp separation between abstract signals and actual systems comprises the most fundamental difference between people working on active sensing today and those in the time of Appleton, Austin, Berkner, Breit, Macdonald, Ratcliffe, Tuve, and Watson.

But that is not the whole story. At the signal processing conference, I also noticed a plenary speech on recent advances in radar imaging of building interiors, obviously relating to the aftermath of 11 September 2001. The speaker called the subject "interdisciplinary," involving signal processing and electromagnetism, and advocated tighter integration of the two areas for indoor radar imaging.[2] Such a statement points exactly to the ambivalent relationship be-

2. Amin, "Recent advances in radar imaging of building interiors" (2010).

tween active sensing and propagation studies. While the claim about interdisciplinarity acknowledges the wide separation between signal processing and research on the physics of the signals, the call for joining (or rejoining?) both fields nonetheless shows the indispensability of signal physics to signal processing. Like *Cassini*'s radar imaging of Titan, indoor radar imaging involves unknown, uncertain, or unspecified propagation conditions not easily reducible to ready-made signal models without further investigation. Nine decades later, we still cannot remove propagation studies from active sensing. The kind of problems that bothered Appleton, Austin, Berkner, Breit, Macdonald, Ratcliffe, Tuve, and Watson still haunt today's scientists and engineers.

# BIBLIOGRAPHY

## ARCHIVES AND MANUSCRIPT COLLECTIONS


American Radio Relay League Archives, Newington, CT.

Arnold Sommerfeld Papers (NL 056), Deutsches Museum Archives, München (Munich), Germany.

AVIA Records, Public Record Office, Kew, London, United Kingdom.

Bureau of Ships Records (RG 19), U.S. National Archives I, Washington, DC.

DSIR Records, Public Record Office, Kew, London, United Kingdom.

Edward Appleton Papers (H37), Special Collections, University of Edinburgh Libraries, Edinburgh, United Kingdom.

General Collection, Philips Research Public Relations Department, Eindhoven, The Netherlands.

George Neville Watson Papers, Special Collections, University of Birmingham Libraries, Birmingham, United Kingdom.

Hiram Percy Maxim Papers, RG 69:12, Connecticut State Library, Hartford, CT.

Historical Archives, Naval Research Laboratory, Anacostia, Washington, DC.

Historical Records, Rutherford Appleton Laboratory (Space Science Department), Didcot, England.

Ionospheric Section Records, 1927–59, Department of Terrestrial Magnetism Archives, Carnegie Institution, Washington, DC.

Jonathan Zenneck Papers (NL 053), Deutsches Museum Archives, München (Munich), Germany.

Joseph Larmor Papers, Special Collection, St John's College Library, Cambridge University, Cambridge, United Kingdom.

Merle Tuve Papers, Library of Congress, Washington, DC.

Oliver Heaviside Papers (UK0108 SC MSS), Institution of Electrical Engineers Archives, London, United Kingdom.
Radio Pioneers, Columbia University Oral History Collection, New York.
Rayleigh Papers, Rare Book Collection, Air Force Research Laboratory, Hanscom Air Force Base, Lexington, MA.
Service Historique de l'Armée de Terre, Paris, France.
Special Collection, Libraries of Radio France, Paris, France.
Special Collection, University of Aberdeen Libraries, Aberdeen, United Kingdom.

## PUBLISHED SOURCES

Aitken, Hugh G. J., Syntony and Spark: The Origins of Radio, New York: Wiley, 1976.
——, The Continuous Wave: Technology and American Radio, 1900–1932, Princeton, NJ: Princeton University Press, 1985.
Allison, David K., New Eye for the Navy: The Origin of Radar at the Naval Research Laboratory, Washington, DC: Naval Research Laboratory, 1981.
Amato, Ivan, Pushing the Horizon: Seventy-five Years of High Stakes Science and Technology at the Naval Research Laboratory, Washington, DC: U.S. Government Printing Office, 1997.
Amin, Moeness, "Recent Advances in Radar Imaging of Building Interiors," Plenary Session 4 (19 March 2010), 2010 IEEE International Conference on Acoustics, Speech, and Signal Processing (ICASSP), Dallas, TX.
Amoudry, Michel, Le Général Ferrié: Naissance des Transmissions et de la Radiodiffusion, Grenoble: Presse Universitaire de Grenoble, 1993.
Anduaga, Aitor, Wireless and Empire: Geopolitics, Radio Industry, and Ionosphere in the British Empire, 1918–1939, Oxford: Oxford University Press, 2009.
Anonymous, "Wireless signals across the ocean," New York Times (15 Dec. 1901).
——, "First of a chain of seven which will girdle the earth and keep the Department at Washington in direct touch with vessels of our fleet wherever they may be," New York Times (20 Oct. 1912).
——, "Who's who in amateur wireless—Fred Schnell," QST, 3:4 (Nov. 1919), 29.
——, "Presentation of prizes to the First Prize winner in the transatlantic amateur tests," Wireless World and Radio Review, 10:1 (1 April 1922), 10–12.
——, "History of the Wireless Society of London," Wireless World and Radio Review, 11:8 (25 Nov. 1922), 257–63.
——, "Who's who in amateur wireless—Léon Deloy," QST, 6:5 (Dec. 1922), 61, 63, 66.
Appleton, Edward V., "Geophysical influences on the transmission of wireless waves," Proceedings of the Physical Society of London, 37 (1924–5), 16D–22D.
——, "On the diurnal variation of ultra-short wave wireless transmissions," Proceedings of the Cambridge Philosophical Society, 23 (1925–7), 155–61.

———, "The existence of more than one ionized layer in the upper atmosphere," Nature, 120 (3 Sept. 1927), 330.

———, "The influence of the earth's magnetic field on wireless transmission," Proceedings of the URSI (Oct. 1927), 2–3.

———, "On some measurements of the equivalent height of the atmospheric ionized layer," Proceedings of the Royal Society of London, 126 (1930), 542–69.

———, "Wireless studies of the ionosphere," Proceedings of the Institution of Electrical Engineers, 71 (1932), 642–50.

———, Science and the Nation: The B.B.C. Reith Lectures for 1956, Edinburgh: Edinburgh University Press, 1957.

———, "The ionosphere," in Nobel Lectures, Physics, 1942–1962 (Amsterdam: Elsevier Publishing Company, 1964), 79-86.

Appleton, Edward V., and M. A. F. Barnett, "Local reflection of wireless waves from the upper atmosphere," Nature, 115 (7 March 1925), 333–4.

———, "On some direct evidence for downward atmospheric reflection of electric rays," Proceedings of the Royal Society of London, 109 (1925), 621–41.

———, "On wireless interference phenomenon between ground waves and waves deviated by the upper atmosphere," Proceedings of the Royal Society of London, 113 (1926–7), 450–8.

Appleton, Edward V., and Geoffery Builder, "The ionosphere as a doubly-refracting medium," Proceedings of the Physical Society, 45 (1933), 208–20.

Appleton, Edward V., and E. C. Childs, "On some radio-frequency properties of ionized air," Philosophical Magazine, 10 (1930), 969–94.

Appleton, Edward V., and Alfred L. Green, "On some short-wave equivalent height measurements of the ionized regions of the upper atmosphere," Proceedings of the Royal Society of London, 128 (1930), 159–78.

Appleton, Edward V., and John Ratcliffe, "On the nature of wireless signal variation—I," Proceedings of the Royal Society of London, 115 (1927), 291–304.

———, "On the nature of wireless signal variation—II," Proceedings of the Royal Society of London, 115 (1927), 305–17.

———, "On a method of determining the state of polarization of downcoming wireless waves," Proceedings of the Royal Society of London, 117 (1928), 576–88.

———, "Some simultaneous observations on downcoming wireless waves," Proceedings of the Royal Society of London, 128 (1930), 133–58.

Arabatzis, Theodore, Representing Electrons: A Biographical Approach to Theoretical Entities, Chicago: University of Chicago Press, 2006.

Austin, Louis W., "Some quantitative experiments in long-distance radiotelegraphy," Bulletin of the Bureau of Standards, 7:3 (1911), 315–63.

———, "The work of the U.S. Naval Radio-Telegraphic Laboratory," Journal of the American Society of Naval Engineers, 24 (1912), 62–172.

———, "Quantitative experiments in radiotelegraphic transmission," Bulletin of the Bureau of Standards, 11 (1914), 69–86.

Baigrie, Brian, Electricity and Magnetism: A Historical Perspective, Westport, CT: Greenwood, 2007.
Baker, W. J., A History of the Marconi Company, London: Methuen & Co., 1970.
Baker, William G., and Alfred L. Green, "The limiting polarization of downcoming radio waves traveling obliquely to the earth's magnetic field," Proceedings of the Institute of Radio Engineers, 21:8 (1933), 1103–31.
Baños, Alfredo, Dipole Radiation in the Presence of a Conducting Half-Space, Oxford: Pergamon, 1966.
Barton, Edwin H., and Walter B. Kilby, "The effect of ionization of air on electrical oscillations and its bearing on long-distance wireless telegraphy," Philosophical Magazine, 26 (1913), 567–8.
Bazerman, Charles, The Languages of Edison's Light, Cambridge, MA: MIT Press, 1999.
Bellini, E., and A. Tosi, "A directive system of wireless telegraphy," Electrical Engineering, 2 (1907), 771–5, 3 (1908), 348–51.
Bergmann, Ludwig, and Walter Düring, "Experimentelle Untersuchungen der Veränderung der Dielektrizitätskonstanten eines sehr verdünnten Gases durch Elektronen," Annalen der Physik, 1 (1929), 1041–68.
Booker, Henry G., and Lloyd V. Berkner, "Constitution of the ionosphere and the Lorentz polarization correction," Nature, 141 (26 March 1938), 562–3.
Bowhill, S. A., "Investigations of the ionosphere by space techniques," Journal of Atmospheric and Terrestrial Physics, 36:12 (1974), 2235–43.
Braun, K. Ferdinand, "On directed wireless telegraphy," Electrician, 57 (1906), 222–4, 244–8.
Breit, Gregory, and Merle A. Tuve, "A radio method of estimating the height of the conducting layer," Nature, 116 (1925), 357.
———, "A test of the existence of the conducting layer," Physical Review, 28 (1926), 554–75.
Breit, Gregory, Merle A. Tuve, and Odd Dahl, "Effective heights of the Kennelly-Heaviside layer in December, 1927 and January, 1928," Proceedings of the Institute of Radio Engineers, 16 (1928), 1236–9.
Bremmer, H., Terrestrial Radio Waves: Theory of Propagation, New York: Elsevier, 1949.
Brittain, James E., "Scanning the past: Albert Hoyt Taylor," Proceedings of the IEEE, 82:6 (June 1994), 958.
Bromberger, Sylvain, On What We Know We Don't Know: Explanation, Theory, Linguistics, and How Questions Shape Them, Chicago: University of Chicago Press, 1992.
Buchwald, Jed Z., From Maxwell to Microphysics: Aspects of Electromagnetic Theory in the Last Quarter of the Nineteenth Century, Chicago: University of Chicago Press, 1985.

———, The Rise of Wave Theory of Light: Optical Theory and Experiment in the Early Nineteenth Century, Chicago: University of Chicago Press, 1989.

———, The Creation of Scientific Effects: Heinrich Hertz and Electric Waves, Chicago: University of Chicago Press, 1994.

Budden, Kenneth George, The Propagation of Radio Waves: The Theory of Radio Waves of Low Power in the Ionosphere and Magnetosphere, Cambridge: Cambridge University Press, 1985.

———, "John Ashworth Ratcliffe," Biographical Memoirs of Fellows of the Royal Society, 34 (1988), 671–711.

Bussey, Gordon, Marconi's Atlantic Leap, Coventry, England: Marconi Communications, 2000.

Carey, Charles W., Jr., "Tuve, Merle Anthony," American National Biography, 22 (1999), 46–8.

Cartwright, Nancy, How the Laws of Physics Lie, Oxford: Oxford University Press, 1983.

Casimir, H. B. G., "Introduction," in H. Bremmer and C. J. Bouwkamp (eds.), Balthasar van der Pol: Selected Scientific Papers (Amsterdam: North-Holland Publishing Company, 1960), vi–viii.

Chang, Hasok, Inventing Temperature: Measurement and Scientific Progress, New York: Oxford University Press, 2004.

Charkravartty, Anjan, A Metaphysics for Scientific Realism: Knowing the Unobservable, Cambridge: Cambridge University Press, 2007.

Child, Clement D., "Discharge from hot CaO," American Physics Society Review, 32 (1911), 492–511.

Child, Maurice, "5WS, the successful transatlantic transmitting station of the Radio Society of Great Britain" (discussion), Wireless World and Radio Review, 11:26 (31 March 1923), 877.

Cochrane, Rexmond C., Measures for Progress: A History of the National Bureau of Standards, Washington, DC: U.S. National Bureau of Standards, 1966.

Collin, Robert E., "Hertzian dipole radiating over a lossy earth or sea: Some early and late 20th-century controversies," IEEE Antennas and Propagation Magazine, 46:2 (2004), 64–79.

Cornell, Thomas, "Merle A. Tuve and his program of nuclear studies at the Department of Terrestrial Magnetism: the early career of a modern American physicist," PhD dissertation, Johns Hopkins University, Baltimore, 1986.

Coursey, Philip R., "Report on receptions by British amateurs in the trans-Atlantic tests, December, 1921," QST, 5:10 (May 1922), 23–7.

———, "On heterodynes," Wireless World and Radio Review, 10:6 (6 May 1922), 161–3.

———, "The transatlantic communication test," Wireless World and Radio Review, 11:6 (11 Nov. 1922), 185–8.

——, "The transatlantic tests: arrangements for transmission from this country," Wireless World and Radio Review, 11:11 (16 Dec. 1922), 379–81.

——, "'5WS,' the successful transatlantic transmitting station of the Radio Society of Great Britain," Wireless World and Radio Review, 11:24 (17 March 1923), 785–8; 11:25 (24 March 1923), 826–30.

Dahl, Odd, and Louis A. Gebhardt, "Measurements of the effective heights of the conducting layer and the disturbances of August 19, 1927," Proceedings of the Institute of Radio Engineers, 16 (1928), 290–6.

Darrigol, Olivier, From C-Numbers to Q-Numbers: The Classical Analogy in the History of Quantum Theory, Berkeley: University of California Press, 1992.

——, Electrodynamics from Ampère to Einstein, Oxford: Oxford University Press, 2000.

Darwin, Charles G., "The optical constants of matter," Transactions of the Cambridge Philosophical Society, 23 (1923-8), 137–67.

——, "Refraction of ionized media," Nature, 133 (13 Jan. 1934), 62.

——, "The refractive index of an ionized medium," Proceedings of the Royal Society of London, 146 (1934), 17–46.

——, "Douglas Rayner Hartree," Biographical Memoirs of Fellows of the Royal Society, 4 (1958), 103–16.

De Forest, Lee, "Absorption (?) of undamped waves," Electrician, 69 (1912), 369–70.

Dellinger, John H., Laurens E. Whittemore, and S. Kruse, "A study of radio signal fading," Scientific Papers of the Bureau of Standards, 19:476 (1923), 193–230.

Deloy, Léon, "A letter from France," QST, 4:2 (Sept. 1920), 52.

——, "My impressions of American amateur radio," QST, 7:5 (Dec. 1923), 17.

——, "Première communication transatlantique bilatérale entre postes d'amateurs," l'Onde électrique, 2 (1923), 678–83.

——, "Communications transatlantiques sur ondes de 100 metres," l'Onde électrique, 3 (1924), 38-42.

DeSoto, Clinton B., Two Hundred Meters and Down: The Story of Amateur Radio, Hartford, CT: American Radio Relay League, 1936.

Douglas, Susan J., "Technological innovation and organizational change: the navy's adoption of radio, 1899–1919," in Merritt Roe Smith (ed.), Military Enterprise and Technological Change: Perspectives on the American Experience (Cambridge, MA: MIT Press, 1985), 117–73.

——, Inventing American Broadcasting: 1899–1922, Baltimore: Johns Hopkins University Press, 1987.

Duchet-Suchaux, G., "Gutton (Camille-Antoine-Marie)," in M. Prevost, Roman D'Amat, and H. Tribout de Morembert, Dictionnaire de biographie française, Paris-VI, Librairie (Paris), 17 (1989), 371.

Duddell, William, and J. E. Taylor, "Wireless telegraphy measurements," Electrician, 55 (1905), 258–61, 299–302, 349–51.

Duhem, Pierre, The Aim and Structure of Physical Theory, Princeton, NJ: Princeton University Press, 1982.

Dunlap, Orrin E., Marconi, the Man and His Wireless, New York: MacMillan, 1937.

Eccles, William H., "On the diurnal variations of the electric waves occurring in nature, and on the propagation of electric waves round the bend of the earth," Proceedings of the Royal Society of London, 87 (1912), 79–99.

———, Wireless Telegraphy and Telephony: A Handbook of Formulae, Data, and Information, New York: D. Van Nostrand, 1918.

Eccles, William H., and H. Morris Airey, "Note on the electrical waves occurring in nature," Proceedings of the Royal Society of London, 85 (1911), 145–50.

Eckersley, P. P., "5WS, the successful transatlantic transmitting station of the Radio Society of Great Britain" (discussion), Wireless World and Radio Review, 11:26 (31 March 1923), 876.

Eckersley, Thomas L., "Refraction of electric waves," Radio Review, 1 (1920), 421-428.

———, "The effect of the Heaviside layer on the apparent direction of electromagnetic waves," Radio Review, 2 (1921), 60-65, 231-248.

———, "On the connection between the ray theory of electric waves and dynamics," Proceedings of the Royal Society of London, 132 (1931), 83-98.

———, "Radio transmission problems treated by phase integral methods," Proceedings of the Royal Society of London, 136 (1932), 499-527.

———, "Long-wave transmission treated by phase integral methods," Proceedings of the Royal Society of London, 137 (1932), 158-173.

———, "Studies in radio transmission," Journal of the Institution of Electrical Engineers, 71 (1932), 405-459.

———, "Musical atmospherics," Nature, 134 (19 January 1935), 104-105.

Eckert, Michael, and Karl Märker, Arnold Sommerfeld, Wissenschaftlicher Briefwechsel, Band 1: 1892–1918, Munich: Deutsches Museum, 2000.

Eddington, Arthur, "Joseph Larmor," Obituary Notices of Fellows of the Royal Society, 4 (1942–4), 197–207.

Edgerton, David, Warfare State: Britain, 1920–1970, Cambridge, England: Cambridge University Press, 2006.

Edwards, Ronald, Co-operative Industrial Research: A Study of the Economic Aspects of the Research Associations Grant-Aided by the Department of Scientific and Industrial Research, London: Sir Issac Pitman & Sons, 1949.

Epstein, Paul S., "Geometrical optics in absorbing media," Proceedings of the National Academy of Sciences, 16 (1930), 37–45.

———, "Reflection of waves in an inhomogeneous absorbing medium," Proceedings of the National Academy of Sciences, 16 (1930), 627–37.

Evans, W. F., History of the Radio Research Board, 1926–1945, Melbourne: Australian Commonwealth Scientific and Industrial Research Organization, 1973.

Fagen, M. D. (ed.), A History of Engineering and Science in the Bell System: The Early Years (1875–1925), vol. 1, New York: Bell Telephone Laboratories, 1975.
Fleming, John A., "On atmospheric reflection and its bending on the transmission of electromagnetic waves round the earth's surface," Proceedings of the Physical Society of London, 26 (1914), 318–33.
———, The Principles of Electric Wave Telegraphy and Telephony, New York: Longmans & Green, 1916.
Fonton, Michaël, "Clouds, sounding balloons, and stratosphere; Teisserenc de Bort: a life in meteorology," Conference paper preprint in From Beaufort to Bjerknes and Beyond: Critical Perspectives on the History of Meteorology (5–9 July 2004, Polling, Germany), International Commission on History of Meteorology.
Försterling, Karl, and Hans Lassen, "Kurzwellenausbreitung im Erdmagnetfeld," Annalen der Physik, 18 (1933), 26–60.
Fuller, Leonard F., "Continuous waves in long-distance radio telegraphy," Transactions of the American Institute of Electrical Engineers, 34 (1915), 809–27.
Galison, Peter, Image and Logic: A Material Culture of Microphysics, Chicago: University of Chicago Press, 1997.
———, Einstein's Clocks and Poincaré's Maps: Empires of Time, New York: W. W. Norton, 2003.
Galison, Peter, and Alexi Assmus, "Artificial clouds, real particles," in David Gooding, Trevor Pinch, and Simon Schaffer (eds.), The Use of Experiment: Studies in the Natural Sciences (Cambridge: Cambridge University Press, 1989), 225–74.
Garraty, John A., and Mark C. Carnes, American National Biography, 14, New York: Oxford University Press, 1999.
Gillispie, Charles C., "Henri Poincaré," in Charles C. Gillispie (ed.), Dictionary of Scientific Biography, 11, New York: Charles Scribner's Sons, 1975, 51–2.
———, "Arnold Sommerfeld," in Charles C. Gillispie (ed.), Dictionary of Scientific Biography, 12, New York: Charles Scribner's Sons, 1975, 526–9.
Gillmor, C. Stewart, "The history of the term 'ionosphere,'" Nature, 262 (29 July 1976), 347–8.
———, "Threshold to space: early studies of the ionosphere," in Paul Hanle and Von del Chamberlain (eds.), Space Science Comes of Age: Perspectives in the History of Space Science (Washington, DC: Smithsonian Institution Press, 1981), 101–14.
———, "Wilhelm Altar, Edward Appleton, and the magneto-ionic theory," Proceedings of the American Philosophical Society, 126:5 (1982), 395–440.
———, "The big story: Tuve, Breit, and the ionospheric sounding, 1923–1928," in Gregory Good (ed.), History of Geophysics, vol. 5: The Earth, the Heavens, and the Carnegie Institution of Washington (Washington, DC: American Geophysical Union, 1994), 133–41.

Godley, Paul F., "Official report on the second transatlantic tests," QST, 5:7 (Feb. 1922), 14–46.

———, "Listening for Europe," QST, 6:5 (Dec. 1922), 33–5.

Goldstein, Sydney, "The influence of the earth's magnetic field on electric transmission in the upper atmosphere," Proceedings of the Royal Society of London, 121 (1928), 261–3.

Good, Gregory, "Vision of a global physics: the Carnegie Institution and the first world magnetic survey," in Gregory Good (ed.), History of Geophysics, vol. 5: The Earth, the Heavens, and the Carnegie Institution of Washington (Washington, DC: American Geophysical Union, 1994), 29–36.

Green, Alfred L., "The polarization of sky waves in the southern hemisphere," Proceedings of the Institute of Radio Engineers, 22:3 (1934), 324–43.

Gutton, Camille, "Sur la décharge électrique à fréquence très élevée," Comptes rendus de l'Académie des Sciences, 178 (1924), 467–70.

———, "Sur les propriétés des gaz ionisés dans les champs électromagnétiques de haute fréquence," Annales de physique, 14 (1930), 7–8.

———, "Sur les propriétés des gaz ionisés dans les champs électromagnétiques de haute fréquence," Comptes rendus de l'Académie des Sciences, 190 (1930), 844–7.

———, "Dix années de T.S.F. 1922–1932: les dix premières années de la Société des Amis de la T.S.F. et de la revue l'Onde électrique," l'Onde électrique, 11 (1932), 397–404.

———, "Propriétés des gaz ionisés dans les champs de haute fréquence," l'Onde électrique, 12 (1933), 63.

Gutton, Camille, Jean Galle, and Henri Joigny, "Sur la réflexion des ondes radiotélégraphiques dans la haute atmosphère," Comptes rendus de l'Academe des Sciences, 199 (1934), 471.

Gutton, Camille, and Henri Gutton, "Sur la décharge électrique en haute fréquence," Comptes rendus de l'Académie des Sciences, 186 (1928), 303–5.

Gutton, Henri, "Sur l'interprétation de résultats expérimentaux relatifs aux propriétés diélectriques des gaz ionisés," l'Onde électrique, 7 (1928), 1–4.

———, "Recherches sur les propriétés diélectriques des gaz ionisés et la décharge en haute fréquence," Annales de physique, 13 (1930), 62–129.

Gutton, Henri, and Jean Clément, "Sur les propriétés diélectriques des gaz ionisés," Comptes rendus de l'Académie des Sciences, 184 (1927), 441–3.

———, "Sur la propagation des ondes électromagnétiques autour de la terre," Comptes rendus de l'Académie des Sciences, 184 (1927), 676–8.

———, "Sur les propriétés diélectriques des gaz ionisés et la propagation des ondes électromagnétiques dans la haute atmosphère," l'Onde électrique, 6 (1927), 137–51.

Hacking, Ian, Representing and Intervening: Introductory Topics in the Philosophy of Natural Science, Cambridge: Cambridge University Press, 1983.

Haines, Catharine, International Women in Science: A Biographical Dictionary to 1950, Santa Barbara, CA: ABC-CLIO, 2001.

Haring, Kristen, Ham Radio's Technical Culture, Cambridge, MA: MIT Press, 2007.

Hartcup, Guy, The War of Invention: Scientific Development, 1914–18, London: Brassey's Defence Publishers, 1988.

Hartree, Douglas R., "The propagation of electromagnetic waves in a stratified medium," Proceedings of the Cambridge Philosophical Society: Mathematical and Physical Sciences, 25 (1929), 97–120.

——, "The propagation of electromagnetic waves in a refracting medium in a magnetic field," Proceedings of the Cambridge Philosophical Society: Mathematical and Physical Sciences, 27 (1931), 143–62.

——, "The dispersion formula for an ionized medium," Nature, 132 (16 Dec. 1933), 929–30.

Heaviside, Oliver, "The theory of electric telegraphy," Encyclopedia Britannica (10th edition), 33 (1902), 215; reprinted in "Kennelly-Heaviside ionized layer—a classic of science," Science News Letter, 17 (18 Jan. 1930), 44.

Hedenus, Michael, Der Komet in der Entladungsröhre: Eugen Goldstein, Wilhelm Foester und die Elektrizität im Weltraum, Berlin: GNT-Verlag, 2007.

Heisenberg, Werner, "Vorwort für die Sommerfeld-Gesamtausgabe," in Sommerfeld: Gesammelte Schriften (1968), 1, I–v.

Hertz, Heinrich, "The forces of electric oscillations, treated according to Maxwell's theory," in Hertz (D. E. Jones trans.), Electric Waves: Researches on the Propagation of Electric Action with Finite Velocity through Space (London: Macmillan, 1900), 137-59; reprinted from Annalen der Physik, 36 (1899), 1–22.

Hevly, Bruce, "Basic research within a military context: the Naval Research Laboratory and the foundations of extreme ultraviolet and X-ray astronomy, 1923–1960," PhD dissertation, Johns Hopkins University, Baltimore, 1987.

——, "Building a Washington network for atmospheric research," in Gregory Good (ed.), History of Geophysics, vol. 5: The Earth, the Heavens, and the Carnegie Institution of Washington (Washington, DC: American Geophysical Union, 1994), 143–8.

Historischen Kommission bei der Bayerischen Akademie der Wissenschaft (ed.), Neue Deutsche Biographie, 13 (Berlin: Duncker & Humblot, 1982), 674.

Hogan, John L., "Quantitative results of recent radio-telegraphic tests between Arlington, Va., and U.S.S. 'Salem,'" Electrician, 63 (1913), 720-3.

Holton, Gerald, Thematic Origins of Scientific Thoughts: Kepler to Einstein, Cambridge, MA: Harvard University Press, 1988.

Home, Roderick W., "To Watherloo and back: the DTM in Australia, 1911–1947," in Gregory Good (ed.), History of Geophysics, vol. 5: The Earth, the Heavens, and the Carnegie Institution of Washington (Washington, DC: American Geophysical Union, 1994), 149–60.

Home, Roderick W. (ed.), Physics in Australia to 1945, Melbourne: Department of the History and Philosophy of Science at the University of Melbourne, 1990.

Hong, Wireless: From Marconi's Black-Box to the Audion, Cambridge, MA: MIT Press, 2001.

Howeth, Linwood S., History of Communications-Electronics in the United States Navy, Washington, DC: Bureau of Ships and Office of Naval History, 1963.

Hulburt, Edward O., "Early theory of the ionosphere," Journal of Atmospheric and Terrestrial Physics, 36 (1974), 2137–40.

Hull, Andrew, "War of words: the public science of the British scientific community and the origins of the Department of Scientific and Industrial Research, 1914–16," British Journal for the History of Science, 32 (1999), 461–81.

Hull, McAllister, "Gregory Breit," Biographical Memoirs: National Academy of Sciences, 74 (1998), 27–56.

Humboldt, Alexander von, Cosmos: A Sketch of a Physical Description of the Universe, Vol. 1, trans. E. C. Otte, Baltimore: Johns Hopkins University Press, 1997.

Hunt, Bruce, The Maxwellians, Ithaca, NY: Cornell University Press, 1991.

Jackson, Henry B., "On the phenomena affecting the transmission of electric waves over the surface of the sea and the earth," Proceedings of the Royal Society of London, 70 (1902), 254–72.

Johns, Adrian, "The great oscillation war: early broadcasting and the politics of popular experiment," Colloquium at the Department of the History of Science, Harvard University, Cambridge, MA, 22 March 2005.

Jolly, W. P., Marconi, New York: Stein and Day, 1972.

Jones-Imhotep, Edward, "Nature, technology, and nation," Journal of Canadian Studies, 38:3 (2004), 5–36.

Jouaust, R., "Le Général Ferrié," l'Onde électrique, 11 (1932), 45–52.

Jungnickel, Christa, and Russell McCormmach, Intellectual Mastery of Nature: Theoretical Physics from Ohm to Einstein, vols. 1–2, Chicago: University of Chicago Press, 1990.

Kaiser, David, Drawing Theories Apart: The Dispersion of Feynman Diagrams in Postwar Physics, Chicago: University of Chicago Press, 2005.

Keen, R., Direction and Position Finding by Wireless, London: Wireless Press, 1922.

Kennelly, Arthur E., "The daylight effect in radio telegraphy," Proceedings of the Institute of Radio Engineers, 1:3 (1913), 1–12.

———, "On the elevation of the electrically-conducting strata of the earth's atmosphere," Electrical World and Engineer, 15 March 1902; reprinted in "Kennelly-Heaviside ionized layer—a classic of science," Science News Letter, 17 (18 Jan. 1930), 45.

Kenrick, George W., "Radio transmission formulae," Physical Review, 31 (1928), 1040–50.

Kevles, Daniel, The Physicists: The History of a Scientific Community in Modern America, Cambridge, MA: Harvard University Press, 1997.

Klein, Ursula, "Paper tools in experimental cultures," Studies in the History and Philosophy of Science, 32 (2001), 265–302.

Kline, Morris, Mathematical Thought from Ancient to Modern Times, Oxford: Oxford University Press, 1972.

Kohler, Robert, Partners in Science: Foundations and Natural Sciences, 1900–1945, Chicago: University of Chicago Press, 1991.

Kruse, S., "Station performance during the Bureau of Standards–ARRL QSS tests of June and July, 1920," QST, 4:2 (Sept. 1920), 11–14.

——, "The Bureau of Standards–ARRL tests of short wave radio signal fading," QST, 4:4 (Nov. 1920), 5–37.

——, "The Bureau of Standards–ARRL tests of short wave radio signal fading (part 2)," QST, 4:5 (Dec. 1920), 13–22.

——, "Exploring 100 meters," QST, 6:8 (March 1923), 12–13.

Kurylo, Friedrich, and Charles Susskind, Ferdinand Braun: A Life of the Nobel Prizewinner and Inventor of the Cathode-Ray Oscilloscope, Cambridge, MA: MIT Press, 1981.

Langmuir, Irving, "Positive ion currents from the positive column of mercury arcs," Science, 52 (1923), 290–1.

Langmuir, Irving, and Harold Mott-Smith, "Studies of electric discharges in gases at low pressures, part I," General Electric Review, 27 (1924), 449–55; part II, 538–48; part III, 616–23.

Laporte, Otto, "Zur Theorie der Ausbreitung elektromagnetischer Wellen auf der Erdkugel," Annalen der Physik, 70 (1923), 595–616.

Larmor, Joseph, "Why wireless electric rays can bend round the earth," Philosophical Magazine, 48 (1924), 1025–36.

Lassen, Hans, "Die täglichen Schwankungen des Ionisationszustandes der Heaviside-Schicht," Elektrische Nachrichten-Technik, 4:4 (1927), 174–9.

——, "Über den Einfluβ des Erdmagnetfeldes auf die Fortpflanzung der elektrischen Wellen der drahtlosen Telegraphie in der Atmosphäre," Elektrische Nachrichten-Technik, 4:8 (1927), 324–34.

Lessing, Lawrence, Man of High Fidelity: Edwin Howard Armstrong, Philadelphia: J. B. Lippincott, 1956.

Lorentz, Hendrik A., The Theory of Electrons and Its Applications to the Phenomena of Light and Radiant Heat, New York: Dover, 1952.

Love, Augustus E. H., "The transmission of electric waves over the surface of the earth," Philosophical Transactions of the Royal Society of London, A215 (1915), 105–31.

Macdonald, Hector M., Electric Waves, Cambridge: Cambridge University Press, 1902.

———, "The bending of electric waves round a conducting obstacle," Proceedings of the Royal Society of London, 71 (1903), 251–8.

———, "The bending of electric waves round a conducting obstacle: amended result," Proceedings of the Royal Society of London, 72 (1904), 59–68.

———, "The transmission of electric waves around the earth's surface," Proceedings of the Royal Society of London, 90 (1914), 50–61.

———, "The transmission of electric waves around the earth's surface," Proceedings of the Royal Society of London, 98 (1920), 216–22, 409–10; 108 (1925), 52–76.

———, "On the determination of the directions of the forces in wireless waves at the earth's surface," Proceedings of the Royal Society of London, 107 (1925), 587–601.

Mahan, Alfred, The Influence of Sea Power upon History, 1660-1783, Boston: Little & Brown, 1903.

March, Hermann William, "Über die Ausbreitung der Wellen der drahtlosen Telegraphie auf der Erdkugel," Annalen der Physik, 37 (1912), 29-50.

Marconi, Degna, My Father Marconi, New York: McGraw Hill, 1962.

Marconi, Guglielmo, "A note on the effect of daylight upon the propagation of electromagnetic impulses over long distances," Proceedings of the Royal Society of London, 70 (1902), 344–7.

———, "On methods whereby the radiation of electric waves may be mainly confined to certain directions, and whereby the receptivity of a receiver may be restricted to electric waves emanating from certain directions," Proceedings of the Royal Society of London, 77 (1906), 413–21.

Martyn, David F., and G. H. Munro, "The Lorentz 'polarization' correction and the behaviour of radio echoes from the ionosphere at frequencies near the gyrofrequency," Nature, 142 (31 Dec. 1938), 1159–60.

Meiβner, Alexander, "Hat das Erdfeld einen Einfluβ auf die Wellenausbreitungsvorgänge?" Elektrische Nachrichten-Technik, 3:9 (1926), 321–4.

Melville, Harry, The Department of Scientific and Industrial Research, London: George Allen & Unwin, 1962.

Millman, S. (ed.), A History of Engineering and Science in the Bell System: Communications Sciences (1925–1980), vol. 4, New York: Bell Telephone Laboratories, 1984.

Milne, E. A., "Augustus Edward Hugh Love," Obituary Notices of Fellows of the Royal Society of London, 3 (1939–41), 467–82.

Mimno, Harry R., "The physics of the ionosphere," Review of Modern Physics, 9:1 (1937), 1–43.

Mindell, David, Between Human and Machine: Feedback, Control, and Computing before Cybernetics, Baltimore: Johns Hopkins University Press, 2002.

Nahin, Paul, Oliver Heaviside, Sage in Solitude: The Life, Work, and Times of an Electrical Genius of the Victorian Age, New York: IEEE Press, 1987.

Nebeker, Frederik, The Dawn of the Electronic Age: Electrical Technologies in the Shaping of the Modern World, 1914-1945, New York: Wiley-IEEE Press, 2009.

Nichols, Harold W., and John C. Schelleng, "Propagation of electric waves over the earth," Bell System Technical Journal, 4 (1925), 215–34.

Nicholson, John W., "On the bending of electric waves round the earth," Philosophical Magazine, 19 (1910), 276–8.

———, "On the bending of electric waves round a large sphere: I," Philosophical Magazine, 19 (1910), 516–37.

———, "On the bending of electric waves round a large sphere: II," Philosophical Magazine, 20 (1910), 157–72.

———, "On the bending of electric waves round a large sphere: III," Philosophical Magazine, 21 (1911), 62–8.

———, "On the bending of electric waves round a large sphere: IV," Philosophical Magazine, 21 (1911), 281–95.

Norton, Kenneth A., "Ionisation of the ionosphere," Nature, 132 (28 Oct. 1933), 676.

Oatley, C. W., "Smith-Rose, Reginald Leslie," in Lord Blake and C. S. Nicholls (eds.), Dictionary of National Biography: 1971-80, Oxford: Oxford University, 1986, 787–8.

Oreskes, Naomi, The Rejection of Continental Drift: Theory and Method in American Earth Science, New York: Oxford University, 1999.

Panofsky, Wolfgang, and Melba Phillips, Classical Electricity and Magnetism, New York: Addison-Wesley, 1962.

Pedersen, Peder Oluf, The Propagation of Radio Waves along the Surface of the Earth and in the Atmosphere, Copenhagen: Danmarks Naturvidenskabelige Samfund, 1927.

Pestre, Dominique, "Studies of the ionosphere and forecasts for radiocommunications: physicists and engineers, the military and national laboratories in France (and Germany) after 1945," History and Technology, 13 (1997), 183–205.

Pierce, George W., Principles of Wireless Telegraphy, New York: McGraw-Hill, 1910.

Poincaré, Henri, "Sur la diffraction des ondes electriques: àpropos d'un article de M. Macdonald," Proceedings of the Royal Society of London, 72 (1904), 42–52.

———, "Sur la diffraction des ondes Hertziennes," Rendiconti del Circolo Matematico di Palermo, 29 (1910), 169–259; reprinted in Henri Poincaré, Œuvres de Henri Poincaré: publiée sous les auspices de l'Académie des Sciences, 10 (Paris: Gauthier-Villars, 1934–56), 94–203.

———, "Sur la diffraction des ondes Hertziennes," Comptes rendus de l'Académie des Sciences, 155 (1912), 795–7.

Ponte, Maurice, "Notice nécrologique sur M. Camille Gutton, Académicien libre," Comptes rendus de l'Académie des Sciences, 257 (1963), 2584–5.

Pyatt, Edward, The National Physical Laboratory: A History, Bristol: Adam Hilger, 1983.

QST editor, "The ARRL QSS test," QST, 3:12 (July 1920), 5–8.

———, "The fading test," QST, 4:6 (Jan. 1921), 12–14.

———, "Transatlantic sending test," QST, 4:7 (Feb. 1921), 20.

———, "Performance of January QSS recorders," QST, 4:10 (May 1921), 14–15.

———, "Godley to England to copy transatlantics," QST, 5:3 (Oct. 1921), 29–32.

———, "2QR's transatlantic claim disproved," QST, 5:6 (Jan. 1922), 8.

———, "Transatlantic tests successful," QST, 5:6 (Jan. 1922), 7.

———, "The story of the transatlantics," QST, 5:7 (Feb. 1922), 7-14.

———, "The European transatlantic results," QST, 5:8 (March 1922), 20.

———, "Failure of the transatlantic test," QST, 5:10 (May 1922), 15–16.

———, "The transatlantic triumph," QST, 6:7 (Feb. 1923), 7–16.

Rankin, R. A., "George Neville Watson," Journal of the London Mathematical Society, 41 (1966), 551–65.

Ratcliffe, John A.———, "The effect of the Lorentz polarization term in ionospheric calculations," Proceedings of the Physical Society, 51:5 (1939), 747–56.

———, The Magneto-Ionic Theory and Its Applications to the Ionosphere: A Monograph, Cambridge: Cambridge University Press, 1959.

———, "Thomas Lydwell Eckersley," Biographical Memoirs of Fellows of the Royal Society, 5 (1959), 69–74.

———, "Edward Victor Appleton," Biographical Memoirs of Fellows of the Royal Society, 12 (1966), 1–21.

———, "William Henry Eccles," Biographical Memoirs of Fellows of the Royal Society of London, 17 (1971), 195–214.

Ratcliffe, John A., and Eric L. C. White, "The effect of the earth's magnetic field on the propagation of short wireless waves," Philosophical Magazine, 16 (1933), 125–44.

Ratcliffe, John A., and F. W. G. White, "The state of polarization of downcoming wireless waves of medium length," Philosophical Magazine, 16 (1933), 423–41.

Rayleigh, Lord (John William Strutt), "On the bending of waves around a spherical obstacle," Proceedings of the Royal Society of London, 72 (1904), 40–41.

———, The Theory of Sound, New York: Dover, 1945.

Reich, Leonard S., The Making of American Industrial Research: Science and Business at GE and Bell, 1876–1926, Cambridge, MA: Cambridge University Press, 1985.

Reinartz, John, "A year's work below forty meters," Radio News (April 1925), 1894–1895, 1983, 1985-1986.

———, "The reflection of short waves," QST, 8:10 (April 1925), 9–12.

Reingold, Nathan, "National science policy in a private foundation: the Carnegie Institution of Washington," in Alexandra Oleson and John Voss (eds.), The Organization of Knowledge in Modern America, 1860–1920 (Baltimore: Johns Hopkins University Press, 1979), 313–41.

———, Science, American Style, New Brunswick, NJ: Rutgers University Press, 1991.

Round, Henry J., "Direction and position finding," Journal of the Institution of Electrical Engineers, 58 (1920), 224–47.

Round, Henry J., Thomas L. Eckersley, K. Tremellen, and F. C. Lunnon, "Report on measurements made on signal strength at great distances during 1922 and 1923," Journal of the Institution of Electrical Engineers, 63 (1925), 933–1011.

Rozwadowski, Helen, Fathoming the Ocean: The Discovery and Exploration of the Deep Sea, Cambridge, MA: Belknap Press of Harvard University Press, 2005.

Russell, Alexander, "The Kennelly-Heaviside layer," Nature, 116:2921 (1925), 609.

Rybczyński, Witold von, "Über die Ausbreitung der Wellen in der drahtlosen Telegraphie auf der Erdkugel," Annalen der Physik, 41 (1913), 191–208.

Rybner, Joergen, "Note sur les expériences relatifs aux propriétés diélectriques des gaz ionisés de MM. Gutton et Clément," l'Onde électrique, 7 (1928), 428–36.

Salpeter, Jakob, "Das Reflexionsvermögen eines ionisierten Gases für elektrische Wellen," Jahrbuch der Drahtlosen Telegraphie und Telephonie, 8 (1914), 247–53.

Sarkar, Tapan, Robert Mailloux, Arthur Oliner, Magdalena Salazar-Palma, and Dipak Sengupta, History of Wireless, New York: Wiley, 2006.

Schaffer, Simon, "Where experiments end: tabletop trials in Victorian astronomy," in Jed Buchwald (ed.), Scientific Practice: Theories and Stories of Doing Physics (Chicago: University of Chicago Press, 1995), 257–99.

Schmucker, George, "Jonathan Zenneck, 1871–1959: Eine Technisch-Wissenschaftliche Biographie," PhD dissertation, University of Stuttgart, 1999.

Schnell, Fred H., "Transatlantic sending tests," QST, 5:2 (Sept. 1921), 12.

———, "Transatlantic sending tests," QST, 5:6 (January 1922), 20.

———, "The ARRL transatlantics, 1922," QST, 6:3 (Oct. 1922), 11–12.

———, "Arrangements for 1922 transatlantics," QST, 6:4 (Nov. 1922), 22–3.

———, "The transatlantic finals," QST, 6:5 (Dec. 1922), 8–10.

———, "The fourth transatlantic tests," QST, 7:5 (Dec. 1923), 9–11.

Schuster, Arthur, "The diurnal variation of terrestrial magnetism," Philosophical Transactions of the Royal Society of London A, 180 (1889), 467–518.

Schweber, Sylvan, QED and the Men Who Made It: Dyson, Feynman, Schwinger, and Tomonaga, Princeton, NJ: Princeton University Press, 1994.

Sen, H. K., and A. A. Wyller, "On the generalization of the Appleton-Hartree magnetoionic formulas," Journal of Geophysical Research, 65 (1960), 3931–50.

Servos, John W., "To explore the borderland: the foundation of the Geophysical Laboratory of the Carnegie Institution of Washington," Historical Studies in the Physical Sciences, 14:1 (1983), 147–85.

Seth, Suman, "Crafting the quantum: Arnold Sommerfeld and the older quantum theory," Studies in History and Philosophy of Science A, 39:3 (2008), 335–48.

——, Crafting the Quantum: Arnold Sommerfeld and the Practice of Theory, 1890-1926, Cambridge, MA: MIT Press, 2010.

Shapin, Steven, and Simon Schaffer, Leviathan and the Air Pump: Hobbes, Boyle, and the Experimental Life, Princeton, NJ: Princeton University Press, 1989.

Sibum, H. Otto, "Science and the changing sense of reality circa 1900," Studies in History and Philosophy of Science A, 39:3 (2008), 295–7.

Siegel, Daniel, Innovation in Maxwell's Electromagnetic Theory: Molecular Vortices, Displacement Current, and Light, Cambridge: Cambridge University Press, 1992.

Smith, Crosbie, and M. Norton Wise, Energy and Empire: A Biographical Study of Lord Kelvin, Cambridge: Cambridge University Press, 1989.

Smith, George E., "The methodology of the Principia," in I. Bernard Cohen and George E. Smith (eds.), The Cambridge Companion to Newton (Cambridge: Cambridge University Press, 2002), 138–73.

Smith, J. O., "Variation of strength of amateur station signals," QST, 3:9 (April 1920), 17.

——, "The ARRL QSS test," QST, 3:11 (June 1920), 5–6.

Smith-Rose, Reginald L., A Discussion of the Practical Systems of Direction-Finding by Reception, Radio Research Special Report 1, London: Department of Scientific and Industrial Research, 1923.

——, A Study of Radio Direction-Finding, Radio Research Special Report 5, London: Department of Scientific and Industrial Research, 1927.

Smith-Rose, Reginald L., and R. H. Barfield, "On the determination of the direction of the forces in wireless waves at the earth's surface," Proceedings of the Royal Society of London, 107 (1925), 587–601.

——, "An investigation of wireless waves arising from the upper atmosphere," Proceedings of the Royal Society of London, 110 (1926), 580–614.

——, "The cause and elimination of night errors in radio direction-finding," Journal of the Institution of Electrical Engineers, 64 (1926), 831–43.

Sommerfeld, Arnold, "Über die Fortpflanzung elektrodynamischer Wellen längs eines Drahtes," Annalen der Physik, 67 (1899), 233–90.

——, "Über die Ausbreitung der Wellen in der drahtlosen Telegraphie," Annalen der Physik, 28 (1909), 665–736.

——, "Autobiographische Skizze," in Arnold Sommerfeld: Gesammelte Schriften, 4, Braunschweig: F. Vieweg, 1968, 673–82.

Staley, Richard, Einstein's Generation: The Origins of the Relativistic Revolution, Chicago: University of Chicago Press, 2008.

Stewart, Balfour, "Meteorology—terrestrial magnetism," Encyclopedia Britannica, 1875–89, ninth edition, vol. 16, 181–4.

Stobbe, H. (ed.), J. C. Poggendorff's Biographie-Literarisches Handwörterbuch, vol. 6, Berlin: Verlag Chemie, 1931.

Strutt, John William. See Rayleigh, Lord.

Stumpers, F. L. H. M., "Some notes on the correspondence between Edward Appleton and Balth. van der Pol," Philips Research Reports, 30 (1975), 344–56.

Swann, William F. G., "The penetrating radiation and its bearing upon the earth's magnetic field," Eos, 2 (1921), 65–73.

Taylor, Albert Hoyt, "An investigation of transmission on the higher radio frequencies," Proceedings of the Institute of Radio Engineers, 13 (1925), 677–83.

———, The First Twenty-five Years of the Naval Research Laboratory, Washington, DC: Navy Department, 1948.

———, Radio Reminiscences: A Half Century, Washington, DC: U.S. Naval Research Laboratory, 1960.

Taylor, Albert Hoyt, and Albert S. Blatterman, "Variations in nocturnal transmission," Proceedings of the Institute of Radio Engineers, 4 (1916), 131–55.

Taylor, Albert Hoyt, and Edward O. Hulburt, "The propagation of radio waves over the earth," Physical Review, 27 (1926), 189–215.

Taylor, Mary, "The Appleton-Hartree formula and dispersion curves for the propagation of electromagnetic waves through an ionized medium in the presence of an external magnetic field. Part 1: curves for zero absorption," Proceedings of the Physical Society of London, 45 (1933), 245–65.

———, "The Appleton-Hartree formula and dispersion curves for the propagation of electromagnetic waves through an ionized medium in the presence of an external magnetic field. Part 2: curves with collisional friction," Proceedings of the Physical Society of London, 46 (1934), 408–19.

Thompson, G. P., "Charles Galton Darwin," Biographical Memoirs of Fellows of the Royal Society, 9 (1963), 69–85.

Tissot, Camille, "Note on the use of the bolometer as a detector of electric waves," Electrician, 56 (1906), 848–9.

Tonks, Lewi, "The high frequency behavior of a plasma," Physical Review, 37 (1931), 1459.

———, "Ionisation density and critical frequency," Nature, 132 (15 July 1933), 101.

———, "Ionisation density and critical frequency," Nature, 132 (4 Nov. 1933), 710–11.

Tonks, Lewi, and Irving Lagmuir, "Oscillations in ionized gases," Physical Review, 33 (1929), 195–210.

Tuve, Merle A., "Early days of pulse radio at the Carnegie Institution," Journal of Atmospheric and Terrestrial Physics, 36:12 (1974), 2079–80.

Tuve, Merle A., and Gregory Breit, "Note on a radio method of estimating the height of the conducting layer," Terrestrial Magnetism and Atmospheric Electricity, 30 (1925), 15–16.

Tuve, Merle A., and Odd Dahl, "A transmitter modulating device for the study of the Kennelly-Heaviside layer by the echo method," Proceedings of the Institute of Radio Engineers, 16 (1928), 794–8.

U.S. Department of Commerce (Bureau of Navigation), Radio Laws and Regulations of the United States, Washington, DC: Washington Government Printing Office, 1914.

Van der Pol, Balthasar, "On the energy transmission in wireless telegraphy," Yearbook of Wireless Telegraphy and Telephony (1918), 858–76.

———, "On the propagation of electromagnetic waves round the earth," Philosophical Magazine, 38 (1919), 365–80.

———, "De Invloed van een Geioniseerd Gas op het Voortschrijden van Electromagnetische Golven en Toepassingen Daarvan op het Gebied der Draadlooze Telegraphie en bij Metingen aan Glimlichtontladingen," PhD dissertation, University of Utrecht, 1920.

Varcoe, Ian, Organizing for Science in Britain: A Case Study, Oxford: Oxford University Press, 1974.

Veblen, Oswald, "Jules Henri Poincaré," Proceedings of the American Philosophical Society, 51 (1912), iii–ix.

Villard, Oswald G., Jr, "The ionospheric sounder and its place in the history of radio science," Radio Science, 11:11 (1976), 850–1.

Von Oettingen, Arthur (ed.), J. C. Poggendorff's Biographisch-Literarisches Handwörterbuch, vol. 4, Leipzig: J. A. Barth, 1904.

Wagner, Karl Willy, "Arthur Edwin Kennelly, zu seinem 70. Geburtstage," Elektrische Nachrichten Technik, 8:12 (1931), 1.

Warner, Kenneth B., "Two-way tests with Europe," QST, 6:8 (March 1923), 13–15.

———, "The progress of transatlantic amateur communications," QST, 7:7 (Feb. 1924), 15–16.

———, "Transatlantic tests report," QST, 7:8 (March 1924), 32–4.

Warwick, Andrew, Masters of Theory: Cambridge and the Rise of Mathematical Physics, Chicago: University of Chicago Press, 2003.

Watson, George N., "The diffraction of electric waves by the earth," Proceedings of the Royal Society of London, 95 (1918–19), 83–99.

———, "The transmission of electric waves round the earth," Proceedings of the Royal Society of London, 95 (1918–19), 546–63.

———, A Treatise of the Theory of Bessel Functions, Cambridge: Cambridge University Press, 1952.

Weaver, J. R. H. (ed.), "Henry Bradwardine Jackson," Dictionary of National Biography: 1922–1930 (London: Oxford University Press, 1930), 448–50.

Webster, Arthur Gordon, "Henri Poincaré as a mathematical physicist," Science, 38 (1913), 901–8.

Weinmeister, P. (ed.), J. C. Poggendorff's Biographisch-Literarisches Handwörterbuch, vol. 5, Berlin: Verlag Chemie, 1922.

Wells, H. W., and L. V. Berkner, Ionospheric Research at Huancayo Observatory, Peru, January 1938–June 1946, Washington, DC: Carnegie Institution of Washington, 1947.

Wenstrom, William H., "Milestones in meteorology," Scientific Monthly, 51:3 (1940), 230.

Whittaker, Edmund Taylor, "Hector Munro Macdonald," Obituary Notices of Fellows of the Royal Society of London, 1 (1935), 551–8.

Whittaker, J. M., "George Neville Watson," Biographical Memoirs of Fellows of the Royal Society, 12 (1966), 521–30.

Wilson, William, "John William Nicholson," Biographical Memoirs of Fellows of the Royal Society of London, 2 (1956), 209–14.

Wolters, Timothy S., "Managing a sea of information: shipboard command and control in the United States Navy, 1899–1945," PhD dissertation, MIT, Cambridge, MA, 2003.

Yavetz, Ido, From Obscurity to Enigma: The Work of Oliver Heaviside, 1872–1889, Berlin: Birkhäuser Verlag, 1995.

Yeang, Chen-Pang, "Between users and developers: American amateurs and radio technology in the early twentieth century," unpublished manuscript, 2007.

———, "From mechanical objectivity to instrumentalizing theory: Inventing radio ionospheric sounders," Historical Studies in the Natural Sciences, 42:3 (2012), 190-234.

Yochelson, Ellis, "Andrew Carnegie and Charles Doolittle Walcott: the origin and early years of the Carnegie Institution of Washington," in Gregory Good (ed.), History of Geophysics, vol. 5: The Earth, the Heavens, and the Carnegie Institution of Washington (Washington, DC: American Geophysical Union, 1994), 1–19.

Yoder, H. S., Jr., "Development and promotion of the initial program for the Geophysical Laboratory," in Gregory Good (ed.), History of Geophysics, vol. 5: The Earth, the Heavens, and the Carnegie Institution of Washington (Washington, DC: American Geophysical Union, 1994), 21–8.

Zenneck, Jonathan, "Über die Fortpflanzung ebener elektromagnetischer Wellen längs einer ebenen Leiterfläche und ihre Beziehung zur drahtlosen Telegraphie," Annalen der Physik, 23 (1907), 846–66.

——— (A. E. Seelig, trans.), Wireless Telegraphy, New York: McGraw-Hill, 1915.

———, Erinnerungen Einer Physikers, Munich: Author-printed, 1961.

## WORLD WIDE WEB SOURCES

Cassini Equinox Mission News Release, http://saturn.jpl.nasa.gov/news/news releases/ (last access on 3 December 2012).

MacTutor History of Mathematics Archive, "Sydney Goldstein," http://www-history.mcs.st-and.ac.uk/Mathematicians/Goldstein.html (last access on 3 December 2012).

Nobel Prize Website (Physics 1947, Edward Victor Appleton), http://nobelprize.org/nobel_prizes/physics/laureates/1947/ (last access on 3 December 2012).

# INDEX